T0185699

The Theoretical Biologist's Toolbox
Quantitative Methods for Ecology and Evolutionary Biology

Mathematical modeling is widely used in ecology and evolutionary biology and it is a topic that many biologists find difficult to grasp. In this new textbook Marc Mangel provides a no-nonsense introduction to the skills needed to understand the principles of theoretical and mathematical biology. Fundamental theories and applications are introduced using numerous examples from current biological research, complete with illustrations to highlight key points. Exercises are also included throughout the text to show how theory can be applied and to test knowledge gained so far. Suitable for advanced undergraduate or introductory graduate courses in theoretical and mathematical biology, this book forms an essential resource for anyone wanting to gain an understanding of theoretical ecology and evolution.

MARC MANGEL is Professor of Mathematical Biology and Fellow of Stevenson College at the University of California, Santa Cruz campus.

The Theoretical Biologist's Toolbox

Quantitative Methods for
Ecology and Evolutionary Biology

Marc Mangel
Department of Applied Mathematics
and Statistics
University of California, Santa Cruz

Shaftesbury Road, Cambridge CB2 8EA, United Kingdom

One Liberty Plaza, 20th Floor, New York, NY 10006, USA

477 Williamstown Road, Port Melbourne, VIC 3207, Australia

314–321, 3rd Floor, Plot 3, Splendor Forum, Jasola District Centre, New Delhi – 110025, India

103 Penang Road, #05–06/07, Visioncrest Commercial, Singapore 238467

Cambridge University Press is part of Cambridge University Press & Assessment, a department of the University of Cambridge.

We share the University's mission to contribute to society through the pursuit of education, learning and research at the highest international levels of excellence.

www.cambridge.org
Information on this title: www.cambridge.org/9780521537483

First published 2006

A catalogue record for this publication is available from the British Library

ISBN   978-0-521-83045-4   Hardback
ISBN   978-0-521-53748-3   Paperback

To all of my teachers, but especially Susan Mangel.

Contents

Preface: Bill Mote, Youngblood Hawke, and Mel Brooks

I conceived of the courses that led to this book on sabbatical in 1999–2000, during my time as the Mote Eminent Scholar at Florida State University and the Mote Marine Laboratory (a chair generously funded by William R. Mote, who was a good friend of science). While at FSU, I worked on a problem of life histories in fluctuating environments with Joe Travis and we needed to construct log-normal random variables of specified means and variances. I did the calculation during my time spent at Mote Marine Laboratory in Sarasota and, while doing the calculation, realized that although this was something pretty easy and important in ecology and evolutionary biology, it was also something difficult to find in the standard textbooks on probability or statistics. It was then that I decided to offer a six-quarter graduate sequence in quantitative methods, starting the following fall. I advertised the course initially as "Quantitative tricks that I've learned which can help you" but mainly as "The Voyage of Quantitative Methods," "The Voyage Continues," etc. This book is the result of that course.

There is an approximate "Part I" and "Part II" structure. In the first three chapters, I develop some basic ideas about modeling (Chapter 1), differential equations (Chapter 2), and probability (Chapter 3). The remainder of the book involves the particular applications that interested me and the students at the time of the course: the evolutionary ecology of parasitoids (Chapter 4), the population biology of disease (Chapter 5), some problems of sustainable fisheries (Chapter 6), and the basics and application of stochastic population theory in ecology, evolutionary biology and biodemography (Chapters 7 and 8).

Herman Wouk's character Youngblood Hawke (Wouk 1962) bursts on the writing scene and produces masterful stories until he literally has nothing left to tell and burns himself out. The stories were somewhere between the ether and the inside of his head and he had to get them out. Much the same is true for music. Bill Monroe (Smith 2000) and Bob Dylan (Sounes 2001) reported that their songs were already present, either in the air or in their heads and that they could not rest until the songs were on paper. Mozart said that he was more transcribing music that was in his head than composing it. In other words, they all had a story to tell and could not rest until it was told. Mel Brooks, the American director and producer, once wrote "I do what I do because I have to get it out. I'm just lucky it wasn't an urge to be a pickpocket."

I too have a story to get out, but mine is about theoretical biology, and once I began writing this book, I could not rest until it was down on paper. Unlike a novel, however, you'll not likely read this book in a weekend or before bed. But I hope that you will read it. Indeed, it took me two years of once a week meetings plus one quarter of twice a week meetings with classes to tell the story (in Chapter 1, I offer some guidelines on how to use the book), so I expect that this volume will be a long-term companion rather than a quick read. And I hope that you will make it so. Like my other books (Mangel 1985, Mangel and Clark 1988, Hilborn and Mangel 1997, Clark and Mangel 2000), my goal is to bring people – keen undergraduates, graduate students, post-docs, and perhaps even a faculty colleague or two – to a skill level in theoretical biology where they will be able to read the primary literature and conduct their own research. I do this by developing tools and showing how they can be used. Suzanne Alonzo, a student of Bob Warner's, post-doc with me and now on the faculty at Yale University, once told me that she carried Mangel and Clark (1988) everywhere she went for the first two years of graduate school. In large part, I write this book for the future Suzannes.

Before writing this story, I told most of it as a six quarter graduate seminar on quantitative methods in ecology and evolutionary biology. These students, much like the reader for whom I write, were keen to learn quantitative methods and wanted to get to the heart of the matter – applying such methods to interesting questions in ecology and evolutionary biology – as quickly as possible. I promised the students that if they stuck with it, they would be able to read and understand almost anything in the literature of theoretical biology. And a number of them did stick through it: Katriona Dlugosch, Will Satterthwaite, Angie Shelton, Chris Wilcox, and Nick Wolf (who, although not a student earned a special certificate of quantitude). Other students were able to attend only part of the series: Nick Bader, Joan Brunkard, Ammon Corl, Eric Danner, EJ Dick, Bret Eldred, Samantha Forde, Cindy Hartway, Cynthia Hays, Becky Hufft, Teresa Ish, Rachel Johnson, Matt Kauffman, Suzanne Langridge, Doug Plante, Jacob Pollock, and Amy Ritter. Faculty and NMFS/SCL colleagues Brent Haddad, Karen Holl, Alec MacCall, Ingrid Parker, and Steve Ralston attended part of the series too (Brent made five of the six terms!). To everyone, I am very thankful for quizzical looks and good questions that helped me to clarify the exposition of generally difficult material.

Over the years, theoretical biology has taken various hits (see, for example, Lander (2004)), but writing at the turn of the millennium,

Sidney Brenner (Brenner 1999) said that there is simply no better description and we should use it. Today, of course, computational biology is much in vogue (I sometimes succumb to calling myself a computational biologist, rather than a theoretical or mathematical biologist) and usually refers to bioinformatics, genomics, etc. Although these are not the motivational material for this book, readers interested in such subjects will profit from reading it. The power of mathematical methods is that they let us approach apparently disparate problems with the same kind of machinery, and many of the tools for ecology and evolutionary biology are the same ones as for bioinformatics, genomics, and systems biology.

I have tried to make this book fun to read, motivated by Mike Rosenzweig's writing in his wonderful book on species diversity (Rosenzweig 1995). There he asserted – and I concur – that because a book deals with a scientific topic in a technical (rather than popular) way, it does not have to be thick and hard to read (not everyone agrees with this, by the way). I have also tried to make it relatively short, by pointing out connections to the literature, rather than going into more detail on additional topics. I apologize to colleagues whose work should have been listed in the Connections section at the end of each chapter, but is not.

For the use of various photos, I thank Luke Baton, Paulette Bierzychudek, Kathy Beverton, Leon Blaustein, Ian Fleming, James Gathany, Peter Hudson, Jay Rosenheim, Bob Lalonde, and Lisa Ranford-Cartwright. Their contributions make the book both more interesting to read and more fun to look at. Permissions to reprint figures were kindly granted by a number of presses and individuals; thank you.

Nicole Rager, a graduate of the Science Illustration Program at UC Santa Cruz and now at the NSF, helped with many of the figures, and Katy Doctor, now in graduate school at the University of Washington, aided in preparation of the final draft, particularly with the bibliography and key words for indexing.

Alan Crowden commissioned this book for Cambridge University Press. His continued enthusiasm for the project helped spur me on. For comments on the entire manuscript, I thank Emma Ådahl, Anders Brodin, Tracy S. Feldman, Helen Ivarsson, Lena Månsson, Jacob Johansson, Niclas Jonzen, Herbie Lee, Jörgen Ripa, Joshua Uebelherr, and Eric Ward. For comments on particular chapters, I thank Per Lundberg and Kate Siegfried (Chapter 1), Leah Johnson (Chapter 2), Dan Merl (Chapter 3), Nick Wolf (Chapter 4), Hamish McCallum, Aand Patil, Andi Stephens (Chapter 5), Yasmin Lucero (Chapter 6), and Steve Munch (Chapters 7 and 8). The members of my research group (Kate, Leah, Dan, Nick, Anand, Andi, Yasmin, and Steve) undertook to check all of the equations and do all of the exercises, thus finding bloopers of various sizes, which I have corrected. Beverley Lawrence is the best copy-editor with whom I have ever worked; she deserves great thanks for helping to clarify matters in a number of places. I shall miss her early morning email messages.

In our kind of science, it is generally difficult to separate graduate instruction and research, since every time one returns to old material, one sees it in new ways. I thank the National Science Foundation, National Marine Fisheries Service, and US Department of Agriculture, which together have continuously supported my research efforts in a 26 year career at the University of California, which is a great place to work.

At the end of *The Glory* (Wouk 1994), the fifth of five novels about his generation of destruction and resurgence, Herman Wouk wrote "The task is done, and I turn with a lightened spirit to fresh beckoning tasks" (p. 685). I feel much the same way.

Have a good voyage.

Permissions

Figure 2.1. Reprinted from Washburn, A.R. (1981). *Search and Detection*. Military Applications Section, Operations Research Society of America, Arlington, VA., with permission of the author.

Figure 2.21a. Reprinted from *Ecology*, volume 71, W.H. Settle and L.T. Wilson, Invasion by the variegated leafhopper and biotic interactions: parasitism, competition, and apparent competition, pp. 1461–1470. Copyright 1990, with permission of the Ecological Society of America.

Figure 3.12. Reprinted from *Statistics*, third edition, by David Freeman, Robert Pisani and Roger Purves. Copyright 1998, 1991, 1991, 1978 by W.W. Norton & Company, Inc. Used by permission of W.W. Norton & Company, Inc.

Figure 4.11. Reprinted from May, Robert M., *Stability and Complexity in Model Ecosystems*. Copyright 1973 Princeton University Press, 2001 renewed PUP. Reprinted by permission of Princeton University Press.

Figure 4.15. Reprinted from *Theoretical Population Biology*, volume 42, M. Mangel and B.D. Roitberg, Behavioral stabilization of host-parasitoid population dynamics, pp. 308–320, Figure 3 (p. 318). Copyright 1992, with permission from Elsevier.

Figure 5.3. Reprinted from *Proceedings of the Royal Society of London*, Series A, volume 115, W.O. Kermack and A.G. McKendrick, A contribution to the mathematical theory of epidemics, pp. 700–721, Figure 1. Copyright 1927, with permission of The Royal Society.

Figure 5.9. Reprinted from *Ecology Letters*, volume 4, J.C. Koella and O. Restif, Coevolution of parasite virulence and host life history, pp. 207–214, Figure 2 (p. 209). Copyright 2001, with permission Blackwell Publishing.

Figure 5.12. From *Infectious Diseases of Humans: Dynamics and Control* by R.M. Anderson and R.M. May, Figure 14.25 (p. 408). Copyright 1991 Oxford University Press. By permission of Oxford University Press.

Chapter 1
Four examples and a metaphor

Robert Peters (Peters 1991) – who (like Robert MacArthur) tragically died much too young – told us that theory is going beyond the data. I thoroughly subscribe to this definition, and it shades my perspective on theoretical biology (Figure 1.1). That is, theoretical biology begins with the natural world, which we want to understand. By thinking about observations of the world, we conceive an idea about how it works. This is theory, and may already lead to predictions, which can then flow back into our observations of the world. Theory can be formalized using mathematical models that describe appropriate variables and processes. The analysis of such models then provides another level of predictions which we take back to the world (from which new observations may flow). In some cases, analysis may be insufficient and we implement the models using computers through programming (software engineering). These programs may then provide another level of prediction, which can flow back to the models or to the natural world. Thus, in biology there can be many kinds of theory. Indeed, without a doubt the greatest theoretician of biology was Charles Darwin, who went beyond the data by amassing an enormous amount of information on artificial selection and then using it to make inferences about natural selection. (Second place could be disputed, but I vote for Francis Crick.) Does one have to be a great naturalist to be a theoretical biologist? No, but the more you know about nature – broadly defined (my friend Tim Moerland at Florida State University talks with his students about the ecology of the cell (Moerland 1995)) – the better off you'll be. (There are some people who will say that the converse is true, and I expect that they won't like this book.) The same is true, of course, for being able to

1

Figure 1.1. Theoretical biology begins with the natural world, which we want to understand. By thinking about observations of the world, we begin to conceive an idea about how it works. This is theory, and may already lead to predictions, which can then flow back into our observations of the world. The idea about how the world works can also be formalized using mathematical models that describe appropriate variables and processes. The analysis of such models then provides another level of predictions which we can take back to the world (from which new observations may flow). In some cases, analysis may be insufficient and we choose to implement our models using computers through programming (software engineering). These programs then provide another level of prediction, which can also flow back to the models or to the natural world.

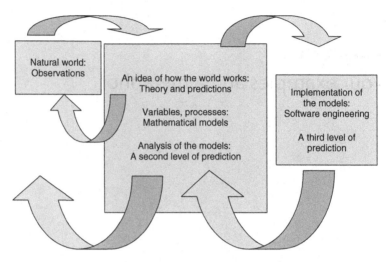

develop models and implementing them on the computer (although, I will tell you flat out right now that I am not a very good programmer – just sufficient to get the job done). This book is about the middle of those three boxes in Figure 1.1 and the objective here is to get you to be good at converting an idea to a model and analyzing the model (we will discuss below what it means to be good at this, in the same way as what it means to be good at opera).

On January 15, 2003, just as I started to write this book, I attended a celebration in honor of the 80th birthday of Professor Joseph B. Keller. Keller is one of the premier applied mathematicians of the twentieth century. I first met him in the early 1970s, when I was a graduate student. At that time, among other things, he was working on mathematics applied to sports (see, for example, Keller (1974)). Joe is fond of saying that when mathematics interacts with science, the interaction is fruitful if mathematics gives something to science and the science gives something to mathematics in return. In the case of sports, he said that what mathematics gained was the concept of the warm-up. As with athletics, before embarking on sustained and difficult mathematical exercise, it is wise to warm-up with easier things. Most of this chapter is warm-up. We shall consider four examples, arising in behavioral and evolutionary ecology, that use algebra, plane geometry, calculus, and a tiny bit of advanced calculus. After that, we will turn to two metaphors about this material, and how it can be learned and used.

Foraging in patchy environments

Some classic results in behavioral ecology (Stephens and Krebs 1986, Mangel and Clark 1988, Clark and Mangel 2000) are obtained in the

(a)

(b)

(c)

Figure 1.2. Two stars of foraging experiments are (a) the great tit, *Parus major*, and (b) the common starling *Sturnus vulgaris* (compliments of Alex Kacelnik, University of Oxford). (c) Foraging seabirds on New Brighton Beach, California, face diet choice and patch leaving problems.

study of organisms foraging for food in a patchy environment (Figure 1.2). In one extreme, the food might be distributed as individual items (e.g. worms or nuts) spread over the foraging habitat. In another, the food might be concentrated in patches, with no food between the patches. We begin with the former case.

The two prey diet choice problem (algebra)

We begin by assuming that there are only two kinds of prey items (as you will see, the ideas are easily generalized), which are indexed by $i = 1, 2$. These prey are characterized by the net energy gain E_i from consuming a single prey item of type i, the time h_i that it takes to handle (capture and consume) a single prey item of type i, and the rate λ_i at which prey items of type i are encountered. The profitability of a single prey item is E_i/h_i since it measures the rate at which energy is accumulated when a single prey item is consumed; we will assume that prey type 1 is more profitable than prey type 2. Consider a long period of time T in which the only thing that the forager does is look for prey items. We ask: what is the best way to consume prey? Since I know the answer that is coming, we will consider only two cases (but you might want to think about alternatives as you read along). Either the forager eats whatever it encounters (is said to generalize) or it only eats prey type 1, rejecting prey type 2 whenever this type is encountered (is said to specialize). Since the flow of energy to organisms is a fundamental biological consideration, we will assume that the overall rate of energy acquisition is a proxy for Darwinian fitness (i.e. a proxy for the long term number of descendants).

In such a case, the total time period can be divided into time spent searching, S, and time spent handling prey, H. We begin by calculating the rate of energy acquisition when the forager specializes. In search time S, the number of prey items encountered will be $\lambda_1 S$ and the time required to handle these prey items is $H = h_1(\lambda_1 S)$. According to our assumption, the only things that the forager does is search and handle prey items, so that $T = S + H$ or

$$T = S + h_1 \lambda_1 S = S(1 + \lambda_1 h_1) \tag{1.1}$$

We now solve this equation for the time spent searching, as a fraction of the total time available and obtain

$$S = \frac{T}{1 + \lambda_1 h_1} \tag{1.2}$$

Since the number of prey items encountered is $\lambda_1 S$ and each item provides net energy E_1, the total energy from specializing is $E_1 \lambda_1 S$, and the rate of acquisition of energy will be the total accumulated energy divided by T. Thus, the rate of gain of energy from specializing is

$$R_s = \frac{E_1 \lambda_1}{1 + h_1 \lambda_1} \tag{1.3}$$

An aside: the importance of exercises

Consistent with the notion of mathematics in sport, you are developing a set of skills by reading this book. The only way to get better at skills is by practice. Throughout the book, I give exercises – these are basically steps of analysis that I leave for you to do, rather than doing them here. You should do them. As you will see when reading this book, there is hardly ever a case in which I write "it can be shown" – the point of this material is to learn how to show it. So, take the exercises as they come – in general they should require no more than a few sheets of paper – and really make an effort to do them. To give you an idea of the difficulty of exercises, I parenthetically indicate whether they are easy (E), of medium difficulty (M), or hard (H).

Exercise 1.1 (E)

Repeat the process that we followed above, for the case in which the forager generalizes and thus eats either prey item upon encounter. Show that the rate of flow of energy when generalizing is

$$R_g = \frac{E_1 \lambda_1 + E_2 \lambda_2}{1 + h_1 \lambda_1 + h_2 \lambda_2} \tag{1.4}$$

We are now in a position to predict the best option: the forager is predicted to specialize when the flow of energy from specializing is greater than the flow of energy from generalizing. This will occur when $R_s > R_g$.

Exercise 1.2 (E)

Show that $R_s > R_g$ implies that

$$\lambda_1 > \frac{E_2}{E_1 h_2 - E_2 h_1} \tag{1.5}$$

Equation (1.5) defines a "switching value" for the encounter rate with the more profitable prey item, since as λ_1 increases from below to above this value, the behavior switches from generalizing to specializing. Equation (1.5) has two important implications. First, we predict that the foraging behavior is "knife-edge" – that there will be no partial preferences. (To some extent, this is a result of the assumptions. So if you are uncomfortable with this conclusion, repeat the analysis thus far in which the forager chooses prey type 2 a certain fraction of the time, p, upon encounter and compute the rate R_p associated with this assumption.) Second, the behavior is determined solely by the encounter rate with the more profitable prey item since the encounter rate with the less profitable prey item does not appear in the expression for the switching value.

Neither of these could have been predicted a priori.

Over the years, there have been many tests of this model, and much disagreement about what these tests mean (more on that below). My opinion is that the model is an excellent starting point, given the simple assumptions (more on these below, too).

The marginal value theorem (plane geometry)

We now turn to the second foraging model, in which the world is assumed to consist of a large number of identical and exhaustible patches containing only one kind of food with the same travel time between them

(a)

(b)

(c)

(d)

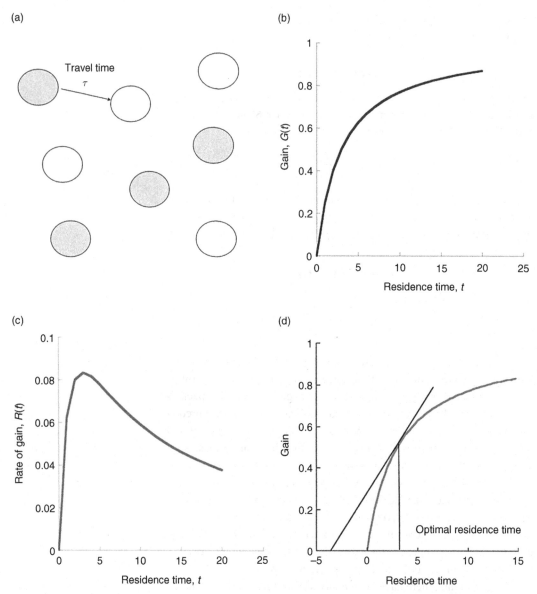

Figure 1.3. (a) A schematic of the situation for which the marginal value theorem applies. Patches of food (represented here in metaphor by filled or empty patches) are exhaustible (but there is a very large number of them) and separated by travel time τ. (b) An example of a gain curve (here I used the function $G(t) = t/(t+3)$, and (c) the resulting rate of gain of energy from this gain curve when the travel time $\tau = 3$. (d) The marginal value construction using a tangent line.

(Figure 1.3a). The question is different: the choice that the forager faces is how long to stay in the patch. We will call this the patch residence time, and denote it by t. The energetic value of food removed by the forager when the residence time is t is denoted by $G(t)$. Clearly $G(0) = 0$ (since nothing can be gained when no time is spent in the patch). Since the patch is exhaustible, $G(t)$ must plateau as t increases. Time for a pause.

Exercise 1.3 (E)

One of the biggest difficulties in this kind of work is getting intuition about functional forms of equations for use in models and learning how to pick them appropriately. Colin Clark and I talk about this a bit in our book (Clark and Mangel 2000). Two possible forms for the gain function are $G(t) = at/(b + t)$ and $G(t) = at^2/(b + t^2)$. Take some time before reading on and either sketch these functions or pick values for a and b and graph them. Think about what the differences in the shapes mean. Also note that I used the same constants (a and b) in the expressions, but they clearly must have different meanings. Think about this and remember that we will be measuring gain in energy units (e.g. kilocalories) and time in some natural unit (e.g. minutes). What does this imply for the units of a and b, in each expression?

Back to work. Suppose that the travel time between the patches is τ. The problem that the forager faces is the choice of residence in the patch – how long to stay (alternatively, should I stay or should I go now?). To predict the patch residence time, we proceed as follows.

Envision a foraging cycle that consists of arrival at a patch, residence (and foraging) for time t and then travel to the next patch, after which the process begins again. The total time associated with one feeding cycle is thus $t + \tau$ and the gain from that cycle is $G(t)$, so that the rate of gain is $R(t) = G(t)/(t + \tau)$. In Figure 1.3, I also show an example of a gain function (panel b) and the rate of gain function (panel c). Because the gain function reaches a plateau, the rate of gain has a peak. For residence times to the left of the peak, the forager is leaving too soon and for residence times to the right of the peak the forager is remaining too long to optimize the rate of gain of energy.

The question is then: how do we find the location of the peak, given the gain function and a travel time? One could, of course, recognize that $R(t)$ is a function of time, depending upon the constant τ and use calculus to find the residence time that maximizes $R(t)$, but I promised plane geometry in this warm-up. We now proceed to repeat a remarkable construction done by Eric Charnov (Charnov 1976). We begin by recognizing that $R(t)$ can be written as

$$R(t) = \frac{G(t)}{t + \tau} = \frac{G(t) - 0}{t - (-\tau)} \tag{1.6}$$

and that the right hand side can be interpreted as the slope of the line that joins the point $(t, G(t))$ on the gain curve with the point $(-\tau, 0)$ on the abscissa (x-axis). In general (Figure 1.3d), the line between $(-\tau, 0)$ and the curve will intersect the curve twice, but as the slope of the line increases the points of intersection come closer together, until they meld when the line is tangent to the curve. From this point of tangency, we can read down the optimal residence time. Charnov called this the marginal value theorem, because of analogies in economics. It allows us to predict residence times in a wide variety of situations (see the Connections at the end of this chapter for more details).

Egg size in Atlantic salmon and parent–offspring conflict (calculus)

We now come to an example of great generality – predicting the size of propagules of reproducing individuals – done in the context of a specific system, the Atlantic salmon *Salmo salar* L. (Einum and Fleming 2000). As with most but not all fish, female Atlantic salmon lay eggs and the resources they deposit in an egg will support the offspring in the initial period after hatching, as it develops the skills needed for feeding itself (Figure 1.4). In general, larger eggs will improve the chances of off-spring survival, but at a somewhat decreasing effect. We will let x denote the mass of a single egg and $S(x)$ the survival of an offspring through the critical period of time (Einum and Fleming used both 28 and 107 days with similar results) when egg mass is x. Einum and Fleming chose to model $S(x)$ by

$$S(x) = 1 - \left(\frac{x_{min}}{x}\right)^a \tag{1.7}$$

where $x_{min} = 0.0676$ g and $a = 1.5066$ are parameters fit to the data. We will define $c = (x_{min})^a$ so that $S(x) = 1 - cx^{-a}$, understanding that $S(x) = 0$ for values of x less than the minimum size. This function is shown in Figure 1.5a; it is an increasing function of egg mass, but has a decreasing slope. Even so, from the offspring perspective, larger eggs are better.

However, the perspective of the mother is different because she has a finite amount of gonads to convert into eggs (in the experiments of Einum and Fleming, the average female gonadal mass was 450 g). Given gonadal mass g, a mother who produces eggs of mass x will make g/x eggs, so that her reproductive success (defined as the expected number of eggs surviving the critical period) will be

$$R(g,x) = \frac{g}{x} S(x) = \frac{g}{x}(1 - cx^{-a}) \tag{1.8}$$

(a)

(b)

(c)

Figure 1.4. (a) Eggs, (b) a nest, and (c) a juvenile Atlantic salmon – stars of the computation of Einum and Fleming on optimal egg size. Photos complements of Ian Fleming and Neil Metcalfe.

and we can find the optimal egg size by setting the derivative of $R(g, x)$ with respect to x equal to 0 and solving for x.

Exercise 1.4 (M)

Show that the optimal egg size based on Eq. (1.8) is $x_{\mathrm{opt}} = \{c(a + 1)\}^{1/a}$ and for the values from Einum and Fleming that this is 0.1244 g. For comparison, the observed egg size in their experiments was about 0.12 g.

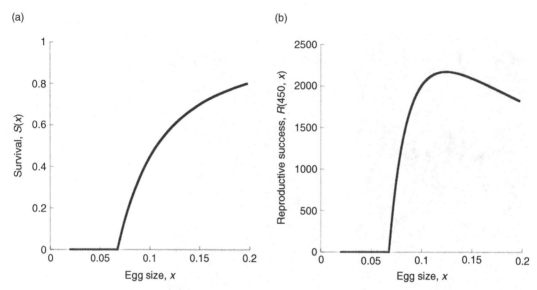

Figure 1.5. (a) Offspring survival as a function of egg mass for Atlantic salmon. (b) Female reproductive success for an individual with 450 g of gonads.

In Figure 1.5b, I show $R(450, x)$ as a function of x; we see the peak very clearly. We also see a source of parent–offspring conflict: from the perspective of the mother, an intermediate egg size is best – individual offspring have a smaller chance of survival, but she is able to make more of them. Since she is making the eggs, this is a case of parent–offspring conflict that the mother wins with certainty.

A calculation similar to this one was done by Heath *et al.* (2003), in their study of the evolution of egg size in Atlantic salmon.

Extraordinary sex ratio (more calculus)

We now turn to one of the most important contributions to evolutionary biology (and ecology) in the last half of the twentieth century; this is the thinking by W. D. Hamilton leading to understanding extraordinary sex ratios. There are two starting points. The first is the argument by R. A. Fisher that sex ratio should generally be about 50:50 (Fisher 1930): imagine a population in which the sex ratio is biased, say towards males. Then an individual carrying genes that will lead to more daughters will have higher long term representation in the population, hence bringing the sex ratio back into balance. The same argument applies if the sex ratio is biased towards females. The second starting point is the observation that in many species of insects, especially the parasitic wasps (you'll see some pictures of these animals in Chapter 4), the sex ratio is highly biased towards females, in apparent contradiction to Fisher's argument.

The parasitic wasps are wonderfully interesting animals and understanding a bit about their biology is essential to the arguments that follow. If you find this brief description interesting, there is no better place to look for more than in the marvelous book by Charles Godfray (Godfray 1994). In general, the genetic system is haplo-diploid, in which males emerge from unfertilized eggs and females emerge from fertilized eggs. Eggs are laid on or in the eggs, larvae or adults of other insects; the parasitoid eggs hatch, offspring burrow into the host if necessary, and use the host for the resources necessary to complete development. Upon completing development, offspring emerge from the wreck that was once the host, mate and fly off to seek other hosts and the process repeats itself. In general, more than one, and sometimes many females will lay their eggs at a single host. Our goal is to understand the properties of this reproductive system that lead to sex ratios that can be highly female biased.

Hamilton's approach (Hamilton 1967) gave us the idea of an "unbeatable" or non-invadable sex ratio, from which many developments in evolutionary biology flowed. The paper is republished in a book that is well worth owning (Hamilton 1995) because in addition to containing 15 classic papers in evolutionary ecology, each paper is preceded by an essay that Hamilton wrote about the paper, putting it in context.

Imagine a population that consists of $N+1$ individuals, who are identical in every way except that N of them (called "normal" individuals) make a fraction of sons r^* and one of them (called the "mutant" individual) makes a fraction of sons r. We will say that the normal sex ratio r^* is unbeatable if the best thing that the mutant can do is to adopt the same strategy herself. (This is an approximate definition of an Evolutionarily Stable Strategy (ESS), but misses a few caveats – see Connections). To find r^*, we will compute the fitness of the mutant given both r and r^*, then choose the mutant strategy appropriately.

In general, fitness is measured by the long term number of descendants (or more specifically the genes carried by them). As a proxy for fitness, we will use the number of grand offspring produced by the mutant female (grand offspring are a convenient proxy in this case because once the female oviposits and leaves a host, there is little that she can do to affect the future representation of her genes).

A female obtains grand offspring from both her daughters and her sons. We will assume that all of the daughters of the mutant female are fertilized, that her sons compete with the sons of normal females for matings, and that every female in the population makes E eggs. Then the number of daughters made by the mutant female is $E(1 - r)$ and the number of grand offspring from these daughters is $E^2(1 - r)$. Similarly, the total number of daughters at the host will be $E(1 - r) + NE(1 - r^*)$, so that the number of grand offspring from all daughters is $E^2\{1 - r + N(1 - r^*)\}$. However, the mutant female will be credited with only a fraction of those offspring, according to the fraction of her sons in the population. Since she makes Er sons and the normal individuals make NEr^* sons, the fraction of sons that belong to the mutant is $Er/(Er + NEr^*)$. Consequently, the fitness $W(r, r^*)$, depending upon the sex ratio r that the female uses and the sex ratio r^* that other females use, from both daughters and sons is

$$W(r,\ r^*) = E^2(1 - r) + E^2\{(1 - r) + N(1 - r^*)\}\left[\frac{r}{r + Nr^*}\right] \qquad (1.9)$$

The strategy r^* will be "unbeatable" (or "uninvadable") if the best sex ratio for the mutant to choose is r^*; as a function of r, $W(r, r^*)$ is maximized when $r = r^*$. We thus obtain a procedure for computing the unbeatable sex ratio: (1) take the partial derivative of $W(r, r^*)$ with respect to r; (2) set $r = r^*$ and the derivative equal to 0; and (3) solve for r^*.

Exercise 1.5 (M)

Show that the unbeatable sex ratio is $r^* = N/2(N + 1)$.

Let us interpret this equation. When $N \to \infty$, $r^* \to 1/2$; this is understandable and consistent with Fisherian sex ratios. As the population becomes increasingly large, the assumptions underlying Fisher's argument are met. How about the limit as $N \to 0$? Formally, the limit as $N \to 0$ is $r^* = 0$, but this must be biologically meaningless. When $N = 0$, the mutant female is the only one ovipositing at a host. If she makes no sons, then none of her daughters will be fertilized. How are we to interpret the result? One way is this: if she is the only ovipositing female, then she is predicted to lay enough male eggs to ensure that all of her daughters are fertilized (one son may be enough). To be sure, there are lots of biological details missing here (see Connections), but the basic explanation of extraordinary sex ratios has stood the test of time.

Two metaphors

You should be warmed up now, ready to begin the serious work. Before doing so, I want to share two metaphors about the material in this book.

Black and Decker

Black and Decker is a company that manufactures various kinds of tools. In Figure 1.6, I show some of the tools of my friend Marv Guthrie, retired Director of the Patent and Technology Licensing Office at Massachusetts General Hospital and wood-worker and sculptor. Notice that Marv has a variety of saws, pliers, hammers, screwdrivers and the like. We are to draw three conclusions from this collection. First, one tool cannot serve all needs; that is why there are a variety of saws, pliers, and screwdrivers in his collection. (Indeed, many of you probably know the saying "When the only tool you have is a hammer, everything looks like a nail".) Similarly, we need a variety of tools in ecology and evolutionary biology because one tool cannot solve all the problems that we face.

Second, if you know how to use one kind of screwdriver, then you will almost surely understand how other kinds of screwdrivers are used. Indeed, somebody could show a new kind of screwdriver to you, and you would probably be able to figure it out. Similarly, the goal in this book is not to introduce you to every tool that could be used in ecology and evolutionary biology. Rather, the point is to give you enough understanding of key tools so that you can recognize (and perhaps develop) other ones.

Third, none of us has envisioned all possible uses of any tool – but understanding how a tool is used allows us to see new ways to use it. The same is true for the material in this book: by deeply understanding some of the ways in which these tools are used, you will be able to discover new ways to use them. So, there will be places in the book where I will

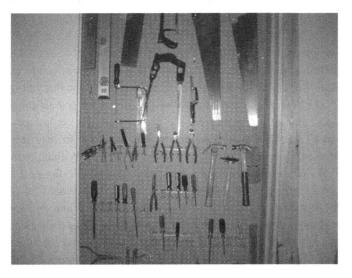

Figure 1.6. The tools of my friend Marv Guthrie; such tools are one metaphor for the material in this book.

set up a situation in which a certain tool could be used, but will not go into detail about it because we've already have sufficient exposure to that tool (sufficient, at least for this book; as with physical tools, the more you use these tools, the better you get at using them).

Fourth, a toolbox does not contain every possible tool. The same is true of this book – a variety of tools are missing. The main tools missing are game theoretical methods and partial differential equation models for structured populations. Knowing what is in here well, however, will help you master those tools when you need them.

There is one tool that I will not discuss in detail but which is equally important: what applies to mathematical methods also applies to writing, once you have used the methods to solve a problem. The famous statistician John Hammersley (Hammersley 1974), writing about the use of statistics in decision-making and about statistical professionalism says that the art of statistical advocacy "resides in one particular tool, which we have not yet mentioned and which we too often ignore in university courses on statistics. The tool is a clear prose style. It is, without any doubt, the most important tool in the statistician's toolbox" (p. 105). Hammersley offers two simple rules towards good prose style: (1) use short words, and (2) use active verbs. During much of the time that I was writing the first few drafts of this book, I read the collected short stories of John Cheever (Cheever 1978) and it occurred to me that writers of short stories face the same problems that we face when writing scientific papers: in the space of 10 or so printed papers, we need to introduce the reader to a world that he or she may not know about and make new ideas substantial to the reader. So, it is probably good to read short stories on a regular basis; the genre is less important. Cheever, I might add, is a master of using simple prose effectively, as is Victor Pritchett (Pritchett 1990a, b).

In his book *On Writing* (King 2000), Stephen King has an entire section called "Toolbox", regarding which he says "I want to suggest that to write to the best of your abilities, it behooves you to construct your own toolbox and then build up enough muscle so that you can carry it with you. Then, instead of looking at a hard job and getting discouraged, you will perhaps seize the correct tool and get immediately to work" (p. 114). King also encourages everyone to read the classic *Elements of Style* (Strunk and White 1979) by William Strunk and E. B. White (of *Charlotte's Web* and *Stuart Little* fame). I heartily concur; if you think that you ever plan to write science – or anything for that matter – you should own Strunk and White and re-read it regularly. One of my favorite authors of fiction, Elizabeth George, has a lovely small book on writing (George 2004) and emphasizes the same when she writes: "that the more you know about your tools, the better you'll be able to use them" (p. 158). She is speaking about the use of words; the concept is more general.

Almost everyone reading this book will be interested in applying mathematics to a problem in the natural world. Skorokhod *et al.* (2002) describe the difference between pure and applied mathematics as this: "This book has its roots in two different areas of mathematics: pure mathematics, where structures are discovered in the context of other mathematical structures and investigated, and applications of mathematics, where mathematical structures are suggested by real-world problems arising in science and engineering, investigated, and then used to address the motivating problem. While there are philosophical differences between applied and pure mathematical scientists, it is often difficult to sort them out." (p. v).

In order to apply mathematics, you must be engaged in the world. And this means that your writing must be of the sort that engages those who are involved in the real world. Some years ago, I co-chaired the strategic planning committee for UC Santa Cruz, sharing the job with a historian, Gail Hershatter, who is a prize winning author (Hershatter 1997). We agreed to split the writing of the first draft of the report evenly and because I had to travel, I sent my half to her before I had seen any of her writing. I did this with trepidation, having heard for so many years about C. P. Snow's two cultures (Snow 1965). Well, I discovered that Gail's writing style (like her thinking style) and mine were completely compatible. She and I talked about this at length and we agreed that there are indeed two cultures, but not those of C. P. Snow. There is the culture of good thinking and good writing, and the culture of bad thinking and bad writing. And as we all know from personal experience, they transcend disciplinary boundaries. As hard as you work on mathematical skills, you need to work on writing skills. This is only done, Stephen King notes, by reading widely and constantly (and, of course, in science we never know from where the next good idea will come – so read especially widely and attend seminars).

Mean Joe Green

The second metaphor involves Mean Joe Green. At first, one might think that I intend Mean Joe Greene, the hall of fame defensive tackle for the Pittsburgh Steelers (played 1968–1981), although he might provide an excellent metaphor too. However, I mean the great composer of opera Giuseppe Verdi (lived 1813–1901; Figure 1.7).

Opera, like the material in this book, can be appreciated at many levels. First, one may just be surrounded by the music and enjoy it, even if one does not know what is happening in the story. Or, one may know

Figure 1.7. The composer G. Verdi, who provides a second metaphor for the material in this book. This portrait is by Giovanni Boldini (1886) and is found in the Galleria Communale d'Arte Moderna in Rome. Reprinted with permission.

the story of the opera but not follow the libretto. One may sit in an easy chair, libretto open and follow the opera. Some of us enjoy participating in community opera. Others aspire to professional operatic careers. And a few of us want to be Verdi. Each of these – including the first – is a valid appreciation of opera.

The material in this book does not come easily. I expect that readers of this book will have different goals. Some will simply desire to be able to read the literature in theoretical biology (and if you stick with it, I promise that you will be able to do so by the end), whereas others will desire different levels of proficiency at research in theoretical biology. This book will deliver for you too.

Regardless of the level at which one appreciates opera, one key observation is true: you cannot say that you've been to the opera unless you have been there. In the context of quantitative methods, working through the details is the only way to be there. From the perspective of the author, it means writing a book that rarely has the phrase "it can be shown" (implying that a particular calculation is too difficult for the reader) and for the reader it means putting the time in to do the problems. All of the exercises given here have been field tested on graduate students at the University of California Santa Cruz and elsewhere. An upper division undergraduate student or a graduate student early in his or her career can master all of these exercises with perseverance – but even the problems marked E may not be easy enough to do quickly in front of the television or in a noisy café. Work through these problems, because they will help you develop intuition. As Richard Courant once noted, if we get the intuition right, the details will follow (for more about Courant, see Reid (1976)). Our goal is to build intuition about biological systems using the tools that mathematics gives to us.

The population biology of disease is one of the topics that we will cover, and Verdi provides a metaphor in another way, too. In a period of about two years, his immediate family (wife, daughter and son) were felled by infectious disease (Greenberg 2001). For more about Verdi and his wonderful music, see Holden (2001), Holoman (1992) or listen to Greenberg (2001).

How to use this book (how I think you got here)

I have written this book for anyone (upper division undergraduates, graduate students, post-docs, and even those beyond) who wants to develop the intuition and skills required for reading the literature in theoretical and mathematical biology and for doing work in this area. Mainly, however, I envision the audience to be upper division and first or second year graduate students in the biological sciences, who want to learn the right kind of mathematics for their interests. In some sense, this is the material that I would like my Ph.D. students to deeply know and understand by the middle of their graduate education. Getting the skills described in this book – like all other skills – is hard but not impossible. As I mentioned above, it requires work (doing the exercises). It also requires returning to the material again and again (so I hope that your copy of this book becomes marked up and well worn); indeed, every time I return to the material, I see it in new and deeper ways and gain new insights. Thus, I hope that colleagues who are already expert in this subject will find new ways of seeing their own problems from reading the book. Siwoff et al. (1990) begin their book with "Flip through these pages, and you'll see a book of numbers. Read it, and you'll realize that this is really a book of ideas. Our milieu is baseball. Numbers are simply our tools" (p. 3). A similar statement applies to this book: we are concerned with ideas in theoretical and mathematical biology and equations are our tools.

Motivated by the style of writing by Mike Rosenzweig in his book on diversity (Rosenzweig 1995), I have tried to make this one fun to read, or at least as much fun as a book on mathematical methods in biology can be. That's why, in part, I include pictures of organisms and biographical material.

I taught all of the material, except the chapter on fisheries, in this book as a six quarter graduate course, meeting once a week for two hours a time. I also taught the material on differential equations and disease in a one quarter formal graduate course meeting three times a week, slightly more than an hour each time; I did the same with the two chapters on stochastic population theory. The chapter on fisheries is based on a one quarter upper division/graduate class that met twice a week for about two hours.

Connections

In an effort to keep this book of manageable size, I had to forgo making it comprehensive. Much of the book is built around current or relatively current literature and questions of interest to me at the timing of writing. Indeed, once we get into the particular applications, you will be treated to a somewhat idiosyncratic collection of examples (that is, stuff which I like very much). It is up to the reader to discover ways that a particular tool may fit into his or her own research program. At the same time, I will end each chapter with a section called Connections that points towards other literature and other ways in which the material is used.

The marginal value theorem

There are probably more than one thousand papers on each of the marginal value theorem, the two prey diet choice problem, parent–offspring conflict, and extraordinary sex ratios. These ideas represent great conceptual advances and have been widely used to study a range of questions from insect oviposition behavior to mate selection; many of the papers add different aspects of biology to the models and investigate the changes in predictions. These theories also helped make behavioral ecology a premier ecological subject in which experiments and theory are linked (in large part because the scale of both theory and observation or experiment match well). At the same time, the ability to make clear and definitive predictions led to a long standing debate about theories and models (Gray 1987, Mitchell and Valone 1990), and what differences between an experimental result and a prediction mean. Some of these philosophical issues are discussed by Hilborn and Mangel (1997) and a very nice, but brief, discussion is found in the introduction of Dyson (1999). The mathematical argument used in the marginal value theorem is an example of a renewal process, since the foraging cycle "renews" itself every time. Renewal processes have a long and rich history in mathematics; Lotka (of Lotka–Volterra fame) worked on them in the context of population growth.

Unbeatable and evolutionarily stable strategies

The notion of an unbeatable strategy leads us directly to the concept of evolutionarily stable strategies and the book by John Maynard Smith (1982) is still an excellent starting point; Hofbauer and Sigmund (1998) and Frank (1998) are also good places to look. Hines (1987) is a more advanced treatment and is a monograph in its own right. In this paper, Hines also notes that differences between the prediction of a model and the observations may be revealing and informative, showing us (1) that the model is inadequate and needs to be improved, (2) the fundamental complexity of biological systems, or (3) an error in the analysis.

On writing and the creative process

In addition to Strunk and White, I suggest that you try to find Robertson Davies's slim volume called *Reading and Writing* (Davies 1992) and get your own copy (and read and re-read it) of William Zinsser's *On Writing Well* (Zinsser 2001) and *Writing to Learn* (Zinnser 1989). You might want to look at Highman (1998), which is specialized about writing for the mathematical sciences, as well. In his book, Davies notes that it is important to read widely – because if you read only the classics, how do you know that you are reading the classics? There is a wonderful, and humourous, piece by Davis and Gregerman (1995) in which this idea is formalized into the quanta of flawedness in a scientific paper (which they call phi) and the quantum of quality (nu). They suggest that all papers should be described as $X{:}Y$, where X is the quanta of phi and Y is the quanta of nu. There is some truth in this humor: whenever you read a paper (or hear a lecture) ask what are the good aspects of it, which you can adapt for your own writing or oral presentations. The interesting thing, of course, is that we all recognize quality but at the same time have difficulty describing it. This is the topic that Prisig (1974) wrestles with in *Zen in the Art of Motorcycle Maintenance*, which is another good addition to your library and is in print in both paperback and hardback editions. In his book, Stephen King also discusses the creative process, which is still a mystery to most of the world (that is – just how do we get ideas). A wonderful place to start learning about this is in the slim book by Jacques Hadamard (1954), who was a first class mathematician and worried about these issues too.

Chapter 2

Topics from ordinary and partial differential equations

We now begin the book proper, with the investigation of various topics from ordinary and partial differential equations. You will need to have calculus skills at your command, but otherwise this chapter is completely self-contained. However, things are also progressively more difficult, so you should expect to have to go through parts of the chapter a number of times. The exercises get harder too.

Predation and random search

We begin by considering mortality from the perspective of the victim. To do so, imagine an animal moving in an environment characterized by a known "rate of predation m" (cf. Lima 2002), by which I mean the following. Suppose that dt is a small increment of time; then

$$\Pr\{\text{focal individual is killed in the next } dt\} \approx m dt \qquad (2.1a)$$

We make this relationship precise by introducing the Landau order symbol $o(dt)$, which represents terms that are higher order powers of dt, in the sense that $\lim_{dt \to 0}[o(dt)/dt] = 0$. (There is also a symbol $O(dt)$, indicating terms that in the limit are proportional to dt, in the sense that $\lim_{dt \to 0}[O(dt)/dt] = A$, where A is a constant.) Then, instead of Eq. (2.1a), we write

$$\Pr\{\text{focal individual is killed in the next } dt\} = m dt + o(dt) \qquad (2.1b)$$

Imagine a long interval of time 0 to t and we ask for the probability $q(t)$ that the organism is alive at time t. The question is only interesting if the organism is alive at time 0, so we set $q(0) = 1$. To survive to time

$t + dt$, the organism must survive from 0 to t and then from t to $t + dt$. Since we multiply probabilities that are conjunctions (more on this in Chapter 3), we are led to the equation

$$q(t + dt) = q(t)(1 - mdt - o(dt)) \tag{2.2}$$

Now, here's a good tip from applied mathematical modeling. Whenever you see a function of $t + dt$ and other terms o(dt), figure out a way to divide by dt and let dt approach 0. In this particular case, we subtract $q(t)$ from both sides and divide by dt to obtain

$$\frac{q(t + dt) - q(t)}{dt} = -mq(t) - q(t)o(dt)/dt = -mq(t) + o(dt)/dt \tag{2.3}$$

since $-q(t)o(dt) = o(dt)$, and now we let dt approach 0 to obtain the differential equation $dq/dt = -mq(t)$. The solution of this equation is an exponential function and the solution that satisfies $q(0) = 1$ is $q(t) = \exp(-mt)$, also sometimes written as $q(t) = e^{-mt}$ (check these claims if you are uncertain about them). We will encounter the three fundamental properties of the exponential distribution in this section and this is the first (that the derivative of the exponential is a constant times the exponential).

Thus, we have learned that a constant rate of predation leads to exponentially declining survival. There are a number of important ideas that flow from this. First, note that when deriving Eq. (2.2), we multiplied the probabilities together. This is done when events are conjunctions, but only when the events are independent (more on this in Chapter 3 on probability ideas). Thus, in deriving Eq. (2.2), we have assumed that survival between time 0 and t and survival between t and $t + dt$ are independent of each other. This means that the focal organism does not learn anything in 0 to t that allows it to better survive and that whatever is attempting to kill it does not learn either. Hence, exponential survival is sometimes called random search.

Second, you might ask "Is the o(dt) really important?" My answer: "Boy is it." Suppose instead of Eq. (2.1) we had written Pr{focal individual is killed in the next dt} $= mdt$ (which I will not grace with an equation number since it is such a silly thing to do). Why is this silly? Well, whatever the value of dt, one can pick a value of m so that $mdt > 1$, but probabilities can never be bigger than 1. What is going on here? To understand what is happening, you must recall the Taylor expansion of the exponential distribution

$$e^x = 1 + x + \frac{x^2}{2!} + \frac{x^3}{3!} + \cdots \tag{2.4}$$

If we apply this definition to survival in a tiny bit of time $q(dt) = \exp(-mdt)$ we see that

$$e^{-mdt} = 1 - mdt + \frac{(-mdt)^2}{2!} + \frac{(-mdt)^3}{3!} + \cdots \tag{2.5}$$

This gives us the probability of surviving the next dt; the probability of being killed is 1 minus the expression in Eq. (2.5), which is exactly $mdt + o(dt)$.

Third, you might ask "how do we know the value of m?" This is another good question. In general, one will have to estimate m from various kinds of survival data. There are cases in which it is possible to compute m from operational parameters. I now describe one of them, due to B. O. Koopman, one of the founders of operations research in the United States of America (Morse and Kimball 1951; Koopman 1980). We think about the survival of the organism not from the perspective of the organism avoiding predation but from the perspective of the searcher. Let's suppose that the search process is confined to a region of area A, that the searcher moves with speed v and can detect the victim within a width W of the search path. Take the time interval $[0, t]$ and divide it into n pieces, so that each interval is length t/n. On one of these small legs the searcher covers a length vt/n and sweeps a search area Wvt/n. If the victim could be anywhere in the region, then the probability that it is detected on any particular leg is the area swept in that time interval divided by A; that is, the probability of detecting the victim on a particular leg is Wvt/nA. The probability of not detecting the victim on one of these legs is thus $1 - (Wvt/nA)$ and the probability of not detecting the victim along the entire path (which is the same as the probability that the victim survives the search) is

$$\text{Prob\{survival\}} = \left(1 - \frac{Wvt}{nA}\right)^n \tag{2.6}$$

The division of the search interval into n time steps is arbitrary, so we will let n go to infinity (thus obtaining a continuous path). Here is where another definition of the exponential function comes in handy:

$$e^x = \lim_{n \to \infty} \left(1 + \frac{x}{n}\right)^n \tag{2.7}$$

so that we see that the limit in Eq. (2.6) is $\exp(-Wvt/A)$ and this tells us that the operational definition of m is $m = Wv/A$. Note that m must be a rate, so that $1/m$ has units of time (indeed, in the next chapter we will see that it is the mean time until death); thus $1/m$ is a characteristic time of the search process.

Perhaps the most remarkable aspect of the formula for random search is that it applies in many situations in which we would not expect it to apply. My favorite example of this involves experiments that Alan Washburn, at the Naval Postgraduate School, conducted in the late 1970s and early 1980s (Washburn 1981). The Postgraduate School provides advanced training (M.S. and Ph.D. degrees) for career officers, many of whom are involved in naval search operations (submarine, surface or air). Alan set out to do an experiment in which a pursuer sought out an evader, played on computer terminals. Both individuals were confined to an square of side L, the evader moved at speed U and the purser at speed $V = 5U$ (so that the evader was approximately stationary compared to the pursuer). The search ended when the pursuer came within a distance $W/2$ of the evader. The search rate is then $m = WV/L^2$ and the mean time to detection about $1/m$.

The main results are shown in Figure 2.1. Here, Alan has plotted the experimental distribution of time to detection, the theoretical prediction based on random search and the theoretical prediction based on exhaustive search (in which the searcher moves through the region in a systematic manner, covering swaths of area until the target is detected.). The differences between panels a and b in Figure 2.1 is that in the former neither the searcher nor evader has any information about the location of the other (except for non-capture), while in the latter panel the evader is given information about the direction towards the searcher. Note how closely the data fit the exponential distribution – including (for panel a) the theoretical prediction of the mean time to detection matching the observation. Now, there is nothing "random" in the search that these highly trained officers were conducting. But when all is said and done, the effect of big brains interacting is to produce the equivalent of a random search. That is pretty cool.

Individual growth and life history invariants

We now turn to another topic of long interest and great importance in evolutionary ecology – characterizing individual growth and its implications for the evolution of life histories. We start the analysis by choosing a measure of the state of the individual. What state should we use? There are many possibilities: weight, length, fat, muscle, structural tissue, and so on – the list could be very large, depending upon the biological complexity that we want to include.

We follow an analysis first done by Ludwig von Bertalanffy; although not the earliest, his 1957 publication in *Quarterly Review of Biology* is the most accessible of his papers (from *JSTOR*, for example). We will assume that the fundamental physiological variable is mass at

Figure 2.1. (a) Experimental results of Alan Washburn for search games played by students at the Naval Postgraduate School under conditions of extremely limited information. (b) Results when the evader knows the direction of the pursuer. Reprinted with permission.

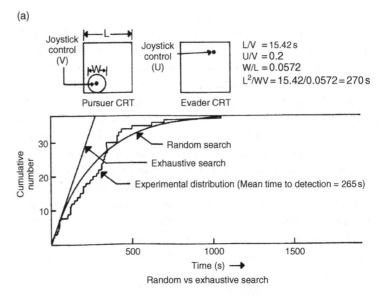

(a)

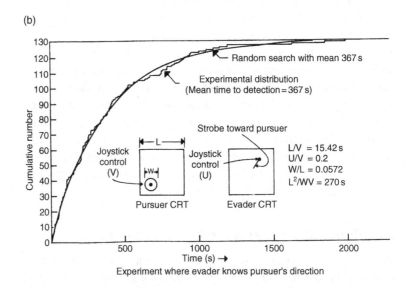

(b)

age, which we denote by $W(t)$ and assume that mass and length are related according to $W(t) = \rho L(t)^3$, where ρ is the density of the organism and the cubic relationship is important (as you will see). How valid is this assumption (i.e. of a spherical or cubical organism)? Well, there are lots of organisms that approximately fit this description if you are willing to forgo a terrestrial, mammalian bias. But bear with the analysis even if you cannot forgo this bias (and also see the nice books by John Harte (1988, 2001) for therapy).

The rate of change of mass is a balance of anabolic and catabolic factors

$$\frac{dW}{dt} = \text{anabolic factors} - \text{catabolic factors} \tag{2.8}$$

We assume that the anabolic factors scale according to surface area, because what an organism encounters in the world will depend roughly on the area in contact with the world. Thus anabolic factors $= \sigma L^2$, where σ is the appropriate scaling parameter. Let us just take a minute and think about the units of σ. Here is one example (if you don't like my choice of units, pick your own): mass has units of kg, time has units of days, so that dW/dt has units of kg/day. Length has units of cm, so that σ must have units of kg/day·cm^2.

We also assume that catabolic factors are due to metabolism, which depends on volume, which is related to mass. Thus catabolic factors $= cL^3$ and I will let you determine the units of c. Combining these we have

$$\frac{dW}{dt} = \sigma L^2 - cL^3 \tag{2.9}$$

Equation (2.9) is pretty useless because W appears on the left hand side but L appears on the right hand side. However, since we have the allometric relationship $W(t) = \rho L(t)^3$

$$\frac{dW}{dt} = 3\rho L^2 \frac{dL}{dt} \tag{2.10}$$

and if we use this equation in Eq. (2.9), we see that

$$3\rho L^2 \frac{dL}{dt} = \sigma L^2 - cL^3 \tag{2.11}$$

so that now if we divide through by $3\rho L^2$, we obtain

$$\frac{dL}{dt} = \frac{\sigma}{3\rho} - \frac{c}{3\rho} L \tag{2.12}$$

and we are now ready to combine parameters.

There are at least two ways of combining parameters here, one of which I like more than the other, which is more common. In the first, we set $q = \sigma/3\rho$ and $k = c/3\rho$, so that Eq. (2.12) simplifies to $dL/dt = q - kL$. This formulation separates the parameters characterizing costs and those characterizing gains. An alternative is to factor $c/3\rho$ from the right hand side of Eq. (2.12), define $L_\infty = \sigma/c$, which we will call asymptotic size, and obtain

$$\frac{dL}{dt} = \frac{c}{3\rho} \left(\frac{\sigma}{c} - L \right) = k(L_\infty - L) \tag{2.13}$$

This is the second form of the von Bertalanffy growth equation. Note that asymptotic size involves a combination of the parameters characterizing cost and growth.

Exercise 2.1 (E)

Check that the units of q, k and asymptotic size are correct.

Equation (2.13) is a first order linear differential equation. It requires one constant of integration for a unique solution and this we obtain by setting initial size $L(0) = L_0$. The solution can be found by at least two methods learned in introductory calculus: the method of the integrating factor or the method of separation of variables.

Exercise 2.2 (M/H)

Show that the solution of Eq. (2.13) with $L(0) = L_0$ is

$$L(t) = L_0 e^{-kt} + L_\infty (1 - e^{-kt}) \qquad (2.14)$$

In the literature you will sometimes find a different way of capturing the initial condition, which is done by writing Eq. (2.14) in terms of a new parameter t_0: $L(t) = L_\infty (1 - e^{-k(t-t_0)})$. It is important to know that these formulations are equivalent. In Figure 2.2a, I show a sample growth curve.

For many organisms, initial size is so small relative to asymptotic size that we can simply ignore initial size in our manipulations of the equations. We will do that here because it makes the analysis much

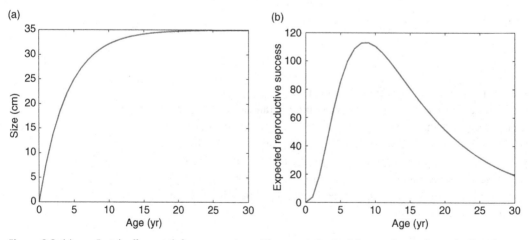

Figure 2.2. (a) von Bertalanffy growth for an organism with asymptotic size 35 cm and growth rate $k = 0.25$/yr. (b) Expected reproductive success, defined by $F(t) = e^{-mt} fL(t)^b$ as a function of age at maturity t.

simpler. Combining our study of mortality and that of individual growth takes us in interesting directions. Suppose that survival to age t is given by the exponential distribution e^{-mt}, where the mortality rate is fixed and that if the organism matures at age t, when length is $L(t)$, then lifetime reproductive output is $fL(t)^b$, where f and b are parameters.

For many fish species, the allometric parameter b is about 3 (Gunderson 1997); for other organisms one can consult Calder (1984) or Peters (1983). The parameter f relates size to offspring number (much as we did in the study of egg size in Atlantic salmon). We now define fitness as expected lifetime reproductive success, the product of surviving to age t and the reproductive success associated with age t. That is $F(t) = e^{-mt}fL(t)^b$. Since survival decreases with age and size asymptotes with age, fitness will have a peak at an intermediate age (Figure 2.2b). It is a standard application in calculus to find the optimal age at maturity.

Exercise 2.3 (M)

Show that the optimal age at maturity, t_{m}, is given by

$$t_{\mathrm{m}} = \frac{1}{k}\log\left(\frac{m+bk}{m}\right) \tag{2.15}$$

In Figure 2.3, I show optimal age at maturity as a function of k for three values of m. We can view these curves in two ways. First, let's fix

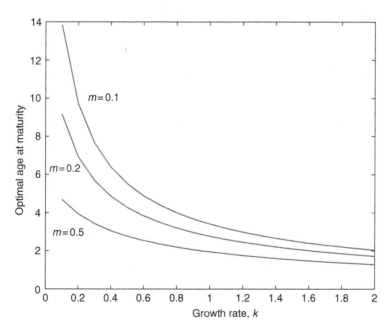

Figure 2.3. Optimal age at maturity, given by Eq. (2.15), as a function of growth rate k, for three values of mortality rate m.

Figure 2.4. Comparison of predicted (by Eq. (2.15)) and inferred age at maturity for different species of *Tilapia*, shown as an inset, and the 1:1 line. Data from Lorenzen (2000).

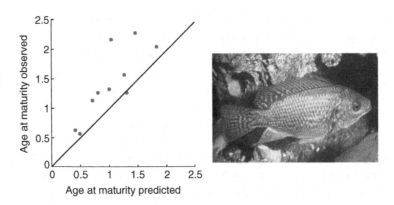

the choice of m and follow one of the curves. The theory then predicts that as growth rate increases, age at maturity declines. If we fix growth rate and take a vertical slice along these three curves, the prediction is that age at maturity declines as mortality decreases. Each of these predictions should make intuitive sense and you should try to work them out for yourself if you are unclear about them. An example of the level of quantitative accuracy of this simple theory is given in Figure 2.4, in which I shown predicted (by Eq. (2.15)) and observed age at maturity for about a dozen species of Tilapia (data from Lorenzen 2000). Fish, like people, mature at different ages, so that when we discuss observed age at maturity, it is really a population concept and the general agreement among fishery scientists is that the age at maturity in a stock is the age at which half of the individuals are mature. Also shown in Figure 2.4 is the 1:1 line; if the theory and data agreed completely, all the points would be on this line. We see, in fact, that not only do the points fall off the line, but there is a slight bias in that when there is a deviation the observed age at maturity is more likely to be greater than the predicted value than less than the predicted value. Once again, we have the thorny issue of the meaning of deviation between a theoretical prediction and an observation (this problem will not go away, not in this book, and not in science). Here, I would offer the following points. First, the agreement, given the relative simplicity of the theory, is pretty remarkable. Second, what alternative theory do we have for predicting age at maturity? That is, if we consider that science consists of different hypotheses competing and arbitrated by the data (Hilborn and Mangel 1997) it makes little practical sense to reject an idea for poor performance when we have no alternative.

Note that both m and k appear in Eq. (2.15) and that there is no way to simplify it. Something remarkable happens, however, when we compute the length at maturity $L(t_m)$, as you should do now.

Exercise 2.4 (E/M)

Show that size at maturity is given by

$$L(t_m) = L_\infty \left(\frac{bk}{m + bk} \right) = L_\infty \left(\frac{b}{b + \frac{m}{k}} \right) \tag{2.16}$$

If you were slick on the way to Eq. (2.15), you actually discovered this before you computed the value of t_m. This equation is remarkable, and the beginning of an enormous amount of evolutionary ecology and here is why. Notice that $L(t_m)/L_\infty$ is the relative size at maturity. Equations (2.15) and (2.16) tell us that although the optimal age at maturity depends upon k and m separately, the relative size at maturity only depends upon their ratio. This is an example of a life history invariant: regardless of the particular values of k and m for different stocks, if their ratio is the same, we predict the same relative size at maturity. This idea is due to the famous fishery scientist Ray Beverton (Figure 2.5) and has been rediscovered many times. Note too that since

$$L(t_m) = L_\infty (1 - \exp(-kt_m)) = L_\infty \left(1 - \exp\left(-\frac{k}{m} m t_m \right) \right)$$

we conclude that if relative size at maturity for two species is the same, then since m/k will be the same (by Eq. (2.16)) that $m t_m$ must be the same.

All of our analysis until this point has been built on the underlying dynamics in Eq. (2.9), in which we assume that gain scales according to area, or according to $W^{2/3}$. For many years, this actually created a problem because whenever experimental measurements were made, the scaling exponent was closer to 3/4 than 2/3. In a series of remarkable papers in the late 1990s, Jim Brown, Ric Charnoff, Brian Enquist, Geoff West, and other colleagues, showed how the 3/4 exponent could be derived by application of scaling laws and fractal analysis. Some representative papers are West *et al.* (1997), Enquist *et al.* (1999), and West *et al.* (2001). They show that it is possible to derive a growth model of the form $dW/dt = aW^{3/4} - bW$ from first principles.

Exercise 2.5 (E)

In the growth equation $dW/dt = aW^{3/4} - bW$, set $W = H^n$, where n is to be determined. Find the equation that $H(t)$ satisfies. What value of n makes it especially simple to solve by putting it into a form similar to the von Bertalanffy equation for length? (See Connections for even more general growth and allometry models.)

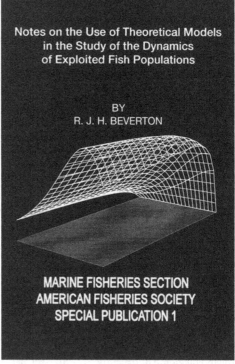

Figure 2.5. Ray Beverton as a young man, delivering his famous lectures that began post-WWII quantitative fishery science, and at the time of his retirement. Photos courtesy of Kathy Beverton.

Population growth in fluctuating environments and measures of fitness

We now come to one of the most misunderstood topics in evolutionary ecology, although Danny Cohen and Richard Lewontin set it straight many years ago (Cohen 1966, Lewontin and Cohen 1969). I include it here because at my university in fall 2002, there was an exchange at a seminar between a member of the audience and the speaker which showed that neither of them understood either the simplicity or the depth of these ideas.

This section will begin in a deceptively simple way, but by the end we will reach deep and sophisticated concepts. So, to begin imagine a population without age structure for which $N(t)$ is population size in year t and $N(0)$ is known exactly. If the per capita growth rate is λ, then the population dynamics are

$$N(t+1) = \lambda N(t) \tag{2.17}$$

from which we conclude, of course, that $N(t) = \lambda^t N(0)$. If the per capita growth rate is less than 1, the population declines, if it is exactly equal to 1 the population is stable, and if it is greater than 1 the population grows. Now let us suppose that the per capita rate of growth varies, first in space and then in time. Because there is no density dependence, the per capita growth rate can also be used as a measure of fitness.

Spatial variation

Suppose that in every year, the environment consists of two kinds of habitats. In the poor habitat the per capita growth rate is λ_1 and in the better habitat it is λ_2. We assume that the fraction of total habitat that is poor is p, so that the fraction of habitat that is good is $1 - p$. Finally, we will assume that the population is uniformly distributed across the entire habitat. At this point, I am sure that you want to raise various objections such as "What if p varies from year to year?", "What if individuals can move from poorer to better locations", etc. To these objections, I simply ask for your patience.

Given these assumptions, in year t the number of individuals experiencing the poor habitat will be $pN(t)$ and the number of individuals experiencing the better habitat will be $(1 - p)N(t)$. Consequently, the population size next year is

$$N(t+1) = (\lambda_1 pN(t) + \lambda_2(1-p)N(t)) = \{p\lambda_1 + (1-p)\lambda_2\}N(t) \tag{2.18}$$

The quantity in curly brackets on the right hand side of this equation is an average. It is the standard kind of average that we are all used to (think about how your grade point average or a batting average is calculated). If we had n different habitat qualities, instead of just two habitat qualities, and let p_i denote the fraction of habitat in which the growth rate is λ_i, then it is clear that what goes in the { } on the right hand side of Eq. (2.18) will be $\sum_{i=1}^{n} p_i \lambda_i$. We call this the arithmetic average. (I am tempted to put "arithmetic average" into bold-face or italics, but Strunk and White (1979) tell me that if I need to do so – to remind you that it is important – then I have not done my job.) Our conclusion thus far: if variation occurs over space, then the arithmetic average is the appropriate description of the growth rate.

Temporal variation

Let us now assume that per capita growth rate varies over time rather than space. That is, with probability p every individual in the population experiences the poorer growth rate in a particular year and with probability $1 - p$ every individual experiences the better growth rate. Let us suppose that t is very big; it will be composed of t_1 years in which the growth rate was poorer and t_2 years in which the growth rate was better. Since there is no density dependence in this model, it does not matter in what order the years happen and we write

$$N(t) = (\lambda_1)^{t_1} (\lambda_2)^{t_2} N(0) \qquad (2.19)$$

If the total time is large, then t_1 and t_2 should be roughly representative of the fraction of years that are poorer or better respectively. That is, we should expect $t_1 \sim pt$ and $t_2 \sim (1 - p)t$. How should you interpret the symbol $\sim$ in the previous sentence? If you are more mathematically inclined, then the law of large numbers allows us to give precise interpretation of what $\sim$ means. If you are less mathematically inclined, this is a case where you can count on your intuition and the world being approximately fair.

Adopting this idea about the good and bad years, Eq. (2.19) becomes

$$N(t) = \lambda_1^{pt} \lambda_2^{(1-p)t} N(0) = \left[\lambda_1^p \lambda_2^{1-p} \right]^t N(0) \qquad (2.20)$$

The quantity in square brackets on the right hand side of this equation is a different kind of average. It is called the geometric mean (or geometric average) and it weights the good and bad years differently than the arithmetic average does. Perhaps the easiest way to see the differences is to think about the extreme case in which the poorer growth rate is 0. According to the arithmetic average, individuals who find themselves in the better habitat will contribute to next year's population and those who find themselves in the poorer habitat will not. On the other hand, if the fluctuations are temporal, then when a poor year occurs, there is no reproduction for the population as a whole and thus the population is gone.

Exercise 2.6 (E/M)

Suppose that λ_1 is less than 1 (so that in poor years, the population declines). Show that the condition for the population to increase using the geometric mean is that $\lambda_2 > \lambda_1^{-p/(1-p)}$. Explore this relationship as λ_1 and p vary by making appropriate graphs. (Do not use three dimensional graphs and recall the advice of the Ecological Detective (Hilborn and Mangel 1997) that you should expect to make 10 times as many graphs for yourself as you would ever show to others.) Compare the results with the corresponding expression making the arithmetic average greater than 1.

If instead of just two kinds of years, we allow n kinds of years, the extension of the square brackets in Eq. (2.20) will be $\prod_{i=1}^{n} \lambda_i^{p_i}$ where the $\prod$ denotes a product (much as $\sum$ denotes a sum, as used above).

Now let us return to Eq. (2.17) for which $N(t) = \lambda^t N(0)$ and recall that the exponential and logarithm are inverse functions, $\lambda = \exp(\log(\lambda))$, which allows us to write $N(t)$ in a different way. In particular we have $N(t) = e^{[\log(\lambda)]t} N(0)$, and if we define $r = \log(\lambda)$, then we have come back to our old friend from introductory ecology $N(t) = e^{rt} N(0)$. That is, if time were continuous, this looks like population growth satisfying $dN/dt = rN$, in which r is the growth rate. But we can actually learn some new things about fluctuating environments from this old friend, because we know that $r = \log(\lambda)$. In Figure 2.6a, I have plotted growth rate as a function of λ and I have shown two particular values of λ that might correspond to good years and poor years. Note that the line segment joining these two points falls below the curve (such a curve is called concave). This means that the growth rate at the arithmetic average of λ is larger than the average value of the growth rates. This phenomenon is called Jensen's inequality.

If we have more than two growth rates, then the expression in square brackets in Eq. (2.20) is replaced by $\prod_{i=1}^{n} \lambda_i^{p_i}$ and if we rewrite this in terms of logarithms we see that

$$N(t) = \exp\left[t \sum_{i=1}^{n} p_i \log(\lambda_i) \right] N(0) \tag{2.21}$$

From this equation, we conclude that the growth rate in a fluctuating environment is $r = \sum_{i=1}^{n} p_i \log(\lambda_i)$, which is the arithmetic average of the logarithm of the per capita growth rates. We thus conclude that for a fluctuating environment, one either applies the geometric mean directly

Figure 2.6. (a) The function $r = \log(\lambda)$ is concave. This implies that fluctuating environments will have lower growth rates than the growth rate associated with the average value of λ. (b) The two color morphs of desert snow *Linanthus parryae* are maintained by fitness differences in fluctuating environments. (c) An example of why this plant is called desert snow. Photos courtesy of Paulette Bierzychudek.

to the per capita growth rates or the arithmetic mean to the logarithm of per capita growth rates.

What about measuring the growth rate of an actual population? Data in a situation such as this one would be population sizes over time $N(0)$, $N(1), \ldots N(t)$ from which we could compute the per capita growth rate as the ratio of population size at two successive years. We would then replace the frequency average by a time average and estimate the growth rate according to

$$r \approx \frac{1}{t}[\log(\lambda(0)) + \log(\lambda(1)) + \cdots + \log(\lambda(t-1))] \qquad (2.22)$$

with the understanding that t is large. Since the sum of logarithms is the logarithm of the product, the term in square brackets in Eq. (2.22) is the same as $\log(\lambda(0)\lambda(1)\lambda(2) \ldots \lambda(t-1))$. But $\lambda(s) = N(s+1)/N(s)$, so that when we evaluate the product of the per capita growth rates, the product is

$$
\begin{aligned}
\log(\lambda(0)\lambda(1)\lambda(2)\ldots\lambda(t-1)) \\
= \log\{(N(1)/N(0))(N(2)/N(1))\ldots(N(t)/N(t-1))\} = \log\{N(t)/N(0)\}
\end{aligned}
$$

However, in a fluctuating environment, the sequence of per capita rates (and thus population sizes) is itself random. Thus, Eq. (2.22) provides the value of r for a specific sequence of population sizes. To allow for others, we take the arithmetic average of Eq. (2.22) and write

$$
r = \lim_{t\to\infty} \frac{1}{t} E\left\{\log\left(\frac{N(t)}{N(0)}\right)\right\} \tag{2.23}
$$

This formula is useful when dealing with data and when using simulation models (for a nice example, see Easterling and Ellner (2000)). A wonderful application of all of these ideas is found in Turelli $et\ al.$ (2001), which deals with the maintenance of color polymorphism in desert snow $Linanthus\ parryae$, a plant (Figure 2.6b, c) that plays an important role in the history of evolutionary biology (Schemske and Bierzychudek 2001). If you stop reading this book now, and choose to read the papers, you will also encounter the "diffusion approximation." We will briefly discuss diffusion approximations in this chapter and then go into them in great detail in the later chapters on stochastic population theory.

Before leaving this section, I want to do one more calculation. It involves a little bit of probability modeling, so you may want to hold off until you've been through the next chapter. Suppose that we do not know the probability distribution of the per capita growth rate, but we do know the mean and variance of λ, which I shall denote by $\bar{\lambda}$ and $\mathrm{Var}(\lambda)$. We begin by a Taylor expansion of $r = \log(\lambda)$ around its mean value, keeping up to second order terms:

$$
\log(\lambda) = \log(\bar{\lambda}) + \frac{1}{\bar{\lambda}}(\lambda - \bar{\lambda}) - \frac{1}{\bar{\lambda}^2}(\lambda - \bar{\lambda})^2 \tag{2.24}
$$

and we now take the expectation of the right hand side. The first term is a constant, so does not change, the second term vanishes because $E\{\lambda\} = \bar{\lambda}$ and the expectation of the quantity in round brackets in the last term is the variance of the per capita growth rate. We thus conclude

$$
r \sim \log(\bar{\lambda}) - \frac{1}{\bar{\lambda}^2}\mathrm{Var}(\lambda) \tag{2.25}
$$

This is a very useful expression for fitness or growth rate in a fluctuating environment. The method is often called Seber's delta method, for G. A. F. Seber who popularized the idea in ecology (Seber 1982). I first learned about it while working in the Operations Evaluation Group of the Center for Naval Analyses (Mangel 1982), so I tend to call it the "method of Navy math." Whatever you call it, the method is handy.

The logistic equation and the discrete logistic map – on the edge of chaos

It is likely true that every reader of this book – and especially any reader who has reached this point – has encountered the logistic equation previously. Even so, by returning to an old friend, we have a good starting point for new kinds of explorations. As in the previous section, we will begin with relatively simple material but end with remarkably sophisticated stuff.

The logistic equation

We allow $N(t)$ to represent population size at time t and assume that it changes according to the dynamics

$$\frac{dN}{dt} = rN\left(1 - \frac{N}{K}\right) \tag{2.26}$$

In this equation, r and K are parameters; K is the population size at which the growth rate of the population is 0. It is commonly called the carrying capacity of the population. When the growth rate is 0, births and deaths are still occurring, but they are exactly balancing each other. The right hand side of Eq. (2.26) is a parabola, with zeros at $N=0$ and $N=K$ and maximum value $rK/4$ when $N=K/2$, which is called the population size that provides maximum net productitivity (MNP); see Figure 2.7a.

In order to understand the parameter r, it is easiest to consider the per capita growth rate of the population

$$\frac{1}{N}\frac{dN}{dt} = r\left(1 - \frac{N}{K}\right) \tag{2.27}$$

Inspection of the right hand side of Eq. (2.27) shows that it is a decreasing function of population size and that its maximum value is r, occurring when $N=0$. Of course, if $N=0$, this is biologically meaningless – there won't be any reproduction if the population size is 0. What we mean, more precisely, is that in the limit of small population size, the per capita growth rate approaches r – so that r is the maximum per capita growth rate.

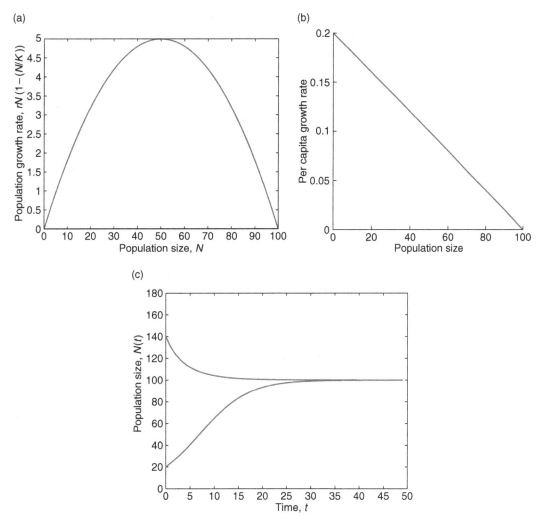

Figure 2.7. An illustration of logistic dynamics when $r = 0.2$ and $K = 100$. (a) Population growth rate as a function of population size. (b) Per capita growth rate as a function of population size. (c) Population size versus time for populations that start above and below the carrying capacity.

The word logistic is derived from the French word logistique, which means to compute. The scientist and mathematician Verhulst wanted to be able to compute the population trajectory of France. He knew that using the exponential growth equation $dN/dt = rN$ would not work because the population grows without bound. This happens because with exponential growth the per capita growth rate is a constant (r). We don't know what Verhulst was thinking, but it might have gone something like this: "I know that a constant per capita growth rate will not be a good representation, and it must be true that per capita growth rate declines as population size increases. Suppose that per capita growth rate falls to zero when the population size is K. What is the simplest way to connect the points $(0, r)$ and $(K, 0)$? Of course – a line. C'est bon." Furthermore, there is only one line that connects the maximum per capita growth rate r when $N = 0$ and per capita growth rate $= 0$ when $N = K$. There are an infinite number of nonlinear ways that we could do it. For example, a per capita growth rate of the form $r(1 - (N/K)^{\alpha})$, for any value of $\alpha > 0$, works equally well to achieve the goal of connecting the maximum and zero per capita growth rates. So, the logistic is not a law of nature, but is a simple and somewhat unique representation of nature. In Figure 2.7b, I show the per capita growth rate for the same parameters as in Figure 2.7a.

Let us now think about the dynamics of a population starting at size $N(0)$ and following logistic growth. If $N(0) > K$, then the growth rate of the population is negative and the population will decline towards K. If $N(0) > 0$ but small, the population will grow, albeit slowly at first, but then as population size increases, the growth rate increases too (even though per capita growth rate is always declining, the product of per capita growth rate and population size increases until $N = K/2$). Once the population size exceeds $K/2$, growth rate begins to slow, ultimately reaching 0 as the population approaches K. We thus expect the picture of population size versus time to be S-shaped or sigmoidal and it is (Figure 2.7c).

Exercise 2.7 (M)

Although Eq. (2.26) is a nonlinear equation, it can be solved exactly (that is how I generated the trajectories in Figure 2.7c) and everyone should do it at least once in his or her career. The exercise is to show that the solution of Eq. (2.26) is $N(t) = [N(0)Ke^{rt}] / [K + N(0)(e^{rt} - 1)]$. To help you along, I offer two hints (the method of partial fractions, if you want to check your calculus text). First, separate the differential equation so that Eq. (2.26) becomes

$$\frac{dN}{N\left(1 - \frac{N}{K}\right)} = r dt$$

Second, recognize that the left hand side of this expression looks like a common denominator, so write

$$\frac{1}{N\left(1 - \frac{N}{K}\right)} = \frac{A}{N} + \frac{B}{\left(1 - \frac{N}{K}\right)}$$

where A and B are constants that you determine by creating the common denominator and simplifying.

The discrete logistic map and the edge of chaos

We now come to what must be one of the most remarkable stories of good luck and good sleuthing in science. To begin this story,

I encourage you to stop reading just now, go to a computer and plot the trajectories for $N(t)$ given by the formula for $N(t)$ in the previous exercise, for a variety of values of r – let r range from 0.4 to about 3.5. After that return to this reading.

Now let us poke around a bit with the logistic equation by recognizing the definition of the derivative as a limiting process. Thus, we could rewrite the logistic equation in the following form:

$$\lim_{dt \to 0} \frac{N(t + dt) - N(t)}{dt} = rN \left(1 - \frac{N}{K} \right) \tag{2.28}$$

This equation, of course, is no different from our starting point. But now let us ignore the limiting process in Eq. (2.28) and simply set $dt = 1$. If we do that Eq. (2.28) becomes a difference equation, which we can write in the form

$$N(t + 1) = N(t) + rN(t) \left(1 - \frac{N(t)}{K} \right) \tag{2.29}$$

This equation is called the logistic map, because it "maps" population size at one time to population size at another time. You may also see it written in the form

$$N(t + 1) = rN(t) \left(1 - \frac{N(t)}{K} \right)$$

which makes it harder to connect to the original differential equation. Note, of course, that Eq. (2.29) is a perfectly good starting point, if we think that the biology operates in discrete time (e.g. insect populations with non-overlapping generations across seasons, or many species of fish in temperate or colder waters).

Although Eq. (2.29) looks like the logistic differential equation, it has a number of properties that are sufficiently different to make us wonder about it. To begin, note that if $N(t) > K$ then the growth term is negative and if r is sufficiently large, not only could $N(t + 1)$ be less than $N(t)$, but it could be negative! One way around this is to use a slightly different form called the Ricker map

$$N(t + 1) = N(t) \exp \left[r \left(1 - \frac{N(t)}{K} \right) \right] \tag{2.30}$$

This equation is commonly used in fishery science for populations with non-overlapping generations (e.g. salmonids) and misused for other kinds of populations. It has a nice intuitive derivation, which goes like this (and to which we will return in Chapter 6). Suppose that maximum per capita reproduction is A, so that in the absence of density dependence $N(t + 1) = AN(t)$, and that density dependence acts in the sense that a focal offspring has probability f of surviving when there is just one adult present. If there are N adults present, the probability that the focal offspring will survive is f^N. Combining these, we obtain $N(t+1) = AN(t)f^{N(t)}$, which surely suggests a good exercise.

Exercise 2.8 (E/M)

Often we set $f^N = e^{-bN}$, so that the Ricker map becomes $N(t+1) = AN(t)e^{-bN(t)}$. First, explain the connection between f and B and the relationship between the parameters A, b and r, K. Second, explain why the Ricker map does not have the nasty property that $N(t)$ can be less than 0. Third, use the Taylor expansion of the exponential function to show how the Ricker and discrete logistic maps are connected.

But now let us return to Eq. (2.29) and explore it. To do this, we begin by simply looking at trajectories. I am going to set $K = 100$, $N(0) = 20$ and show $N(t)$ for a number of different values of r (Figure 2.8). When r is moderate, things behave as we expect: starting at $N(0) = 20$, the population rises gradually towards $K = 100$. However, when $r = 2.0$ (Figure 2.8c), something funny appears to be happening. Instead of settling down nicely at $K = 100$, the population exhibits small oscillations around that value. For r slightly larger ($r = 2.3$, panel d) the oscillations become more pronounced, but still seem to be flipping back and forth across $K = 100$. The behavior becomes even more complicated when r gets larger – now there are multiple population sizes that are consistently visited (Figure 2.8e). When r gets even larger, there appears to be no pattern, just wild and erratic behavior. This behavior is called deterministic chaos. It was discovered more or less accidentally in a number of different ways in the 1960s and 1970s (see Connections).

Before explaining what is happening, I want to present the results in a different way, obtained using the following procedure. I fixed r. However, instead of fixing $N(0)$, I picked it randomly and uniformly (all values equally likely) between 1 and K. I then ran the population dynamics for 500 time steps and plotted the point $(r, N(500))$. I repeated this, with r still fixed, for 50 different starting values, then changed r and began the process over again. The results, called a bifurcation (for branching) diagram, are shown in Figure 2.9. When r is small, there is only one place for $N(500)$ to be – at carrying capacity $K = 100$. However, once we enter the oscillatory regime, $N(500)$ is never K – it is either larger or smaller than K. And as r increases, we see that we jump from 2 values of $N(500)$ to 4 values, then on to 8, 16, 32 and so forth (with the transition regions becoming closer and closer). As r continues to increase, virtually all values can be taken by $N(500)$. You may want to stop reading now, go to your computer and create a spreadsheet that does this same set of calculations.

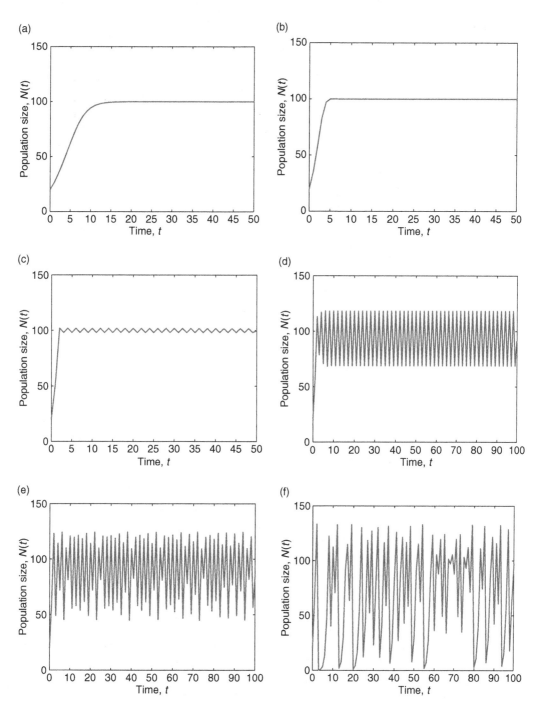

Figure 2.8. Dynamics of the discrete logistic, for varying values of r: (a) $r=0.4$, (b) $r=1.0$, (c) $r=2.0$, (d) $r=2.3$, (e) $r=2.6$, (f) $r=3$.

Figure 2.9. The bifurcation plot of $N(500)$ versus r; see text for details.

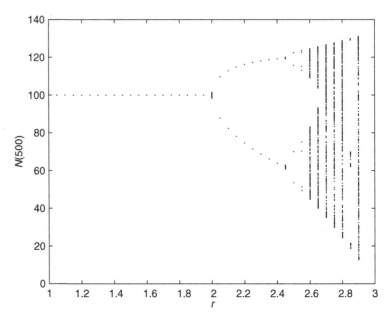

How do we understand what is happening? To begin we rewrite Eq. (2.29) as

$$N(t+1) = (1+r)N(t) - \frac{rN(t)^2}{K}$$

and investigate this as a map relating $N(t+1)$ to $N(t)$. Clearly if $N(t)=0$, then $N(t+1)=0$; also if $N(t)=K(1+r)/r$, then $N(t+1)=N(t)$. In Figure 2.10, I have plotted this function, for three values of r, when $K=100$. I have also plotted the 1:1 line. The three curves and the line intersect at the point $(100, 100)$, or more generally at the point (K, K). Using this figure, we can read off how the population dynamics grow. Let us suppose that $N(0)=50$, and $r=0.4$. We can see then that $N(1)=60$ (by reading where the line $N=50$ intersects the curve). We then go back to the x-axis, for $N(1)=60$, we see that $N(2)=69.6$; we then go back to the x-axis for $N(2)$ and obtain $N(3)$. In this case, it is clear that the dynamics will be squeezed into the small region between the curve and the 1:1 line. This procedure is called cob-webbing.

What happens if $N(0)=50$ and $r=2.3$? Well, then $N(1)=107.5$, but if we take that value back to the x-axis, we see that $N(2)$ is about 89. We have jumped right across the steady state at 100. From $N(2)=89$, we will go to $N(3)$ about 111 and from there to $N(4)$ about 82. The behavior is even more extreme for the case in which $r=3$: starting at $N(0)=50$, we go to 125 and from there to about 31; from 31 to about 95, and so forth.

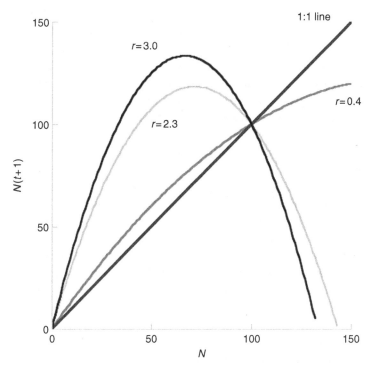

Figure 2.10. Logistic maps for three different values of r, allowing us to understand how simple deterministic dynamics can lead to oscillations and to apparently random trajectories.

This is a very interesting process – one in which simple deterministic dynamics can produce a wide range of behaviors, including oscillations and apparently random trajectories. These kinds of results fall under the general rubric of deterministic chaos (see Connections).

A bit about bifurcations

The results of the previous section suggest that when we encounter a differential or difference equation, we should consider not only the solution, but how the solution depends upon the parameters of the equation. This subject is generally called bifurcation theory (because, as we will see, solutions "branch" as parameters vary). In this section, we will consider the two simplest bifurcations and some of their implications. As we discuss the material, do not try to apply biological interpretations to the equations; I have picked them to make illustrating the main points as simple as possible. At the end of this section, I will do one biological example and in Connections point you towards the literature for other ones.

We begin with the differential equation

$$\frac{dx}{dt} = x^2 - \alpha \tag{2.31}$$

for the variable $x(t)$ depending upon the single parameter α. When we first encounter a differential equation, we may ask "What is the solution of this equation?". The trouble is, the vast majority of differential equations do not have explicit solutions. Given that restriction, a good first question is "What are the steady states, that is for what values of x is dx/dt equal to 0?". This is always a good question, and can often be answered. For the dynamics in Eq. (2.31), the steady states are given by $x_s = \pm\sqrt{\alpha}$. We thus conclude that if $\alpha < 0$ there are no steady states (more precisely, there are no real steady states) and that if $\alpha \geq 0$ there are one (when $\alpha = 0$) or two steady states. We will call these steady solutions branches; there are thus two branches, one of which is positive and one of which is negative. Along these branches, $dx/dt = 0$. What about elsewhere in the plane? Between the branches, α is greater than x^2, so we conclude that $dx/dt < 0$ and that $x(t)$ will decrease, thus moving towards the lower branch. Anywhere else in the plane α is less than x^2, so that $dx/dt > 0$ and $x(t)$ will increase; I have summarized this analysis in Figure 2.11.

Before going on with the analysis, a few stylistic comments. First, note that I have put x on the ordinate and α on the abscissa. Thus, one might say "x is on the y axis, how confusing." However, the labeling of axes is a convention, not a rule, and one just needs to be careful when conducting the analysis (more of this to come with the next bifurcation). Second, I have used $x(t)$ and x interchangeably; this is done for convenience (and for avoiding writing things in a more cumbersome manner). Once again, this is not a problem if one is careful in understanding and presentation.

Returning to the figure, imagine that α is fixed, but x may vary, and that we are at some point along the positive branch. Then $dx/dt = 0$ and

Figure 2.11. The steady states of the differential equation $dx/dt = x^2 - \alpha$, showing the positive and negative branches.

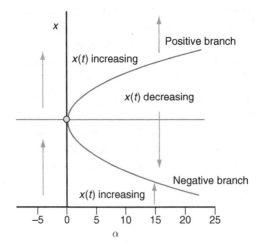

we will stay there forever. However, if we receive a small perturbation off that branch, interesting things happen. If the perturbation (until otherwise notified, all perturbations are small) puts us between the two branches, then $x(t)$ declines and we move towards the negative branch. If the perturbation puts us above the positive branch, then $x(t)$ increases and we move away from the positive branch. So, in either case, a perturbation moves us away from the positive branch. We say that such a branch is dynamically unstable (or just unstable). A similar argument shows that perturbations from the negative branch return to it; we say that the negative branch is stable. What happens when $\alpha = 0$? The differential equation becomes $dx/dt = x^2$, so that $x(t)$ is always increasing. Thus, if $x(0) < 0$, $x(t)$ rises towards 0; however if $x(0) > 0$, $x(t)$ moves away from 0. We say that such a point is marginally stable; we also say that the equation $dx/dt = x^2 - \alpha$ is structurally unstable (these words may appear to be needlessly complex, but think about them and they make sense) when $\alpha = 0$, because small changes of α from the value 0 lead to very different properties of the equation (in this case, either no steady states or two steady states). We also sometimes say that the stable steady state and unstable steady state coalesce and annihilate each other (kind of like matter and antimatter) when $\alpha = 0$.

The next most complicated equation involves two parameters and a cubic in x:

$$\frac{dx}{dt} = -x^3 + \alpha x + \beta \qquad (2.32)$$

where α and β are the parameters of interest. The steady states of this equation satisfy the cubic equation $x^3 - \alpha x - \beta = 0$. We will momentarily discuss geometric solutions of this equation, but now begin with a bit of algebra. A cubic equation has three solutions (by the fundamental theorem of algebra), of which one may be real and two complex, three may be real with two equal, or three may be real and unequal. Which case applies is determined by the value of the discriminant $D(\alpha, \beta) = (\beta^2/4) - (\alpha^3/27)$. (You probably once learned this in high school algebra, but most likely don't remember it. This is a case where I ask that you trust me; of course you can also go and check the formula in a book.) If $D(\alpha, \beta) > 0$, then there is one real solution; if $D(\alpha, \beta) = 0$, then there are three real solutions, two of which are equal; if $D < (\alpha, \beta)$ then there are three real, unequal solutions. Thus, in some sense $D(\alpha, \beta) = 0$ is a boundary. So, we need to think about the shape of $\beta^2 = 4\alpha^3/27$, which is shown in Figure 2.12. This kind of equation (in which the independent variable appears as a 3/2 power) is called a cusp; hence this is called the cusp bifurcation or sometimes the cusp catastrophe (see Connections).

Figure 2.12. A plot of the equation $\beta^2 = 4\alpha^3/27$, which is called a cusp. Along the curves, there are two real solutions of the cubic (and thus three steady states of Eq. (2.32)). Elsewhere, there are either one real solution or three real solutions.

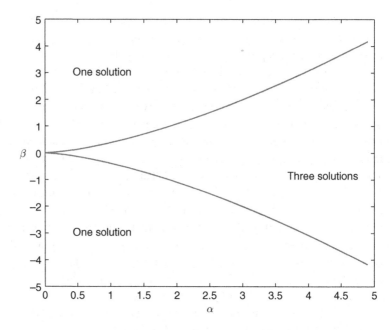

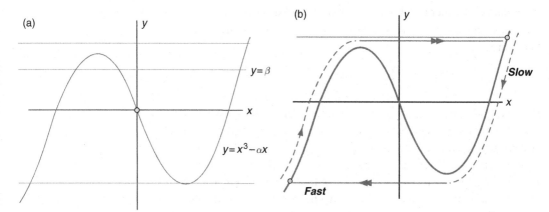

Figure 2.13. (a) The geometric solution of the equation $x^3 - \alpha x = \beta$. (b) When we append dynamics for β, there is no longer a steady state, but both x and β change in time.

Now I want to discuss the solution in a more geometric manner, because learning to think geometrically about these matters is absolutely essential for your understanding of the material. The steady states of the differential equation (2.32) satisfy $x^3 - \alpha x = \beta$. In Figure 2.13a, I plotted the curve $y = x^3 - \alpha x$ and the line (actually a number of lines) $y = \beta$. Since the steady states correspond to values of x where these are equal, we conclude that the steady states are values of x for which the line and the curve intersect. We also see that there may be just one intersection point (on the left hand branch of the curve or on the right hand branch), there may be two intersection points (if the line is tangent to the curve) or three intersection points (if the value of β falls between the local maximum and local minimum of the curve). We thus have a geometric interpretation of the cusp. When the horizontal line is tangent to the curve, the system is once again structurally unstable: at the point of tangency there are two steady states, one of which is marginally stable. However, a small change in either of the parameters leads to a situation in which there are either three or one steady states.

But this really is not the situation that I wanted to consider. Rather, I want to consider the situation in which β varies as well. In particular, let us append the equation $d\beta/dt = -\varepsilon x$, in which ε is a new parameter, to Eq. (2.32). We will assume that ε is small (that is much less than 1), and we know that when ε is set equal to 0 we obtain the cusp bifurcation.

The steady state is now $x = 0$, $\beta = 0$, but the dynamics are very interesting. To be explicit, suppose we start on the right hand branch of the cubic, where the line is above the local maximum, as shown in Figure 2.13b. If ε were 0, the system would stay there. But since ε is not 0, things change. In light of $x > 0$, β will decline (since its derivative is negative). Thus in the next bit of time, the line will lower a little. Furthermore, now the line is slightly below the cubic and since $dx/dt = \beta - (x^3 - \alpha x)$, x declines slightly too. At this new value of x, $d\beta/dt$ is still negative, so that both β and x will continue to decline. We will thus slowly move down along the right hand branch of the cubic (Figure 2.13b). For how long will this go on? Until we reach the local minimum of the cubic at $x = \sqrt{\alpha}$. At this point, β is still declining, but once it does so there is no intersection between the line and the curve for positive values of x. We thus predict a rapid transition from the right hand branch of the cubic to the left hand branch. When we get near the left hand branch, x is negative so that $d\beta/dt$ is positive and β begins to rise. Once again, this happens slowly, along the left hand branch, until the local maximum is crossed, at which point there will be a rapid transition back to the right hand branch of the cubic. In other words, we predict oscillations, and that the oscillations will have a shape that involves a slowly changing component and a rapidly changing component.

In Figure 2.14, I show the numerical solution of the differential equations for the case in which $\alpha = 1$, $\varepsilon = 0.005$ with initial values $x(0) = 2$ and $\beta(0) = x(0)^3 - \alpha x(0)$. Starting at $x(0) = 2$, we see a slow decline along the right hand branch of the cubic, until there is a rapid drop, then a slow rise, and oscillations set in. To help make this point clearer, Figures 2.14b and c show just parts of the trajectory; in Figure 2.14c, we most clearly see the slow and fast parts of the oscillation.

Oscillations such as the ones described here are called "relaxation oscillations" and they arise in many different ecological contexts,

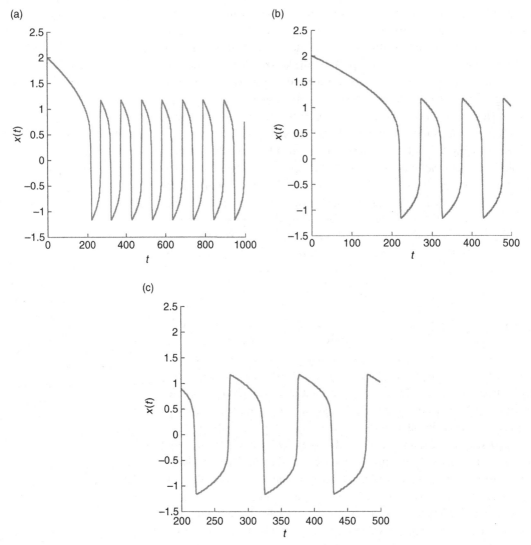

Figure 2.14. The oscillations induced by allowing the parameter β to slowly change, as described in the text. Three panels are shown, with increasingly fine resolution in time, so that we can clearly see the slow and fast parts of the oscillatory system.

typically in relationship to some kind of pest outbreak or plankton bloom (see Connections).

Two dimensional differential equations and the classification of steady states

Many of the models that we encounter in population biology involve two or more differential equations of the form $dx/dt = f(x, y)$ and

$dy/dt = g(x, y)$. Some examples are the Lotka–Volterra predator (P)–prey (V) equations

$$\frac{dV}{dt} = rV\left(1 - \frac{V}{K}\right) - bPV$$

$$\frac{dP}{dt} = cPV - mP$$

(2.33)

the Lotka–Volterra competition equations

$$\frac{dx}{dt} = r_1 x\left(1 - \frac{x + \alpha y}{K_1}\right)$$

$$\frac{dy}{dt} = r_2 y\left(1 - \frac{y + \beta x}{K_2}\right)$$

(2.34)

and equations that could describe a mutualistic interaction (for example between ants and butterflies, see Pierce and Nash (1999) or Pierce *et al.* (2002); for yuccas and moths see Pellmyr (2003))

$$\frac{dA}{dt} = r_a A\left(1 - \frac{A}{K_0 + K_1 B}\right)$$

$$\frac{dB}{dt} = B(r_b A - mB)$$

(2.35)

If these equations are not familiar to you, do not despair, but read on – we shall explicitly consider the first two pairs in what follows.

When considering differential equations such as these in the plane, one can usefully apply a three step procedure (which is generalized to systems of higher dimension): understand the steady states, the qualitative dynamics, and only then the quantitative dynamics. We will approach this procedure slowly, beginning with some very specific examples and then ending with the general case.

We start with a specific example: consider the following system of differential equations for a pair of variables $u(t)$ and $v(t)$ (don't try to ascribe biological meaning to them just now, that will come later on).

$$\frac{du}{dt} = Au$$

$$\frac{dv}{dt} = Dv$$

(2.36)

The choice of the constants A and D, which may be mysterious now, will also become apparent later.

The steady states of this system are the values in the (u, v) plane for which $du/dt = 0$ and $dv/dt = 0$. We can determine by inspection that the only steady state is the origin $(0, 0)$. Furthermore, we can determine by inspection that $u(t)$ and $v(t)$ must be exponential functions of time. Thus, we conclude that $u(t) = u(0)e^{At}$ and that $v(t) = v(0)e^{Dt}$.

(We could also note that $du/dv = Au/Dv$, which integrates to $A\ln(u) = D\ln(v) + \text{constant}$, and which then becomes $u = cv^{A/D}$, where c is a constant. But we are not going to make a big deal out of this because it does not help us except in the special case).

What does help us, however, is to think about the exponential solutions of time in a plane represented by u on the abscissa and v on the ordinate. This is called the phase plane (Figure 2.15). We will distinguish three cases. First suppose that $A > 0$ and $D > 0$. If we start the system at $u(0) = 0$ and $v(0) = 0$, then it stays there forever. However, if we start it anywhere else, both $u(t)$ and $v(t)$ grow in time. We say that points in the u–v plane "flow away from the origin." This is represented by the arrows in Figure 2.15a pointing away from the origin. Note that we are not trying to characterize the shape of those curves that represent the flow away from the origin, just that points move away. We call this an unstable node. Second, suppose that $A < 0$ and $D < 0$. Then everything that we just concluded applies, but in reverse. If initial values are not at the origin, they decline in time; we say that the flow is towards the origin and that this is a stable node (Figure 2.15b). Third, suppose that one of A or D is positive and that the other is negative. For concreteness, I will do the case $A < 0$ and $D > 0$ and let you draw the picture for the other one. Now some interesting things can happen. Note, if we start exactly on the u-axis, we flow towards the origin. If we start exactly on the v-axis, we flow away from the origin. For any starting point with $u(0) \neq 0$ and $v(0) \neq 0$ but close to the origin, we will first flow towards the origin, kind of "along the u-axis" and then flow away from it "along the v-axis." So we see that the u-axis separates the plane into two regions; these are often called domains of attraction and the u-axis is called the separatrix. In this case the origin is called a saddle point (Figure 2.15c), in analogy to real saddles (Figure 2.15d) in which one falls into the middle of the saddle moving along the back of the horse but off the saddle moving laterally to the back of the horse.

This case was nice, but perhaps a bit too simple because the dynamics of u and v were not connected in any way. The next most complicated case would be linear, but with connection. Here we will go back to x and y and write

$$\frac{dx}{dt} = Ax + By$$

$$\frac{dy}{dt} = Cx + Dy$$

(2.37)

and now I hope you understand my choice of A and D in the previous discussion. How could we analyze these equations? We might try to find some combination of x and y so that new variables $u = \alpha x + \beta y$ and $v = \gamma x + \delta y$ satisfy $du/dt = A'u$ and $dv/dt = B'v$ for some constants

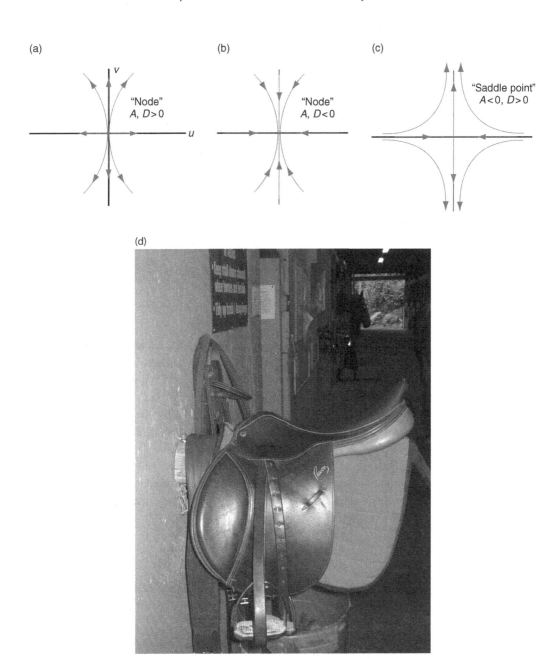

Figure 2.15. The phase plane for the simple dynamical system $du/dt = Au$, $dv/dt = Dv$. If A and D are both greater than 0, the origin is an unstable node (panel a). If A and D are both less than 0, the origin is a stable node (panel b). If one of A or D is positive and the other is negative, the origin is called a saddle point (panel c), in analogy with actual saddles (panel d; compliments of Gabby Roitberg).

A' and B'. Rather than doing that, we will try to generalize what we have already learned.

I will now show two different ways to get to the same answer. The first method is completely independent of anything outside of this book. The second requires that you know a bit of linear algebra. The first method proceeds as follows. We differentiate the first equation in Eq. (2.37) with respect to time to obtain $d^2x/dt^2 = A(dx/dt) + B(dy/dt) = A(dx/dt) + B(Cx + Dy)$. Now we use the first equation in (2.37) once again, by noting that $y = (1/B)[(dx/dt) - Ax]$. Combining these, we obtain a single, second order differential equation for $x(t)$.

Exercise 2.9 (E)

Show that when we combine the last two equations, we obtain

$$\frac{d^2x}{dt^2} - (A + D)\frac{dx}{dt} + (AD - BC)x = 0 \qquad (2.38)$$

Now, before discussing the solution of this equation, let us think about some of its properties. Since this is a second order differential equation, two constants of integration will appear in the solution. These are called the initial conditions. For the original system, we might specify $x(0)$ and $y(0)$ (for example, two population sizes), but for Eq. (2.38) we might specify $x(0)$ and $dx/dt|_{t=0}$ (these are an analogous specification since we know that $y = (1/B)[(dx/dt) - Ax]$). Because of the integration constants, there will be many different solutions of Eq. (2.38). The next exercise, which is called the linear superposition of solutions, will be extremely useful for the rest of the chapter.

Exercise 2.10 (E/M)

Suppose that $x_1(t)$ and $x_2(t)$ are solutions of Eq. (2.38). (That is, each of them satisfies the differential equation.) Show that $X(t) = ax_1(t) + bx_2(t)$, where a and b are constants, is also a solution of Eq. (2.38).

We still have to deal with the matter of finding the solution of Eq. (2.38). We know that a first order linear differential equation of the form $dx/dt = Ax$ has exponential solutions, so let's guess that the solution of Eq. (2.38) has the form $x(t) = x_0 e^{\lambda t}$ where x_0 is a constant (corresponding to the initial value of x) and we need to find λ. If we accept this guess, then the derivatives of $x(t)$ are $dx/dt = \lambda x_0 e^{\lambda t}$ and $d^2x/dt^2 = \lambda^2 x_0 e^{\lambda t}$. When we substitute these forms for $x(t)$ and its two derivatives back into Eq. (2.38), note that both x_0 and $e^{\lambda t}$ will cancel since they appear in all of the terms. We are then left with a quadratic equation for the parameter λ:

$$\lambda^2 - (A+D)\lambda + AD - BC = 0 \tag{2.39}$$

Before interpreting Eq. (2.39), I will show a different way to reach it. For this second method, let us assume that there are certain special initial values of $x(0) = u$ and $y(0) = v$ such that $x(t) = ue^{\lambda t}$ and $y(t) = ve^{\lambda t}$. Note that these are clearly not the u and v with which we started this section. I use them here because in life we are symbol-limited. Given this form for $x(t)$ and $y(t)$ the derivatives are $dx/dt = \lambda ue^{\lambda t} = \lambda x(t)$ and $dy/dt = \lambda ve^{\lambda t} = \lambda y(t)$. For this reason, λ is called an eigenvalue (from the German word "eigen" meaning similar or equivalent) of the differential equations (2.37) because when we take the derivatives of $x(t)$ and $y(t)$ we get back multiples of $x(t)$ and $y(t)$. In a geometrical way, we can think of a vector that joins the origin and the point (u, v); it is called the eigenvector, for much the same reason.

Now we substitute these derivatives into Eq. (2.37). Once again, the exponential terms cancel and when we combine terms we obtain

$$\begin{aligned}(A - \lambda)u + Bv &= 0 \\ Cu + (D - \lambda)v &= 0\end{aligned} \tag{2.40}$$

One solution of these linear algebraic equations is $u = v = 0$. For there to be other solutions, we recall that the determinant of the coefficients of u and v must be equal to 0. That is

$$\begin{vmatrix} A - \lambda & B \\ C & D - \lambda \end{vmatrix} = 0 \tag{2.41}$$

and when we apply the rule for determinants (i.e. that Eq. (2.41) is equivalent to $(A - \lambda)(D - \lambda) - BC = 0$) we obtain the same equation for λ, Eq. (2.39).

Equation (2.39) is a quadratic equation, so that we know there are two solutions, given by the quadratic formula. We will denote these solutions by $\lambda_{1,2}$ and they are

$$\begin{aligned}\lambda_{1,2} &= \frac{(A+D) \pm \sqrt{(A+D)^2 - 4(AD - BC)}}{2} \\ &= \frac{(A+D) \pm \sqrt{(A-D)^2 + 4BC}}{2}\end{aligned} \tag{2.42}$$

where, for convention, we will assume that 1 corresponds to $+$ and 2 to $-$ in the quadratic formula. If we define the discriminant by $\Delta = (A - D)^2 + 4BC$, then we can write that $\lambda_{1,2} = [(A+D) \pm \sqrt{\Delta}]/2$.

We are now able to classify the steady state $(0, 0)$ of the system given in Eq. (2.37). Before doing that, let's have a brief interlude.

Exercise 2.11 (M)

Show that if λ is a solution of Eq. (2.42) and that if we set $u = B$ and $v = \lambda - A$ that Eq. (2.40) is satisfied. Thus, we know how to find the eigenvectors too.

As long as $\Delta \neq 0$, which we will assume in this chapter, the exercises up to this point have allowed us to find the general solution of the system given by Eq. (2.37):

$$x(t) = c_1 Be^{\lambda_1 t} + c_2 Be^{\lambda_2 t}$$
$$y(t) = c_1(\lambda_1 - A)e^{\lambda_1 t} + c_2(\lambda_2 - A)e^{\lambda_2 t} \tag{2.43}$$

Although it is nice to have an explicit form for the solution, what is nicer is that we now know how to classify the steady state.

We begin with the case in which $\Delta > 0$. Then both of the eigenvalues are real. We conclude that if they are both positive, the origin is an unstable node. Since solutions will grow exponentially, whichever eigenvalue is larger will ultimately dominate the behavior of the solution. If both of the eigenvalues are negative, we conclude that the origin is a stable node. If one of the eigenvalues is positive and the other is negative, we conclude that the origin is a saddle point.

When $\Delta < 0$, the eigenvalues are complex numbers, so if we set $q = \sqrt{|\Delta|}$ we can rewrite the eigenvalues as $\lambda_{1,2} = [(A + D) \pm iq]/2$, where $i = \sqrt{-1}$. Consequently, when we compute solutions given by Eq. (2.43), we will need to consider expressions of the form

$$\exp\left(\frac{(A + D)t + iqt}{2}\right) = \exp\left(\frac{(A + D)t}{2}\right)\exp\left(\frac{iqt}{2}\right)$$

From this, we see that if $A + D$, the real part of the eigenvalues, is negative, then whatever else happens solutions will decline in time. If $A + D$ is positive, they will grow in time. The question then becomes how we interpret the exponential of iqt.

For this interpretation, we need a brief reminder. Recall that the solution of the differential equation $d^2x/dt^2 = -kx$ involves sines or cosines. (If you do not recall this, confirm that if $x = \sin(\sqrt{k}t)$ or $x = \cos(\sqrt{k}t)$ then the differential equation is satisfied.) Since this is a linear equation, the general solution must be of the form $c_1 \sin(\sqrt{k}t) + c_2 \cos(\sqrt{k}t)$, where the c_i are constants. Suppose that we had guessed an exponential solution for this equation, i.e. that $x = ce^{\lambda t}$. In this case, the second derivative of $x(t)$ is $c\lambda^2 e^{\lambda t}$ so that we conclude λ must satisfy the equation $\lambda^2 = -k$ or that $\lambda = \pm i\sqrt{k}$. In other words, exponentials involving $\sqrt{-1}$ lead to oscillations. Our solution is at hand. If $\Delta < 0$, we now know that the solutions will oscillate. Such a steady state is called a focus or a spiral point. If $A + D < 0$, the focus is stable and if $A + D > 0$ the focus is unstable.

But, of course, all this work (and it is hard work) only corresponds to the linear system of equations (2.37) and the equations that we actually encounter in population biology are nonlinear. What do we do about this? That is, in general we will have a pair of differential equations of the form

$$\frac{dx}{dt} = f(x,y)$$
$$\frac{dy}{dt} = g(x,y)$$
(2.44)

and let us suppose that the point (x_s, y_s) is a steady state of this system so that $f(x_s, y_s) = g(x_s, y_s) = 0$. We go forward from Eq. (2.44) by linearizing the equations around the steady state. That is, we write $x(t) = x_s + \tilde{x}(t)$ and that $y(t) = y_s + \tilde{y}(t)$ so that $\tilde{x}(t)$ and $\tilde{y}(t)$ measure the deviations from the steady state. Since the steady states are constant, we know that $dx/dt = d\tilde{x}/dt$ and $dy/dt = d\tilde{y}/dt$. Now we will Taylor expand $f(x, y)$ around the steady state and keep only the linear term:

$$f(x,y) \approx f(x_s + \tilde{x}, y_s + \tilde{y})$$
$$= f(x_s, y_s) + \frac{\partial}{\partial x} f(x,y)|_{(x_s, y_s)} \tilde{x} + \frac{\partial}{\partial y} f(x,y)|_{(x_s, y_s)} \tilde{y}$$
(2.45)

Now let us consider the three terms in the right hand expression of this equation. The first term on the right hand side is identically zero, because (x_s, y_s) is a steady state. The second term is the partial derivative of $f(x, y)$ with respect to x, evaluated at the steady state. To help simplify what we have to write, we will use subscripts for partial derivatives and, with a slight abuse of notation, replace the second and third terms on the right hand side of Eq. (2.45) by $f_x(x_s, y_s)\tilde{x}$ and $f_y(x_s, y_s)\tilde{y}$. A similar argument shows that $g(x,y) \approx g_x(x_s, y_s)\tilde{x} + g_y(x_s, y_s)\tilde{y}$. The point of all this work is that we can now replace the nonlinear differential equation (2.44) by a linear system that characterizes the deviations from the steady state

$$\frac{d\tilde{x}}{dt} = f_x(x_s, y_s)\tilde{x} + f_y(x_s, y_s)\tilde{y}$$
$$\frac{d\tilde{y}}{dt} = g_x(x_s, y_s)\tilde{x} + g_y(x_s, y_s)\tilde{y}$$
(2.46)

and we now compare Eq. (2.46) with Eq. (2.37) to determine the values of A, B, C, and D (note that they are not arbitrary but must match the various partial derivatives in Eq. (2.46)), from which we can determine the stability characteristics of the steady states.

To help make the preceding more concrete, we will first consider an example, then an exercise. A very simple model for competition between two types or species $x(t)$ and $y(t)$ is

$$\frac{dx}{dt} = x(1 + a - x - ay)$$

$$\frac{dy}{dt} = y(1 + a - y - ax)$$

(2.47)

where a is a parameter, which we assume to be positive. From the form of these equations, we see that the presence of x increases the rate of change of x and that the presence of both x and y decreases the rate of change of x (and vice versa for y). We say that x and y are auto-catalysts for themselves and anti-catalysts for the other type. This thinking underlay the work of Sir F. C. Frank in his study of spontaneous asymmetric synthesis (see Connections); Eq. (2.47) is also a simple analog of the Lotka–Volterra competition equations, in which the competition is symmetric.

We find the steady states of Eq. (2.47) by setting $dx/dt = 0$ and $dy/dt = 0$. For the former, we find that $x = 0$ or $x + ay = 1 + a$. For the latter, we find that $y = 0$ or $y + ax = 1 + a$. Thus, $(1, 1)$ is a steady state. Before conducting an eigenvalue analysis, we use the isoclines (or more properly, the nullclines, lines on which $dx/dt = 0$ or $dy/dt = 0$) of the differential equations to understand properties of the solution. These are shown in Figure 2.16. The steady state $(1, 1)$ can be either a node (if $a < 1$) or a saddle point (if $a > 1$). When $a = 1$, the two isoclines sit on top of each other and the system is structurally unstable. Note also that, because x and y are interchangeable in the two equations, the line $y = x$ is a solution of the equations – points on the line $y = x$ move towards $(1, 1)$, regardless of whether it is a node or a saddle point.

We can now conduct the eigenvalue analysis. In this case $f(x,y) = x(1 + a - x - ay) = x(1 + a) - x^2 - axy$ and $g(x, y) = y(1 + a) - y^2 - axy$. The partial derivatives are thus $f_x = 1 + a - 2x - ay$, $f_y = -ax$, $g_x = -ay$, and $g_y = 1 + a - 2y - ax$, and we evaluate these at $(1, 1)$ in order to obtain A, B, C, and D, so that $A = 1 + a - 2 - a = -1$, $B = -a$, $C = -a$, and $D = 1 + a - 2 - a = -1$. We substitute this into Eq. (2.42) and find that $\lambda_{1,2} = -1 \pm \sqrt{a}$. From the eigenvalue analysis, we reach the same conclusion as from the phase plane analysis – that $(1, 1)$ is either a stable node or saddle point, depending upon the value of a. Thus, in this case, the eigenvalue analysis told us little that we could not understand from the phase plane. Here's an example where it tells us much more.

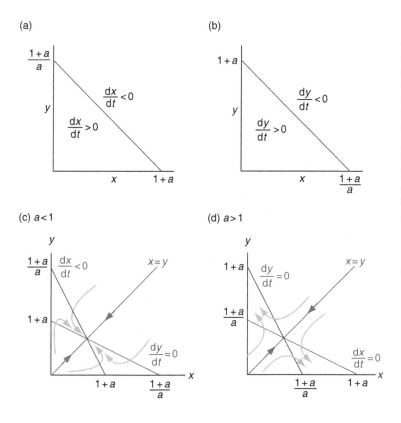

Figure 2.16. The isocline analysis of the equations for spontaneous asymmetric synthesis/symmetric competition. In panels (a) and (b), I show the separate isoclines for $dx/dt = 0$ and $dy/dt = 0$ and the flow of points in the phase plane. When these are put together, the resulting phase plane shows either a stable node at $(1,1)$ (panel c) or a saddle point (panel d).

Exercise 2.12 (M → H)

Consider the following predator (P)–prey (V) system

$$\frac{dV}{dt} = rV\left(1 - \frac{V}{K}\right) - bPV \qquad \frac{dP}{dt} = cPV - mP$$

Assume that the biomass of each is measured in numbers of individuals (but treated as a continuous variable) and time is measured in years. It might be helpful for what follows to think of rabbits and foxes as the victims and predators. It might also be helpful, especially for parts (a) and (b), to convert to per capita growth rates. (a) What are the units of all the parameters? (b) Interpret the biology of both predator and prey. What must be true about the relationship between b and c if the system is mammalian predators such as rabbits and foxes? (c) Conduct an isocline analysis. Note: there are two cases, depending upon the relationship between K and m/c. Be sure to get both of them and carefully think about what each means. (d) Classify the steady states of the system according to their eigenvalues. What does the eigenvalue calculation tell you that the isocline analysis did not? Once again, there are two cases that require careful interpretation. (e) What happens to the eigenvalues as $K \to \infty$?

What does this mean for the dynamics of the system? If you want more after this, do the same kind of analysis for the mutualism described in Eq. (2.35). We will return to this exercise, in a different context, in Chapter 6.

Diffusion as a random walk

We close this chapter with a discussion of diffusion – first a model of how the process takes place (this section), then diffusion with linear population growth (next section), and finally diffusion with nonlinear population growth. This material is merely an introduction (although it does help cement many of the ideas we've discussed thus far) and Chapters 7 and 8 will be an extensive study of diffusion processes and their applications to stochastic population theory. We will also use occasional references to rules of probability (material from Chapter 3).

We envision a "random walker" whose position is denoted by $X(t)$. Exactly what is intended by the word "walking" does not matter at this point – $X(t)$ could equally be the position of an individual in physical space, the frequency of a genotype, the size of a population, or the price of a stock. We assume that the walk takes place on a one dimensional lattice of "sites" (Figure 2.17) that are spaced distance Δ apart. We will ultimately let Δ shrink to 0, but not just yet. The sites will be indexed by the letter i and we thus measure distance by $x = i\,\Delta$; for concreteness, we refer to $X(t)$ as the position of the walker at time t. The walker will make moves ("jumps") at a fixed time interval τ, which will also be allowed to shrink to 0 later in this section. These jumps are characterized by a transition function $\rho(s)$, which is the probability that if the walker is currently at the site i, it will next be at the site j a distance s away. For example, if the walker were to move by flipping a fair coin (e.g. moving left by one step if heads comes up, right by one step if tails comes up), then $\rho(-1) = \rho(1) = 1/2$ and $\rho(s) = 0$ for any other value of s. If the walker moved by reaching into a bag and pulling one of two dice, marked left or right, and then rolling the second die to determine how far to move, then it could move left 1, 2, 3, 4, 5, or 6 sites each with probability 1/12 and right the same amounts with the same probability.

Given this framework, we define $p(x,\ t) = \text{Probability}\{X(t) = x\}$. This seems sensible enough, but there is actually a subtlety to it. When we talk about the walker "being at a site," we actually mean within the vicinity of the site. That is, when we say the walker is at

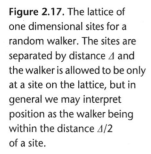

Figure 2.17. The lattice of one dimensional sites for a random walker. The sites are separated by distance Δ and the walker is allowed to be only at a site on the lattice, but in general we may interpret position as the walker being within the distance $\Delta/2$ of a site.

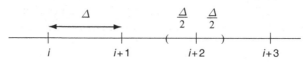

site $i+2$ in Figure 2.17, we could interpret this to mean that the walker is within $\Delta/2$ to the left or $\Delta/2$ to the right of site $i+2$. Thus, perhaps a more precise way to think about $p(x, t)$ is that $p(x, t)$ Δ is the probability that the walker is within $\pm\Delta/2$ of the site $i=x/\Delta$. This may seem like a lot of pedantry, and to some extent it is for just now, but thinking carefully about what these things mean will be enormously helpful later on, when we let the distance between sites shrink towards 0.

We will now derive an equation for $p(x, t)$. How can the walker be at spatial point x (site i) at time $t+\tau$? There are many ways, but they all boil down to this argument: the walker had to be at spatial point y at time t and take a jump of size $|x - y|$. We thus define j by $y=x - \Delta s$. Assuming that the walker's location at time t and the size of the jump are independent of each other allows us to multiply probabilities (you will be reminded of these rules in Chapter 3) so that we have

$$p(x, t + \tau) = \sum_{s} \rho(s)p(x - \Delta s, t) \qquad (2.48)$$

We Taylor expand this equation around the point (x, t) to obtain

$$p(x, t) + p_t\tau + o(\tau) = \sum_{s} \rho(s)\left[p(x, t) - \Delta s p_x + \frac{\Delta^2}{2}s^2 p_{xx} + o(\Delta^2)\right] \qquad (2.49)$$

where subscripts denote partial derivatives. We simplify this expression by thinking about the sums. For example, we know that $\sum_s \rho(s) = 1$ because a jump of some size (including 0) must occur. Furthermore, since $p(x, t)$ does not depend upon j, we know that $\sum_s \rho(s)p(x, t) = p(x, t)\sum_s \rho(s)$. This takes care of the first term in square brackets on the right hand side of Eq. (2.49). The fourth term will similarly simplify to $o(\Delta^2)$. The second and third terms require a bit more thought. Factoring the things that do not depend upon s out of the second term allows us to rewrite it as $-\Delta p_x \sum_s \rho(s)s$. We recognize the sum as the average jump size: that is s is the size of the jump from site j to site i and $\rho(s)$ is the chance of making this jump. We will denote this average by the symbol m_1 (for first moment). Exactly the same kind of argument will apply to the third term on the right hand side of Eq. (2.49), except that we will have the square of the jump size; hence we will use the symbol m_2 for that summation. We put all this together by subtracting $p(x, t)$ from both sides of Eq. (2.49), divide by τ and obtain

$$p_t + \frac{o(\Delta)}{\tau} = -\frac{\Delta}{\tau}m_1 p_x + \frac{\Delta^2}{2\tau}m_2 p_{xx} + \frac{o(\Delta^2)}{\tau} \qquad (2.50)$$

Now we let $\tau \to 0$, $\Delta \to 0$ and assume that as this happens $(m_1\Delta)/\tau \to v$, $(m_2\Delta^2)/\tau \to \sigma^2$, $o(\Delta^2)/\tau \to 0$, which is the definition of the quantities v and σ^2. When we do this, the resulting equation is

$$p_t = -vp_x + \frac{\sigma^2}{2}p_{xx} \tag{2.51}$$

With this equation, we have much to talk about (but little to reminisce, as we will by Chapters 7 and 8). First, let us consider these new parameters. Since Δ has units of distance, τ has units of time, and both m_1 and m_2 are pure numbers (averages of the jump size), we see that v has units of distance/time – it is a velocity. On the other hand, σ^2 has units of (length)^2/time; these units make it a diffusion coefficient. Hence, Eq. (2.51) is called a diffusion equation.

We need to discuss one subtlety of this limit. In particular, how is $p(x, t)$ interpreted now that a set of discrete sites has become a continuum (as $\Delta \to 0$)? Recall, that we earlier agreed to think of $p(x, t)$ as the probability that the walker was within $\pm\Delta/2$ of the spatial point x (site i) at time t. In the limit, we interpret $p(x, t)\mathrm{d}x$ as the probability that the walker is within $\mathrm{d}x$ of the spatial point x at time t. Since the walker must be some place, we obtain a normalization condition

$$\int_{-\infty}^{\infty} p(x, t)\mathrm{d}x = 1 \tag{2.52}$$

Equation (2.51) also involves one time derivative and two spatial derivatives. This means that there are three conditions needed to completely specify the solution. One is an initial condition: we specify $p(x, 0)$ – the chance of initially finding the walker at spatial point x. The choices about boundary conditions depend upon the nature of the spatial region.

First, suppose that this region is unbounded. It is reasonable to expect, then, that the chance that the walker can reach $\pm\infty$ in any finite time is 0. We then specify that $p(x, t) \to 0$ as $|x| \to \pm\infty$. Sometimes we can get a solution of this equation virtually for free, as in the following exercise.

Exercise 2.13 (M)

Show that

$$p(x, t) = \frac{1}{\sqrt{2\pi\sigma^2 t}}\exp\left[\frac{-(x - vt)^2}{2\sigma^2 t}\right] \tag{2.53}$$

is a solution of the diffusion equation and satisfies the boundary conditions for an unbounded domain.

The function $p(x, t)$ defined in Eq. (2.53) is called the Gaussian or normal distribution with mean vt and variance $\sigma^2 t$. If you did the exercise, you know that it satisfies the boundary conditions and the differential equation. But how do we know that it satisfies the normalization condition Eq. (2.52)? The following exercise helps with that.

Exercise 2.14 (M/H)

First show that setting $u = (x - vt)/\sqrt{\sigma^2 t}$ means that the normalization condition is equivalent to showing that $\int_{-\infty}^{\infty} 1/\sqrt{2\pi} \exp(-u^2/2)du = 1$. Second, to show that this is true, consider the double integral $\int_{-\infty}^{\infty} \int_{-\infty}^{\infty} \exp(-u^2/2) \exp(-w^2/2)dudw$, switch to polar coordinates in which $r = \sqrt{u^2 + w^2}$ and evaluate the resulting integral to show that the integral is 2π.

But we are not done with the solution given by Eq. (2.53). In Figure 2.18, I have plotted four Gaussian distributions with mean 0 (i.e. $v = 0$ so that the original walk is unbiased in either direction) and $\sigma^2 t = 0.1, 0.5, 1$, or 3. As the variance decreases, the curves become more peaked and centered around the origin. Now the area under each of this curves is 1 (because of the normalization constant). Let us think about the limit of $p(x, t)$, for the case in which $v = 0$, as t approaches 0. The function we are considering is thus $(1/\sqrt{2\pi\sigma^2 t}) \exp(-x^2/2\sigma^2 t)$. Now, if $x \neq 0$, as t approaches 0, the reciprocal of the square root goes to infinity, but the exponential function goes to 0 and since exponentials

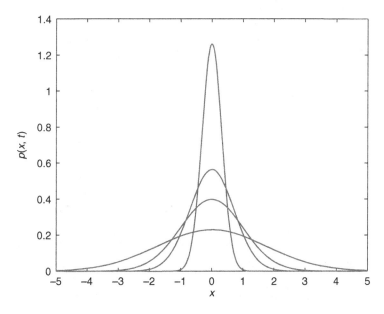

Figure 2.18. Gaussian distributions with mean 0 and $\sigma^2 t = 0.1, 0.5, 1$, or 3.

decline faster than any algebraic function (something that you need to remember from calculus), the product goes to 0. However, if $x = 0$, the exponential term is always equal to 1, but the reciprocal square root still goes to infinity as t approaches 0. What do we conclude from this discussion? When $v = 0$, as $t \to 0$ $p(x, t)$ approaches a function $\delta(x)$ with these properties

$$\int_{-\infty}^{\infty} \delta(x)\mathrm{d}x = 1$$

(2.54)

$$\delta(x) = \begin{matrix} 0 & \text{if } x \neq 0 \\ \infty & \text{if } x = 0 \end{matrix}$$

In physics, a function with these properties is called the Dirac delta function, named after Paul Dirac (see Connections); in applied mathematics it is called a generalized function, so named by Sir James Lighthill (Lighthill 1958), although at the time he was just M. J. Lighthill (also see Connections). We will use generalized functions considerably when we deal with stochastic population theory in Chapters 7 and 8.

For the time being, this takes care of the infinite domain. What about a bounded region? That is, suppose that the range that the walker can take is $0 \leq x \leq L$. Once again, we need to apply two spatial conditions. We might have one at $x = 0$ and one at $x = L$, or both at one of the boundaries. For example, if the walker disappears (is absorbed) at $x = 0$ or $x = L$, we would have the boundary condition $p(0, t) = p(L, t) = 0$ because there is no chance of finding the walker at those points. Suppose, on the other hand, we knew that the walker was always constrained to stay within $(0, L)$. We then have the condition that $\int_0^L p(x, t)\mathrm{d}x = 1$ and if we differentiate this equation with respect to time, we obtain $\int_0^L p_t(x, t) = 0$. We now use Eq. (2.51) to conclude that $\int_0^L [-vp_x + (\sigma^2/2)p_{xx}]\mathrm{d}x = 0$, which we integrate to obtain

$$\left[(-vp) + \frac{\sigma^2}{2}p_x \right]\Bigg|_{x=0}^{L} = -vp(L, t) + \frac{\sigma^2}{2}p_x(L, t) - \left[-vp(0, t) + \frac{\sigma^2}{2}p_x(0, t) \right]$$

If we want what happens at $x = 0$ and $x = L$ to be independent of each other, the way to do it is to require that each of the terms on the right hand side are 0, so that we obtain the boundary conditions $-vp(L, t) + (\sigma^2/2)p_x(L, t) = 0$ and $-vp(0, t) + (\sigma^2/2)p_x(0, t) = 0$. In the literature these are often called "no flux" (from chemical analogies) or "reflecting" boundary conditions. The latter makes sense: the walker is constrained to stay between 0 and L, so that when it comes up against $x = 0$ or $x = L$, it must be reflected back into the region, much as a ball bouncing around in a room will bounce off the walls of the room.

Before moving on, I want to introduce one more concept, which we will use in a slightly different way in the next section, but which you will encounter frequently in the literature. To illustrate these ideas, let us continue with the case of reflecting boundary conditions, and for simplicity set $v = 0$. Our problem is then to solve $p_t = (\sigma^2/2)p_{xx}$, given some initial condition $p(x, t) = p_0(x)$ and subject to the boundary conditions that $p_x(x, 0) = p_x(L, 0) = 0$. To do this, let us guess that $p(x, t)$ is the product of a function of time and a function of space. How do we know to make this guess? Well, first, generations of other scientists and mathematicians have tried it and found that it worked. So, we have history on our side. Second, new things are often discovered by good guessing. In this book, of course, I am not going to take you down too many blind roads (that is, bad guesses). The movie *The Color of Money* begins with an off-screen voice describing 9 ball pool and continues "which is to say that in 9 ball luck plays a part. . . but for some players, luck itself is an art." The same is true of applying mathematical methods to understand scientific questions. We need good guesses and good luck that the guess is correct, but sometimes we create our own luck through experience and thought. The Czech chess instructor Jan Amos Komensky once said, regarding chess, "Through play, knowledge" (Pandolfini, 1989, p. xix). It works here too.

Accepting this guess, which is called the method of separation of variables, means that $p(x, t) = T(t)S(x)$, where $T(t)$ is a function depending only upon time and $S(x)$ is a function depending only upon space. The diffusion equation then becomes $T_t(t)S(x) = (\sigma^2/2)T(t)S_{xx}(x)$, where I still use subscripts to denote derivatives, although these are now ordinary (not partial) derivatives. Dividing both sides by $T(t)S(x)$ we obtain

$$\frac{T_t(t)}{T(t)} = \frac{\sigma^2}{2}\frac{S_{xx}(x)}{S(x)} \tag{2.55}$$

Now, the left hand side of Eq. (2.55) depends only upon time and the right hand side depends only upon space. What does this mean? It means that they both had better be independent of both time and space – each side should be constant. The left hand side also implies that $T(t)$ must be an exponential function. We do not want the solution of the diffusion equation to grow without bound in time, because that makes no sense, so the constant must be negative (or at least not positive). A way of writing a non-positive number is $-n^2$, where $n = 0, 1, 2$, etc. Then we know that $T(t) \sim \exp(-n^2t)$. The equation that $S(x)$ satisfies will then become $S_{xx}(x) = -(2n^2/\sigma^2)S(x)$. Since the second derivative of $S(x)$ is a negative number times $S(x)$, we know that $S(x)$ must involve sines or cosines. The diffusion equation is a linear equation, so Exercise 2.10 (on the linear combination of solutions) tell us that the most general form of the solution for the equation that $S(x)$ satisfies will be a mixture of sines and cosines. In particular, we can write

$$S(x) = \sum_n A_n \sin\left(x\sqrt{\frac{2n^2}{\sigma^2}}\right) + B_n \cos\left(x\sqrt{\frac{2n^2}{\sigma^2}}\right) \qquad (2.56)$$

where the A_n and B_n are constants, which we must somehow determine. The way this is done is explained in the next section.

Diffusion and exponential population growth

We now consider how diffusion and population growth interact. That is, instead of simply exponential population growth in time or diffusion in space, we consider population size $N(x, t)$ depending upon both spatial point x and time t. This population will be characterized by the equation

$$N_t = \frac{\sigma}{2} N_{xx} + rN \qquad (2.57)$$

where I have suppressed the dependence of N on x and t. As before, we will require an initial condition and boundary conditions. We will assume that $N(x, 0)$ is specified and that the population is confined to a region $[0, L]$. In that case the appropriate boundary conditions are $N_x(0, t) = N_x(L, t) = 0$.

Before doing any mathematics, let us spend time thinking about Eq. (2.57). We begin with a profile of population size in time, $N(x, 0)$. One such a profile (made up by me) is shown in Figure 2.19a. If we were to describe this profile, we might say that there is a cline of increasing population size, with some small deviations from what looks to be a straight line. It is those deviations that we are interested in learning about, so to focus on them we define the average population size by $\bar{N} = (1/L) \int_0^L N(x, 0) dx$ and the deviation $n(x, 0)$ by $n(x, 0) = N(x, 0) - \bar{N}$. In Figure 2.19b, I show the scaled value of $n(x, 0)$, scaled by the average (that is, I am plotting $n(x, 0)/\bar{N}$).

We know that if the population started out completely homogeneous in space with initial value $\bar{N}$, then its size at any later time would be $\bar{N}e^{rt}$, so let us define $n(x, t) = N(x, t) - \bar{N}e^{rt}$. We already know $n(x, 0)$ and since the boundary conditions for $N(x, t)$ involve derivatives, we have the same boundary conditions for $n(x, t)$. Regarding the equation that $n(x, t)$ satisfies, see Exercise 2.15.

Exercise 2.15 (E)

Show that $n(x, t)$ satisfies $n_t = (\sigma^2/2)n_{xx} + rn$.

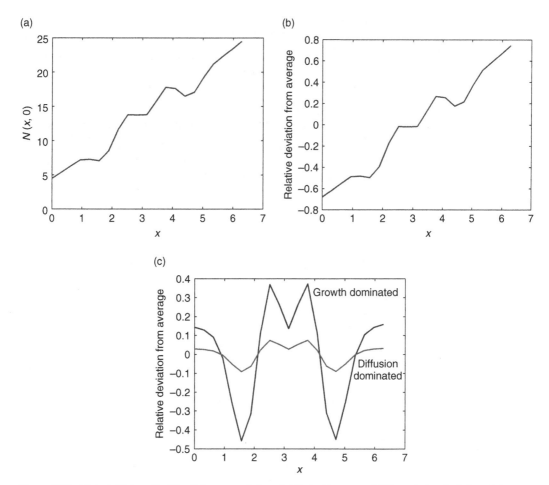

Figure 2.19. (a) An initial profile $N(x, 0)$ for a population distributed in space. (b) The relative deviation of the initial profile from its average. (c) The relative deviation may expand or shrink, depending on the relative strength of growth and diffusion. Our job is to figure out which is which.

We now know the equation that $n(x, t)$ must satisfy. But how do we find the solution? One method would be to make the substitution $n(x, t) = w(x, t)e^{rt}$, in which case we find that $w(x, t)$ satisfies the diffusion equation $w_t = (\sigma^2/2)w_{xx}$ (you can consider this an optional exercise). One could say that we've reduced this problem to the previous one, which could be solved by separation of variables. But let's go on, to see what new insights can be gained.

Another route is the following. Given the result in Eq. (2.56), let us guess that $n(x, t)$ can be represented by a mixture of sines and cosines, in which case we might guess a form such as $n(x, t) = A(t)\sin(wx) + B(t)\cos(wx)$ where the coefficients $A(t)$ and $B(t)$ and the frequency w need to be determined. Let's think about the boundary conditions, which apply to $n_x(x, t) = A(t)w\cos(wx) - B(t)w\sin(wx)$. Since $\cos(0) = 1$ and $\sin(0) = 0$, we can satisfy the boundary condition at $x = 0$ by picking $A(t) = 0$ for all time. Then to satisfy the boundary condition at $x = L$, we must have $\sin(wx) = 0$. This will be true if $wL = 0$, $wL = \pi$, $wL = 2\pi$, etc. Now, if $wL = 0$, w must be 0, in which case $\cos(wx) = 1$, independent of space. But we have already taken account of the spatially independent aspects of the solution, so we can ignore $w = 0$. We thus conclude that $wL = k\pi$, for $k = 1, 2, 3$, and so forth. Because of our result on the linear superposition of solutions, this means that the solution must be

$$n(x, t) = \sum_{k=1} B_k(t) \cos\left(\frac{k\pi}{L}x\right) \tag{2.58}$$

But we still do not know the values of the different $B_k(t)$, which we call the amplitude of the kth mode. We will find them in two steps. First, we will derive an equation for each $B_k(t)$. Second, we will find the initial value $B_k(0)$; together these will tell us the entire solution. For the first step, we take the partial derivatives of $n(x, t)$ given by Eq. (2.58)

$$n_t = \sum_{k=1} \frac{d}{dt} B_k(t) \cos\left(\frac{k\pi}{L}x\right) \qquad n_{xx} = -\sum_{k=1} B_k(t) \left(\frac{k\pi}{L}\right)^2 \cos\left(\frac{k\pi}{L}x\right)$$

where I have now used d/dt to denote the time derivative of each of the $B_k(t)$. If we now substitute these back into the equation for $n(x, t)$, we obtain

$$\sum_{k=1} \frac{d}{dt} B_k(t)\cos\left(\frac{k\pi}{L}x\right) = -\frac{\sigma^2}{2} \sum_{k=1} B_k(t) \left(\frac{k\pi}{L}\right)^2 \cos\left(\frac{k\pi}{L}x\right)$$
$$+ r \sum_{k=1} B_k(t)\cos\left(\frac{k\pi}{L}x\right)$$

This equation will be satisfied if we choose the coefficients $B_k(t)$ to satisfy

$$\frac{d}{dt}B_k(t) = \left[r - \frac{\sigma^2}{2}\left(\frac{k\pi}{L}\right)^2\right]B_k(t) \tag{2.59}$$

Equation (2.59) provides us with the main intuition about the interaction of diffusion and population growth. We see from this equation that $B_k(t)$ is an exponential function of time. It is exponentially growing if $r > (\sigma^2/2)(k\pi/L)^2$, constant if equality holds, and exponentially declining if $r < (\sigma^2/2)(k\pi/L)^2$. We have thus derived a very precise relationship between the rate of population growth and the rate of diffusion. This relationship tells us how the amplitude of the kth mode grows or declines. The result has nice intuitive appeal: if diffusion is very strong (so that $r < (\sigma^2/2)(k\pi/L)^2$ even when $k = 1$) then all of the modes will decline and initial fluctuations in $n(x, 0)$ will smear out over time. On the other hand, if the diffusion coefficient is not too big, then for some values of k we will have $r > (\sigma^2/2)(k\pi/L)^2$ and the amplitude of those modes will grow in time; for other values of k the amplitudes will decline (and there may be one value of k where exact equality holds, in which case the amplitude will remain constant). In this case, the more slowly varying amplitudes will be accentuated and small deviations in $n(x, 0)$ will be enhanced in time (Figure 2.19c).

To understand what is happening, and to find that pesky value of $B_k(0)$, it is helpful to think geometrically about the cosine function (we will get mathematical details in a minute). In Figure 2.20a, I have plotted $y = \cos(\pi x/L)$ and $y = \cos(6\pi x/L)$ for $L = 10$ (the choices $k = 1$, $k = 6$, and $L = 10$ are arbitrary, and you might want to make your own similar plots with different values of k). Notice the shape of the plot for $k = 1$: the curve starts at 1 when $x = 0$, smoothly decreases, passing through 0 when $x = 5$ and reaches -1 when $x = 10$. The curve is symmetric around the line $y = 0$ when $x = 5$: each value of $x < 5$ has a certain value of $y = \cos(\pi x/L)$ and there is a value of $x > 5$ with exactly the opposite value. Thus, for example, the integral of $y = \cos(\pi x/L)$ from $x = 0$ to $x = L$ will be 0 (you could do this, of course, by simply integrating, but understanding the geometry is important). The same is true of the curve $y = \cos(6\pi x/L)$, which fluctuates much more between $x = 0$ and $x = L$ but is also symmetrical. In Figure 2.20b, I have plotted the product $y = \cos(\pi x/L)\cos(6\pi x/L)$, which is also symmetric around the line $y = 0$ at $x = 5$. So, the integral of this product will also be 0. The situation is different, however, if we consider the squares $y = \cos^2(\pi x/L)$ or $y = \cos^2(6\pi x/L)$. In this case, of course, there are no negative values. The more slowly fluctuating $\cos^2(\pi x/L)$ has much less frenetic changes, but a remarkable fact from calculus is that their integrals are the same.

To be completely general, I report results for both sine and cosine. Suppose that j and k are any two integers greater than or equal to 1. If $j \neq k$

$$\int\limits_0^L \sin\left(\frac{j\pi x}{L}\right)\sin\left(\frac{k\pi x}{L}\right)dx = \int\limits_0^L \cos\left(\frac{j\pi x}{L}\right)\cos\left(\frac{k\pi x}{L}\right)dx = 0$$

and if $j = k$

$$\int\limits_0^L \sin^2\left(\frac{k\pi x}{L}\right)dx = \int\limits_0^L \cos^2\left(\frac{k\pi x}{L}\right)dx = \frac{L}{2}$$

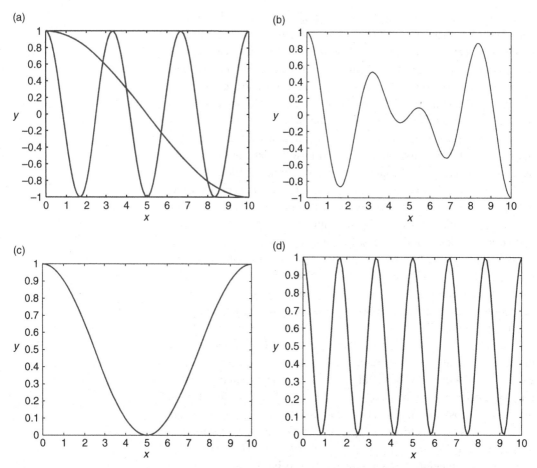

Figure 2.20. The key to finding how the amplitudes of different modes are initially set is understanding the cosine function. In these examples, $L=10$ and $k=1$ or $k=6$. (a) The functions $y=\cos(\pi x/L)$ or $y=\cos(6\pi x/L)$. (b) The product $\cos(\pi x/L)\cos(6\pi x/L)$. (c) The square $\cos^2(\pi x/L)$. (d) The square $\cos^2(6\pi x/L)$.

These expressions tell us how to find the initial value of the amplitude of each mode. That is, recall that we know $n(x, 0)$. But we have also represented $n(x, 0)$ by $\sum_{k=1}B_k(0)\cos(k\pi x/L)$. If we multiply both $n(x, 0)$ and its series representation by $\cos(k\pi x/L)$, integrate between 0 and L, and take advantage of the relationships between the integrals of cosine, we will obtain

$$\int_0^L n(x,0)\cos\left(\frac{k\pi x}{L}\right)dx = B_k(0)\frac{L}{2}$$

and this is the initial condition to go along with the differential equation (2.59). The $B_k(t)$ are called Fourier coefficients and the series representation for $n(x, t)$ called a Fourier series (see Connections). This completes the story and we now know how diffusion and population growth interact when population growth is exponential. The story is quite different when population growth is logistic, as we will see in the next section.

Diffusion and logistic population growth: invasions, the Fisher equation, and traveling waves

We conclude this chapter with a short introduction to a complicated topic, and one that comes the closest to pure mathematics yet – we are going to show that a solution to a question exists, but we are not going to actually find the solution. By way of motivation, we begin with the empirical phenomenon.

In Figure 2.21a, I show the spatial distribution of the variegated leafhopper (VLH, *Erythroneura variabilis*) which is a pest of grapes in California (Settle and Wilson 1990), during an invasion in which *E. variabilis* more or less replaced a congener, the grape leafhopper *E. elegantula*. Note that in 1985, the proportion of VLH was 1 for distances less than about 3 km and dropped to 50% at about 5 km. However, in 1986 these respective distances are about 7 km

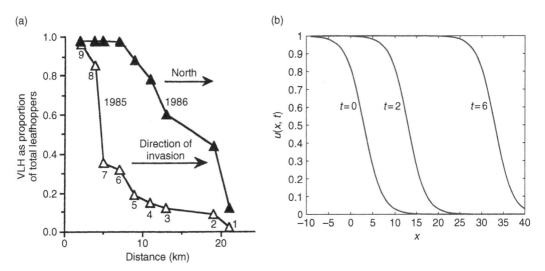

Figure 2.21. (a) The invasion pattern of a leafhopper in California (after Settle and Wilson 1990; with permission). Note that 1986 pattern is similar to that of 1985, only shifted to the right. We call this a traveling wave (of invasion). (b) A caricature of the traveling wave of invasion for three different times. Our goal is to understand how diffusion and logistic population growth combine to move the initial profile to the right.

and 20 km; it is as if the entire 1985 graph had shifted to the right.
Figure 2.21b is an idealized example of this phenomenon. The abscissa
is space and the ordinate is a function $u(x, t)$, which will become explicit
in a moment, shown at three different times – the subsequent times have
the same shape, but translated to the right. This kind of spatial–temporal
behavior is called a traveling wave.

R. A. Fisher thought a lot about this question in the context of the
spread of an advantageous allele. That is, imagine a single locus with
two alleles, a and A. Assume that A is more fit, but that the population is
initially mainly a; thus the fitness of the genotype AA is greater than that
of the genotype aa and heterozygotes are somewhere in between.
Suppose that we denote the frequency of A by $u(t)$. Then, in the absence
of spatial effects, the dynamics of $u(t)$ are

$$\frac{du}{dt} = su(1 - u) \tag{2.60}$$

where $s > 0$ is the selection coefficient, a function of the fitnesses of the
different genotypes AA, Aa and aa. As long as $u(0) > 0$, so that some of
the advantageous allele is present, we see that $u(t)$ will grow logistically
towards 1.

Fisher modified the dynamics given by Eq. (2.60) to include
space by assuming that there was undirected diffusion in space that
accompanied the logistic growth in time. Hence the resulting equation
would be

$$u_t = \frac{\sigma^2}{2} u_{xx} + su(1 - u) \tag{2.61}$$

We will consider an infinite spatial domain, but defer for a bit the
discussion of boundary conditions. For the initial condition, we assume
$u(x, 0)$ similar to the profile in Figure 2.21b.

To simplify Eq. (2.61) (and to show how exactly the same equation
arises in the discussion of invading organisms, rather than invading
genes), we will begin by scaling variables. First, divide both sides of
the equation by s. The left hand side is now $(1/s)(\partial u/\partial t)$, so that if
we defined a new time variable by $t' = st$, the left hand side would be
$\partial u/\partial t'$. After division by s, the first term on the right hand side of
Eq. (2.61) will be $(\sigma^2/2s)(\partial^2 u/\partial x^2)$, so that if we define a new space
like variable by $y = \sqrt{(2s/\sigma^2)}x$ the entire equation will become
$u_{t'} = u_{yy} + u(1 - u)$. Understanding that we are using scaled variables,
we can thus just as easily consider the equation

$$u_t = u_{xx} + u(1 - u) \tag{2.62}$$

which is called the Fisher equation.

Exercise 2.16 (E)

The model for logistic population growth and non-directed diffusion of an invading organism would be $N_t = (\sigma^2/2)N_{xx} + rN[1 - (N/K)]$. What scalings are needed to convert this to the same form as the Fisher equation (2.62)?

Now a traveling wave, such as shown in Figure 2.21b, keeps its shape as time changes but is displaced. Thus, at some time t, if we want to know the value of $u(x, t)$, we ask for the corresponding value of u at the initial time, but at a spatial point that is moved backwards from x. If the wave is traveling at speed $c > 0$, then to reach the point x at time t, it had to start at $x - ct$ at time 0. Thus $u(x, t)$ is only a function, let's call it U, of the combination $x - ct$, which we will call τ. In symbols, we write that $u(x, t) = U(\tau)$, where $\tau = x - ct$. Then the chain rule tells us $\partial u/\partial t = (dU/d\tau)(\partial\tau/\partial t) = -c(dU/d\tau)$ and $\partial^2 u/\partial x^2 = d^2 U/d\tau^2$. We thus are able to convert Eq. (2.62) from a partial differential equation for $u(x, t)$ to an ordinary differential equation for $U(t)$:

$$-c\frac{dU}{d\tau} = \frac{d^2 U}{d\tau^2} + U(1 - U) \qquad (2.63)$$

Since this is a second order equation, we need two conditions to specify its solution. With reference to Figure 2.21b, recall that we are thinking about an infinite spatial domain but a finite time domain. Also, with reference to that figure, at any time, as $x \to -\infty$, U approaches 1 and as $x \to \infty$, U approaches 0. We thus have the conditions that $U(-\infty) = 1$ and $U(\infty) = 0$ to go along with Eq. (2.63).

We are not going to try to solve Eq. (2.63), but we will succeed in analyzing it. The first step in this analysis is to convert it to a system of ordinary differential equations by introducing W as the derivative of U:

$$\frac{dU}{d\tau} = W$$

$$\frac{dW}{d\tau} = -cW - U(1 - U) \qquad (2.64)$$

The steady states of Eq. (2.64) are $(U, W) = (0, 0)$ and $(1, 0)$. This is very handy, since we know that $U = 1$ corresponds to $\tau \to -\infty$ and $U = 0$ corresponds to $\tau \to \infty$. The isoclines are also easy to compute: the line $W = 0$ (i.e. the U-axis) is the isocline for U and the parabola $W = [-U(1 - U)]/c$ is the isocline for W. These are shown in Figure 2.22a and b respectively. Our next step will be to characterize the steady states.

Exercise 2.17 (E)

Show that the eigenvalues of Eq. (2.64) when linearized around $(1, 0)$ are $\lambda_{1,2} = (-c \pm \sqrt{c^2 + 4})/2$ and when linearized around $(0, 0)$ are $\lambda_{1,2} = (-c \pm \sqrt{c^2 - 4})/2$.

Figure 2.22. Analysis of the traveling wave solution of the Fisher equation. (a) The isocline for U is the line $W = 0$. (b) Isocline for W is the parabola $W = [-U(1 - U)]/c$. (c) The eigenvalue analysis tells us that (1,0) is a saddle point and that if $c > 2$ that (0,0) is a stable node. (d) The traveling wave comes out of the saddle point and moves towards the stable node.

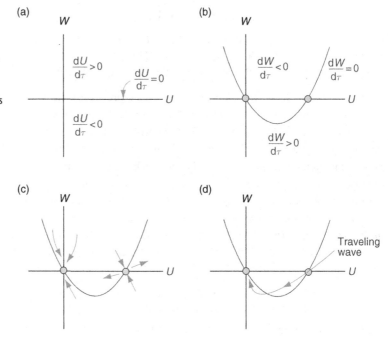

Regardless of the value of c, when we linearize around $(1, 0)$ one eigenvalue will be positive and one will be negative. We thus conclude that $(1, 0)$ is a saddle point; the isocline analysis tells us that one of the directions moving away from $(1, 0)$ moves towards $(0, 0)$, as shown in Figure 2.22c.

Now, the situation around $(0, 0)$ is a little bit more complicated. First, note that if $c < 2$, the eigenvalues will be complex, with a negative real part. This means that $(0, 0)$ will be a stable focus. However, if the origin is a focus, trajectories will spiral around it – which includes visiting values of U that are negative. Now, W can be negative, since it is the derivative of U, but U itself cannot be negative. We thus conclude on a biological basis that $c > 2$, in which case the origin is a stable node. As time increases, trajectories will approach the origin in the direction of the smaller (in absolute value) of the two eigenvalues.

Exercise 2.18 (E)

The condition $c > 2$ pertains to the scaled time and space variables. How does it translate to the original variables, involving the strength of selection (or maximum per capita growth rate for the ecological case) and the diffusion coefficient?

For a traveling wave to exist, a trajectory must come out of the saddle point at $(1, 0)$ and go directly into the stable node at $(0, 0)$ along the smaller of the two eigenvalues. The results of Exercise 2.11 tell us that the vector which joins the point $(1, 0)$ and $(1 - \varepsilon, -\varepsilon[(-c + \sqrt{c^2 + 4})/2])$ is the eigenvector corresponding to the positive eigenvalue at $(1, 0)$. We move to the point $(1 - \varepsilon, -\varepsilon[(-c + \sqrt{c^2 + 4})/2])$ and integrate Eq. (2.64) forward in time, given $c > 2$. In this way, we construct the trajectory that comes out of the saddle point directly into the origin, along its eigenvector (Figure 2.22d).

Connections

Life history invariants

To my knowledge, life history invariants were first made explicit by Ray Beverton and Sidney Holt in the 1950s, both in their studies of fishery management and their work on aging (Beverton and Holt 1957, 1959; Beverton 1992). Life history invariants have been rediscovered many times (and probably will continue to be rediscovered (Roff 1984, 1991)), although Charnov's book (Charnov 1993) will probably help to reduce the frequency of rediscovery. In this context, there often arises confusion between dimensionless parameters and life history invariants. In the context of our work, for example, k/m (or m/k) is a dimensionless variable – because both k and m are rates – but they are not necessarily invariants of the life history. However, their ratio is an invariant for relative size at maturity (as explained in the text). In 1994, Beverton gave a series of lectures at laboratories of the National Marine Fisheries Service across the USA. These were recently transcribed and published; they can be found at http://spo.nwr.noaa.gov//.Beverton Lectures1994/. The papers of Essington *et al.* (2001) and He and Stewart (2001) provide a very nice generalization of some of the ideas we have discussed, with applications to fisheries (which we discuss in Chapter 6).

Population dynamics in fluctuating environments

The papers of Dan Cohen are still classic in this area, and well worth reading. One topic that we did not consider here is the situation in which there is both individual and global variation. Such a situation arises, for example, when plants experience both microsite variation as individuals and environmental variation (e.g. weather) across the population. Rees *et al.* (2000) develop methods for such a mix of microsite and temporal variation. Provine's book about Sewall Wright

(Provine 1986) is well worth the read. Stearns (2000) has a lovely discussion of Daniel Bernoulli's contributions to these ideas (Bernoulli (1738/1954).

The logistic equation

It is important to know the history of one's discipline, and for population biology there is no better starting point than the book by Kingsland (1985). She explains how "logistic" enters English from Verhulst trying to compute ("logistique") the population of France.

The Ricker recruitment function

The Ricker recruitment function is widely used in population biology. Many years ago, Bob Costantino indicated in conversation to me that he thought the original ideas came from R. N. Chapman, an insect ecologist, who argued much in the same way that I did concerning the form of the recruitment function. Citation to Chapman's work (one example is Chapman (1928)) can be found in Costantino and Desharnais (1991), which itself is a rich and informative volume on flour beetles (*Tribolium* spp.). An obituary of Ricker (Beamish 2002), who died in 2001 at age 93, noted that he began as an entomologist and that it was only on a post-doctoral tour of Europe that he became professionally interested in fish population dynamics.

Chaos and complexity

It is hard, of course, to be awake in the twentyfirst century and not be aware of chaos theory and complexity (Steven Spielberg's dinosaur movie guaranteed that). But it is important to recognize that these remarkable properties of nonlinear dynamical systems were appreciated only in the late 1970s (although some of the more mathematical behavior was known long before that). The history by Gleick (1988) is well worth reading. Strogatz (1994) provides a good introduction to more mathematical approaches and Stewart (2000) has written a short turn-of-the-millennium report on the subject.

Bifurcations and catastrophe theory

The solution of the cubic equation, and its associated cusp discriminant, has an interesting history itself. Guedj (2000) provides a fictionalized account of this story in a delightful book. Catastrophe theory, popularized in the 1970s by René Thom and Chris Zeeman, entered our language again from the French word for sudden change ("catastrophe"), and in the mid 1970s the cusp catastrophe attracted lots of mathematicians and physicists who saw ways of explaining many kinds of biological and social phenomena without having to know the details of the application (Kolata 1977). An example of relaxation oscillation in a marine system is given by James *et al.* (2003); one of chaos by Chattopadhyay and Sarkar (2003). The next bifurcation in the one dimensional series is called the swallowtail (the names in use are still the ones picked by René Thom (1972/1975)) and corresponds to the steady states of the differential equation $dx/dt = -x^4 + \alpha x^2 + \beta x + \gamma$. Hernandez and Barradas (2003) put a nice ecological context around bifurcations and catastrophes. Readers of this book interested in conservation biology will surely already know what must be the simplest of the bifurcations (so simple that it is never even named, but see below), which occurs in the Levins patch model (Levins 1969, 1970). In this model, one focuses on the dynamics of the fraction of occupied patches in a metapopulation connected by dispersal. Patches become extinct at a rate m and are colonized in proportion to the number of occupied patches with proportionality constant c. The dynamics thus become

$$dp/dt = cp(1 - p) - mp = (c - m)p - cp^2$$

for which $p = 0$ is always a steady state. There is another steady state when $p = (c - m)/c$, which makes biological sense only if $c > m$. It is easy to see that $p = 0$ is an unstable steady state and that $p = (c - m)/c$ is stable. We thus conclude that as c declines towards m, the two steady states collide and the unstable steady state at $p = 0$ becomes stable. The bifurcation picture in this case is simple and I suggest that you try to draw it. A more complicated version of the Levins model involves the "rescue effect," in which patches can go extinct in some time interval and be colonized in that same time interval. The equation for the dynamics of patches then becomes $dp/dt = cp(1 - p) - [mp/(1 + Ap)]$ where A measures the size of the rescue. By simultaneously sketching $y = cp(1 - p)$ and $y = mp/(1 + Ap)$ you should convince yourself that there may be one, two, or three steady states of this system, depending upon the parameter values. There are, of course, plenty more complicated versions of the Levins model with various applications (Lin 2003). One of my personal favorite examples of situations corresponding to the cusp catastrophe involves work on models of the tuna–dolphin purse seine fishery that Colin Clark and I did when I was a graduate student, working for Colin as a research assistant (Clark and Mangel 1979). In this paper, we develop models for the tuna purse seine fishery and ask the question "What information does fishery related data give us about the status of the stock?". In cases where a cusp bifurcation may

occur, the answer can be "very little." The recently published book of
Bazykin (1998) makes his important work generally available (for many
years, the only English versions were preprints of translated papers). This
book contains a fully complete description of phase planes for predator–
prey, competitive, and mutualistic systems, and a good amount of work
on three species systems. It is well worth looking at. Scheffer *et al.* (2001)
and Scheffer and Carpenter (2003) discuss the role of the cusp catastrophe in ecosystem dynamics.

Differential equations in the phase plane

The case of $\Delta = 0$ requires more mathematics, as do cases in which
periodic orbits (limit cycles) exist surrounding an unstable focus. The
way that one demonstrates the existence of such limit cycles is to show
that the steady state is an unstable focus but that points in the phase plane
far away from the origin move towards it. A variety of good texts at the
next level exist; I suggest that you poke around at a book store and spend
time looking through different ones. A particularly simple example,
which is called the Hopf bifurcation, corresponds to a pair of differential
equations in polar coordinates for angle (θ) and radius from the origin (r):

$$\frac{d\theta}{dt} = c$$

$$\frac{dr}{dt} = a - r^2$$

where $c \neq 0$ and a are constants. In this case, the angular velocity is a
constant c. The dynamics of radius are more interesting. If $a < 0$, then
$dr/dt < 0$ and we see that the origin is a stable focus. However, if $a > 0$,
the origin is unstable and the circle $r = \sqrt{a}$ is stable. The parameter a
passing through 0, a stable focus becoming unstable, and the appearance
of a periodic orbit is called a Hopf bifurcation.

Spontaneous asymmetric synthesis

F. C. Frank (see Enderby (1998) to learn more about Frank) envisioned
the equations of spontaneous asymmetric synthesis as a means for the
evolution of optimal activity in biological compounds. That is, the two
types x and y in our equations are dynamically identical, but suppose
that x rotated light clockwise and y rotated light counter clockwise. We
could imagine a situation in which we go from $a < 1$ to $a > 1$ in
Eq. (2.47); for example if temperature is inversely related to a, then
as a system cools, the dynamics would change from (1, 1) being a stable
node (and a racemic mixture) to (1, 1) being a saddle point (and

optically active). However, because of molecular fluctuations, when this change takes place we do not expect the system to be exactly at (1, 1). Thus small deviations from (1, 1) will become amplified if $a > 1$. This is what Frank meant by spontaneous asymmetric synthesis and when my first paper on this subject appeared (Mangel and Ludwig 1977) he wrote to me suggesting that I had developed methods for a problem that he was unable to solve, which was a great treat for a young scientist. This topic is still of great interest. I encourage you to read Frank's original paper (Frank 1953) and a one dimensional version that predates it by Max Delbrück (Delbrück 1940); some of the more recent work on this subject – which can also point you towards other literature – is found in Pincock *et al.* (1971), Pincock and Wilson (1973), Soai *et al.* (1990), Link *et al.* (1997), Berger *et al.* (2000), Blackmond *et al.* (2001), Siegel (2002) and Singleton and Vo (2002).

More on mutualism

Mutualism has not received the attention it deserves (Wilkinson and Sherrat 2001), in comparison to predation or competition, but that is starting to be rectified (Wilson *et al.* 2003). Margulis and Sagan (2002) argue that ecology has been too dominated by the metaphors of competition and predation and that we should, in fact, have more focus on mutualism and symbiosis. Their argument is an interesting one and the book a worthwhile read, although their hostility to all mathematical methods puts this reader off a bit. Law and Dieckmann (1998) show how models similar to the ones we use here can help us to understand symbiosis.

Diffusion as a random walk

There are many different derivations of the diffusion equation. The one that I used here follows (Hughes 1995). In biology, of course, we trace the notion of diffusion to the botanist Robert Brown, who reported the irregular movement of pollen particles observed under the microscopic (Brown 1828). Hence, diffusion is often called – even by mathematicians – Brownian motion. In his miraculous year of 1905, Einstein published papers on the photoelectric effect (for which he was awarded the Nobel prize), special relativity and Brownian motion (a very nice reprint of this paper is found in Stachel (1998); there is also a Dover edition containing it). The paper on Brownian motion is particularly interesting because at the time he wrote it, there was still discussion about whether the atomic theory of matter was correct. Einstein wondered what the atomic theory of matter would mean for a large particle surrounded by a large number of randomly moving small ones. In answering this question, he derived the solution in Eq. (2.51) and connected the diffusion coefficient to temperature and Boltzmann's constant. Einstein had evidently heard about Brownian motion, but had not read the paper because he wrote "It is possible that the motions to be discussed here are identical with so-called Brownian molecular motion; however, the data available to me on the latter are so imprecise that I could not form a judgment on the question" (p. 85 in Stachel (1998)). The history of diffusion itself is quite interesting. As starting points, I suggest that you look at Wheatley and Augutter (1996) and Narasimhan (1999), which give two very interesting perspectives. In his interesting and provocative essay, Simberloff (1980) notes that 1859 was the year of publication of both *Origin of Species* and of Maxwell's work on the statistical distribution of velocities of particles in a gas; thus beginning the revolutions against determinism in both biology and physics coincide. We will have much more to say about diffusion and the random walk in Chapter 7 and 8.

Dirac delta functions and generalized functions

The book by Lighthill (1958) is a gem and well worth owning. The dedication of this book is wonderful: "To PAUL DIRAC who saw that it must be true, LAURENT SCHWARTZ who proved it, and GEORGE TEMPLE who showed how simple it could be made." Lighthill came to Vancouver in 1976 or 1977 and when my advisor Don Ludwig and I went to hear his lecture, Don said to me "Now, he's a real applied mathematician."

Separation of variables and Fourier series

The computation that we did for the diffusion equation with linear population growth is an example of a Fourier series solution of the diffusion equation. The way that we computed amplitudes of the different modes of cosine is an example of how one finds the Fourier coefficients. The method of Fourier series is an extremely powerful one and is used in many different ways in applied mathematics. A good introductory book on partial differential equations will explain how the method works in general; see, for example, Haberman (1998). In biological systems, we may have different boundary conditions, depending upon the situation (e.g. the size of a cell or a region and the nature of transport across the boundary). Some of my favorite investigations in this area involve the interaction of boundary conditions and the resulting patterns (Keller 2002, Murray 2003).

Linear and nonlinear diffusion

The calculation that we have completed is perhaps the simplest example dealing with pattern formation in biological systems. The classic paper in this area is due to Alan Turing (Turing 1952). An update c. 1990 is offered by Murray (1990) and updated again in Murray (2002). Maynard Smith (1968) offers a nice introduction to this material too; this was one of the first – if not the first – books on mathematical biology that I purchased; Harrison (1993) offers the perspective of a physical chemist turned biologist. The classic reference for linear population growth and diffusion is Skellam (1951); for a broader historical context, see Toft and Mangel (1991).

The Fisher equation, invasion biology and reaction diffusion equations

As with life history invariants, the creation of dimensionless combinations of variables is very useful. A good place to start getting more details about these methods is Lin and Segel (1988 (1974)). One could study the Fisher equation for different kinds of genetic models, such as heterozygote superiority, in which case we replace $su(1 - u)$ in the Fisher equation by a function $f(u)$ with the properties that $f(0) = f(a) = f(1) = 0$ and with the requirement that $u = a$ be a stable steady state, or heterozygote inferiority, in which $u = a$ is an unstable steady state. The literature on reaction–diffusion equations is enormous. And once one begins with systems that involve two variables and two spatial dimensions, the variety of interesting patterns and solutions is nearly endless. The books by Grindrod (1996), Kot (2001), and Murray (2002) are a good place to start learning about these; the paper by Levin and Segel (1985) is a classic. An interesting alternative approach for logistic growth in space and time is offered in the paper by Law *et al.* (2003); Medvinsky *et al.* (2002) and McLeod *et al.* (2002) use such models to understand plankton blooms; by Klein *et al.* (2003) to understand patterns of pollen dispersal, and by Kot *et al.* (1996) to understand invasions.

Chapter 3
Probability and some statistics

In the January 2003 issue of *Trends in Ecology and Evolution*, Andrew Read (Read 2003) reviewed two books on modern statistical methods (Crawley 2002, Grafen and Hails 2002). The title of his review is "Simplicity and serenity in advanced statistics" and begins as follows:

> One of the great intellectual triumphs of the 20th century was the discovery of the generalized linear model (GLM). This provides a single elegant and very powerful framework in which 90% of data analysis can be done. Conceptual unification should make teaching much easier. But, at least in biology, the textbook writers have been slow to get rid of the historical baggage. These two books are a huge leap forward.

A generalized linear model involves a response variable (for example, the number of juvenile fish found in a survey) that is described by a specified probability distribution (for example, the gamma distribution, which we shall discuss in this chapter) in which the parameter (for example, the mean of the distribution) is a linear function of other variables (for example, temperature, time, location, and so on).

The books of Crawley, and Grafen and Hails, are indeed good ones, and worth having in one's library. They feature in this chapter for the following reason. On p. 15 (that is, still within the introductory chapter), Grafen and Hails refer to the t-distribution (citing an appendix of their book). Three pages later, in a lovely geometric interpretation of the meaning of total variation of one's data, they remind the reviewer of the Pythagorean theorem – in much more detail than they spend on t-distribution. Most of us, however, learned the Pythagorean theorem long before we learned about the t-distribution.

If you already understand the *t*-distribution as well as you understand the Pythagorean theorem, you will likely find this chapter a bit redundant (but I encourage you to look through it at least once). On the other hand, if you don't, then this chapter is for you. My objective is to help you gain understanding and intuition about the major distributions used for general linear models, and to help you understand some tricks of computation and application associated with these distributions.

With the advent of generalized linear models, everyone's power to do statistical analysis was made greater. But this also means that one must understand the tools of the trade at a deeper level. Indeed, there are two secrets of statistics that are rarely, if ever, explicitly stated in statistics books, but I will do so here at the appropriate moments.

The material in this chapter is similar to, and indeed the structure of the chapter is similar to, the material in chapter 3 of Hilborn and Mangel (1997). However, regarding that chapter my colleagues Gretchen LeBuhn (San Francisco State University) and Tom Miller (Florida State University) noted its denseness. Here, I have tried lighten the burden. We begin with a review of probability theory.

A short course in abstract probability theory, with one specific application

The fundamentals of probability theory, especially at a conceptual level, are remarkably easy to understand; it is operationalizing them that is difficult. In this section, I review the general concepts in a way that is accessible to readers who are essentially inexperienced in probability theory. There is no way for this material to be presented without it being equation-dense, and the equations are essential, so do not skip over them as you move through the section.

Experiments, events and probability fundamentals

In probability theory, we are concerned with outcomes of "experiments," broadly defined. We let S be all the possible outcomes (often called the sample space) and A, B, etc., particular outcomes that might interest us (Figure 3.1a). We then define the probability that A occurs, denoted by $\Pr\{A\}$, by

$$\Pr\{A\} = \frac{\text{Area of } A}{\text{Area of } S} \qquad (3.1)$$

Figuring out how to measure the Area of A or the Area of S is where the hard work of probability theory occurs, and we will delay that hard work until the next sections. (Actually, in more advanced treatments, we replace the word "Area" with the word "Measure" but the fundamental

Figure 3.1. (a) The general set up of theoretical probability consists of a set of all possible outcomes S, and the events A, B, etc., within it. (b) Two helpful metaphors for discrete and continuous random variables: the fair die and a ruler on which a needle is dropped, constrained to fall between 1 cm and 6 cm. (c) The set up for understanding Bayes's theorem.

notion remains the same). Let us now explore the implications of this definition.

In Figure 3.1a, I show a schematic of S and two events in it, A and B. To help make the discussion in this chapter a bit more concrete, in Figure 3.1b, I show a die and a ruler. With a standard and fair die, the set of outcomes is 1, 2, 3, 4, 5, or 6, each with equal proportion. If we attribute an "area" of 1 unit to each, then the "area" of S is 6 and the probability of a 3, for example, then becomes 1/6. With the ruler, if we "randomly" drop a needle, constraining it to fall between 1 cm and 6 cm, the set of outcomes is any number between 1 and 6. In this case, the "area" of S might be 6 cm, and an event might be something like the needle falls between 1.5 cm and 2.5 cm, with an "area" of 1 cm, so that the probability that the needle falls in the range 1.5–2.5 cm is 1 cm/6 cm $= 1/6$.

Suppose we now ask the question: what is the probability that either A or B occurs. To apply the definition in Eq. (3.1), we need the total area of the events A and B (see Figure 3.1a). This is Area of A + Area of B − overlap area (because otherwise we count that area twice). The overlap area represents the event that both A and B occur, we denote this probability by

$$\Pr\{A, B\} = \frac{\text{Area common to } A \text{ and } B}{\text{Area of } S} \qquad (3.2)$$

so that if we want the probability of A or B occurring we have

$$\Pr\{A \text{ or } B\} = \Pr\{A\} + \Pr\{B\} - \Pr\{A, B\} \qquad (3.3)$$

and we note that if A and B share no common area (we say that they are mutually exclusive events) then the probability of either A or B is the sum of the probabilities of each (as in the case of the die).

Now suppose we are told that B has occurred. We may then ask, what is the probability that A has also occurred? The answer to this question is called the conditional probability of A given B and is denoted by $\Pr\{A|B\}$. If we know that B has occurred, the collection of all possible outcomes is no longer S, but is B. Applying the definition in Eq. (3.1) to this situation (Figure 3.1a) we must have

$$\Pr\{A|B\} = \frac{\text{Area common to } A \text{ and } B}{\text{Area of } B} \qquad (3.4)$$

and if we divide numerator and denominator by the area of S, the right hand side of Eq. (3.4) involves $\Pr\{A, B\}$ in the numerator and $\Pr\{B\}$ in the denominator. We thus have shown that

$$\Pr\{A|B\} = \frac{\Pr\{A, B\}}{\Pr\{B\}} \qquad (3.5)$$

This definition turns out to be extremely important, for a number of reasons. First, suppose we know that whether A occurs or not does not depend upon B occurring. In that case, we say that A is independent of B and write that $\Pr\{A|B\} = \Pr\{A\}$ because knowing that B has occurred does not affect the probability of A occurring. Thus, if A is independent of B, we conclude that $\Pr\{A, B\} = \Pr\{A\}\Pr\{B\}$ (by multiplying both sides of Eq. (3.5) by $\Pr\{B\}$). Second, note that A and B are fully interchangeable in the argument that I have just made, so that if B is independent of A, $\Pr\{B|A\} = \Pr\{B\}$ and following the same line of reasoning we determine that $\Pr\{B, A\} = \Pr\{B\}\Pr\{A\}$. Since the order in which we write A and B does not matter when they both occur, we conclude then that if A and B are independent events

$$\Pr\{A, B\} = \Pr\{A\}\Pr\{B\} \qquad (3.6)$$

Let us now rewrite Eq. (3.5) in its most general form as

$$\Pr\{A, B\} = \Pr\{A|B\}\Pr\{B\} = \Pr\{B|A\}\Pr\{A\} \qquad (3.7)$$

and manipulate the middle and right hand expression to conclude that

$$\Pr\{B|A\} = \frac{\Pr\{A|B\}\Pr\{B\}}{\Pr\{A\}} \qquad (3.8)$$

Equation 3.8 is called Bayes's Theorem, after the Reverend Thomas Bayes (see Connections). Bayes's Theorem becomes especially useful when there are multiple possible events $B_1, B_2, \ldots B_n$ which themselves are mutually exclusive. Now, $\Pr\{A\} = \sum_{i=1}^{n} \Pr\{A, B_i\}$ because the B_i are mutually exclusive (this is called the law of total probability). Suppose now that the B_i may depend upon the event A (as in

Figure 3.1c; it always helps to draw pictures when thinking about this material). We then are interested in the conditional probability $\Pr\{B_i|A\}$. The generalization of Eq. (3.8) is

$$\Pr\{B_i|A\} = \frac{\Pr\{A|B_i\}\Pr\{B_i\}}{\sum\limits_{j=1}^{n}\Pr\{A|B_j\}\Pr\{B_j\}} \tag{3.9}$$

Note that when writing Eq. (3.9), I used a different index (j) for the summation in the denominator. This is helpful to do, because it reminds us that the denominator is independent of the numerator and the left hand side of the equation.

Conditional probability is a tricky subject. In *The Ecological Detective* (Hilborn and Mangel 1997), we discuss two examples that are somewhat counterintuitive and I encourage you to look at them (pp. 43–47).

Random variables, distribution and density functions

A random variable is a variable that can take more than one value, with the different values determined by probabilities. Random variables come in two varieties: discrete random variables and continuous random variables. Discrete random variables, like the die, can have only discrete values. Typical discrete random variables include offspring numbers, food items found by a forager, the number of individuals carrying a specific gene, adults surviving from one year to the next. In general, we denote a random variable by upper case, as in Z or X, and a particular value that it takes by lower case, as in z or x. For the discrete random variable Z that can take a set of values $\{z_k\}$ we introduce probabilities p_k defined by $\Pr\{Z=z_k\}=p_k$. Each of the p_k must be greater than 0, none of them can be greater than 1, and they must sum to 1. For example, for the fair die, Z would represent the outcome of 1 throw; we then set $z_k=k$ for $k=1$ to 6 and $p_k=1/6$.

Exercise 3.1 (E)

What are the associated z_k and p_k when the fair die is thrown twice and the results summed?

A continuous random variable, like the needle falling on the ruler, takes values over the range of interest, rather than discrete specific values. Typical continuous random variables include weight, time, length, gene frequencies, or ages. Things are a bit more complicated now, because we can no longer speak of the probability that $Z=z$, because a continuous variable cannot take any specific value (the area of a point on a line is 0; in general we say that the measure of any specific value for a continuous random variable is 0). Two approaches are taken. First, we might ask for the probability that Z is less than or equal to a particular z. This is given by the probability distribution function (or just distribution function) for Z and usually denoted by an upper case letter such as $F(z)$ or $G(z)$ and we write:

$$\Pr\{Z \le z\} = F(z) \qquad (3.10)$$

In the case of the ruler, for example, $F(z) = 0$ if $z < 1$, $F(z) = z/6$ if z falls between 1 and 6, and $F(z) = 1$ if $z > 6$. We can create a distribution function for discrete random variables too, but the distribution function has jumps in it.

Exercise 3.2 (E)

What is the distribution function for the sum of two rolls of the fair die?

We can also ask for the probability that a continuous random variable falls in a given interval (as in the 1.5 cm to 2.5 cm example mentioned above). In general, we ask for the probability that Z falls between z and $z + \Delta z$, where Δz is understood to be small. Because of the definition in Eq. (3.10), we have

$$\Pr\{z \le Z \le z + \Delta z\} = F(z + \Delta z) - F(z) \qquad (3.11)$$

which is illustrated graphically in Figure 3.2. Now, if Δz is small, our immediate reaction is to Taylor expand the right hand side of Eq. 3.11 and write

$$\Pr\{z \le Z \le z + \Delta z\} = [F(z) + F'(z)\Delta z + o(\Delta z)] - F(z)$$
$$= F'(z)\Delta z + o(\Delta z) \qquad (3.12)$$

where we generally use $f(z)$ to denote the derivative $F'(z)$ and call $f(z)$ the probability density function. The analogue of the probability density function when we deal with data is the frequency histogram that we might draw, for example, of sizes of animals in a population.

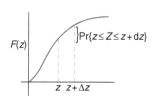

The exponential distribution

We have already encountered a probability distribution function, in Chapter 2 in the study of predation. Recall from there, the random variable of interest was the time of death, which we now call T, of an organism subject to a constant rate of predation m. There we showed that

$$\Pr\{T \le t\} = 1 - e^{-mt} \qquad (3.13)$$

Figure 3.2. The probability that a continuous random variable falls in the interval $[z, z + \Delta z]$ is given by $F(z + \Delta z) - F(z)$ since $F(z)$ is the probability that Z is less than or equal to z and $F(z + \Delta z)$ is the probability that Z is less than or equal to $z + \Delta z$. When we subtract, what remains is the probability that $z \le Z \le z + \Delta z$.

and this is called the exponential (or sometimes, negative exponential) distribution function with parameter m. We immediately see that $f(t) = me^{-mt}$ by taking the derivative, so that the probability that the time of death falls between t and $t + dt$ is $me^{-mt}dt + o(dt)$.

We can combine all of the things discussed thus far with the following question: suppose that the organism has survived to time t; what is the probability that it survives to time $t + s$? We apply the rules of conditional probability

$$\Pr\{\text{survive to time } t + s | \text{survive to time } t\} = $$
$$\frac{\Pr\{\text{survive to time } t + s, \text{survive to time } t\}}{\Pr\{\text{survive to time } t\}}$$

The probability of surviving to time t is the same as the probability that $T > t$, so that the denominator is e^{-mt}. For the numerator, we recognize that the probability of surviving to time $t + s$ and surviving to time t is the same as surviving to time $t + s$, and that this is the same as the probability that $T > t + s$. Thus, the numerator is $e^{-m(t+s)}$. Combining these we conclude that

$$\Pr\{\text{survive to } t + s | \text{survive to } t\} = \frac{e^{-m(t+s)}}{e^{-mt}} = e^{-ms} \qquad (3.14)$$

so that the conditional probability of surviving to $t + s$, given survival to t is the same as the probability of surviving s time units. This is called the memoryless property of the exponential distribution, since what matters is the size of the time interval in question (here from t to $t + s$, an interval of length s) and not the starting point. One way to think about it is that there is no learning by either the predator (how to find the prey) or the prey (how to avoid the predator). Although this may sound "unrealistic" remember the experiments of Alan Washburn described in Chapter 2 (Figure 2.1) and how well the exponential distribution described the results.

Moments: expectation, variance, standard deviation, and coefficient of variation

We made the analogy between a discrete random variable and the frequency histograms that one might prepare when dealing with data and will continue to do so. For concreteness, suppose that z_k represents the size of plants in the kth category and f_k represents the frequency of plants in that category and that there are n categories. The sample mean (or average size) is defined as $\bar{Z} = \sum_{k=1}^{n} f_k z_k$ and the sample variance (of size), which is the average of the dispersion $(z_k - \bar{Z})^2$ and usually given the symbol σ^2, so that $\sigma^2 = \sum_{k=1}^{n} f_k (z_k - \bar{Z})^2$.

These data-based ideas have nearly exact analogues when we consider discrete random variables, for which we will use $E\{Z\}$ to denote the mean, also called the expectation, and $Var\{Z\}$ to denote the variance and we shift from f_k, representing frequencies of outcomes in the data, to p_k, representing probabilities of outcomes. We thus have the definitions

$$E\{Z\} = \sum_{k=1}^{n} p_k z_k \qquad Var\{Z\} = \sum_{k=1}^{n} p_k (z_k - E\{Z\})^2 \qquad (3.15)$$

For a continuous random variable, we recognize that $f(z)dz$ plays the role of the frequency with which the random variable falls between z and $z + dz$ and that integration plays the role of summation so that we define (leaving out the bounds of integration)

$$E\{Z\} = \int zf(z)dz \qquad Var\{Z\} = \int (z - E\{Z\})^2 f(z)dz \qquad (3.16)$$

Here's a little trick that helps keep the calculus motor running smoothly. In the first expression of Eq. (3.16), we could also write $f(z)$ as $-(d/dz)[1 - F(z)]$, in which case the expectation becomes

$$E\{Z\} = -\int z\left(\frac{d}{dz}(1 - F(z))\right)dz$$

We integrate this expression using integration by parts, of the form $\int udv = uv - \int vdu$ with the obvious choice that $u = z$ and find a new expression for the expectation: $E\{Z\} = \int (1 - F(z))dz$. This equation is handy because sometimes it is easier to integrate $1 - F(z)$ than $zf(z)$. (Try this with the exponential distribution from Eq. (3.13).)

Exercise 3.3 (E)

For a continuous random variable, the variance is $Var\{Z\} = \int (z - E\{Z\})^2 f(z)dz$. Show that an equivalent definition of variance is $Var\{Z\} = E\{Z^2\} - (E\{Z\})^2$ where we define $E\{Z^2\} = \int z^2 f(z)dz$.

In this exercise, we have defined the second moment $E\{Z^2\}$ of Z. This definition generalizes for any function $g(z)$ in the discrete and continuous cases according to

$$E\{g(Z)\} = \sum_{k=1}^{n} p_k g(z_k) \qquad E\{g(Z)\} = \int g(z)f(z)dz \qquad (3.17)$$

In biology, we usually deal with random variables that have units. For that reason, the mean and variance are not commensurate, since the mean will have units that are the same as the units of the random variable but variance will have units that are squared values of the units of the random variable. Consequently, it is common to use the standard deviation defined by

$$SD(Z) = \sqrt{Var(Z)} \qquad (3.18)$$

since the standard deviation will have the same units as the mean. Thus, a non-dimensional measure of variability is the ratio of the standard deviation to the mean and is called the coefficient of variation

$$CV\{Z\} = \frac{SD(Z)}{E\{Z\}} \qquad (3.19)$$

Exercise 3.4 (E, and fun)

Three series of data are shown below:

Series A: 45, 32, 12, 23, 26, 27, 39
Series B: 1401, 1388, 1368, 1379, 1382, 1383, 1395
Series C: 225, 160, 50, 115, 130, 135, 195

Ask at least two of your friends to, by inspection, identify the most variable and least variable series. Also ask them why they gave the answer that they did. Now compute the mean, variance, and coefficient of variation of each series. How do the results of these calculations shed light on the responses?

We are now in a position to discuss and understand a variety of other probability distributions that are components of your toolkit.

The binomial distribution: discrete trials and discrete outcomes

We use the binomial distribution to describe a situation in which the experiment or observation is discrete (for example, the number of Steller sea lions *Eumatopias jubatus* who produce offspring, with one pup per mother per year) and the outcome is discrete (for example, the number of offspring produced). The key variable underlying a single trial is the probability p of a successful outcome. A single trial is called a Bernoulli trial, named after the famous probabilist Daniel Bernoulli (see Connections in both Chapter 2 and here). If we let X_i denote the outcome of the ith trial, with a 1 indicating a success and a 0 indicating a failure then we write

$$X_i = \begin{array}{l} 1 \text{ with probability } p \\ 0 \text{ with probability } 1-p \end{array} \qquad (3.20)$$

Virtually all computer operating systems now provide random numbers that are uniformly distributed between 0 and 1; for a uniform random number between 0 and 1, the probability density is $f(z) = 1$ if $0 \le z \le 1$ and is 0 otherwise. To simulate the single Bernoulli trial, we specify p, allow the computer to draw a uniform random number U and if

$U < p$ we consider the trial a success; otherwise we consider it to be a failure.

The binomial distribution arises when we have N Bernoulli trials. The number of successes in the N trials is

$$K = \sum_{i=1}^{N} X_i \tag{3.21}$$

This equation also tells us a good way to simulate a binomial distribution, as the sum of N Bernoulli trials.

The number of successes in N trials can range from $K = 0$ to $K = N$, so we are interested in the probability that $K = k$. This probability is given by the binomial distribution

$$\Pr\{K = k\} = \binom{N}{k} p^k (1-p)^{N-k} \tag{3.22}$$

In this equation $\binom{N}{k}$ is called the binomial coefficient and represents the number of different ways that we can get k successes in N trials. It is read "N choose k" and is given by $\binom{N}{k} = N!/k!(N-k)!$, where $N!$ is the factorial function.

We can explore the binomial distribution through analytical and numerical means. We begin with the analytical approach. First, let us note that when $k = 0$, Eq. (3.22) simplifies since the binomial coefficient is 1 and $p^0 = 1$:

$$\Pr\{K = 0\} = (1-p)^N \tag{3.23}$$

This is also the beginning of a way to calculate the terms of the binomial distribution, which we can now write out in a slightly different form as

$$\begin{aligned}
\Pr\{K = k\} &= \frac{N!}{k!(N-k)!} p^k (1-p)^{N-k} \\
&= \frac{N!(N-(k-1))}{k(k-1)!(N-(k-1))!} p^{k-1} p \frac{(1-p)^{N-(k-1)}}{1-p}
\end{aligned} \tag{3.24}$$

To be sure, the right hand side of Eq. (3.24) is a kind of mathematical trick and most readers will not have seen in advance that this is the way to proceed. That is fine, part of learning how to use the tools is to apprentice with a skilled craft person and watch what he or she does and thus learn how to do it oneself. Note that some of the terms on the right hand side of Eq. (3.24) comprise the probability that $K = k - 1$. When we combine those terms and examine what remains, we see that

$$\Pr\{K = k\} = \frac{N-k+1}{k} \frac{p}{1-p} \Pr\{K = k - 1\} \tag{3.25}$$

Equation (3.25) is an iterative relationship between the probability that $K=k-1$ and the probability that $K=k$. From Eq. (3.23), we know explicitly the probability that $K=0$. Starting with this probability, we can compute all of the other probabilities using Eq. (3.25). We will use this method in the numerical examples discussed below.

Although Eq. (3.24) seems to be based on a bit of a trick, here's an insight that is not: when we examine the outcome of N trials, something must happen. That is $\sum_{k=0}^{N} \Pr\{K = k\} = 1$. We can use this observation to find the mean and variance of the random variable K. The expected value of K is

$$
\begin{aligned}
\mathrm{E}\{K\} &= \sum_{k=0}^{N} k \Pr\{K = k\} = \sum_{k=0}^{N} k \binom{N}{k} p^k (1-p)^{N-k} \\
&= \sum_{k=1}^{N} k \binom{N}{k} p^k (1-p)^{N-k}
\end{aligned}
\tag{3.26}
$$

There is nothing tricky about what we have done thus far, but another trick now comes into play. We know how to evaluate the binomial sum from $k=0$, but not from $k=1$. So, we will manipulate terms accordingly by first writing the binomial coefficient explicitly and then factoring out Np from the expression on the right hand side of Eq. (3.26)

$$
\begin{aligned}
\mathrm{E}\{K\} &= \sum_{k=1}^{N} k \frac{N!}{k!(N-k)!} p^k (1-p)^{N-k} \\
&= Np \sum_{k=1}^{N} \frac{(N-1)!}{(k-1)!(N-k)!} p^{k-1} (1-p)^{N-k}
\end{aligned}
\tag{3.27}
$$

and we now set $j=k-1$. When $k=1, j=0$ and when $k=N, j=N-1$. The last expression in Eq. (3.27) becomes a recognizable summation:

$$
\mathrm{E}\{K\} = Np \sum_{j=0}^{N-1} \binom{N-1}{j} p^j (1-p)^{N-1-j}
\tag{3.28}
$$

In fact, the summation on the right hand side of Eq. (3.28) is exactly 1. We thus conclude that $\mathrm{E}\{K\} = Np$.

Exercise 3.5 (M)

Show that $\mathrm{Var}\{K\} = Np(1-p)$.

Next, let us think about the shape of the binomial distribution. That is, since the random variable K takes discrete values from 0 to N, when we plot the probabilities, we can (and will) do it effectively as a histogram and we can ask what the shape of the resulting histograms might look like. As a starting point, you should do an easy exercise that will help you learn to manipulate the binomial coefficients.

Exercise 3.6 (E)

By writing out the binomial probability terms explicitly and simplifying show that

$$\frac{\Pr\{K = k+1\}}{\Pr\{K = k\}} = \frac{(N - k)p}{(k + 1)(1 - p)} \qquad (3.29)$$

The point of Eq. (3.29) is this: when this ratio is larger than 1, the probability that $K = k + 1$ is greater than the probability that $K = k$; in other words – the histogram at $k + 1$ is higher than that at k. The ratio is bigger than 1 when $(N - k)p > (k + 1)(1 - p)$. If we solve this for k, we conclude that the ratio in Eq. (3.29) is greater than 1 when $(N + 1)p > k + 1$. Thus, for values of k less than $(N + 1)p - 1$, the binomial probabilities are increasing and for values of k greater than $(N + 1)p - 1$, the binomial probabilities are decreasing. Equations (3.25) and (3.29) are illustrated in Figure 3.3, which shows the binomial probabilities, calculated using Eq. (3.25), when $N = 15$ for three values of p (0.2, 0.5, or 0.7).

In science, we are equally interested in questions about what things might happen (computing probabilities given N and p) and inference or learning about the system once something has happened. That is, suppose we know that $K = k$, what can we say about N or p? In this case, we no longer think of the probability that $K = k$, given the parameters N and p. Rather, we want to ask questions about N and p, given the data. We begin to do this by recognizing that $\Pr\{K = k\}$ is really $\Pr\{K = k|N, p\}$ and we can also interpret the probability as the likelihood of different values of N and p, given k. We will use the symbol $\tilde{L}$ to denote likelihood. To begin, let us assume that N is known. The experiment we envision thus goes something like this: we conduct N trials, have k successes and want to make an inference about the value of p. We thus write the likelihood of p, given k and N as

$$\tilde{L}(p|k, N) = \binom{N}{k} p^k (1 - p)^{N-k} \qquad (3.30)$$

Note that the right hand side of this equation is exactly what we have been working with until now. But there is a big difference in interpretation: when the binomial distribution is summed over the potential values of k (0 to N), we obtain 1. However, we are now thinking of Eq. (3.30) as a function of p, with k fixed. In this case, the range of p clearly has to be 0 to 1, but there is no requirement that the integral of the likelihood from 0 to 1 is 1 (or any other number). Bayesian statistical methods (see Connections) allow us to both incorporate prior information about potential values of p and convert likelihood into things that we can think of as probabilities.

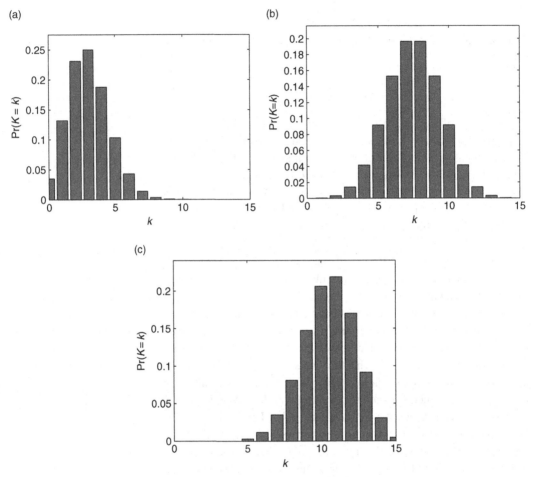

Figure 3.3. The binomial probability distribution when $N=15$ and $p=0.2$ (panel a), $p=0.5$ (panel b), or $p=0.7$ (panel c).

Only the left hand side – the interpretation – differs. For both historical (i.e. mathematical elegance) and computational (i.e. likelihoods often involve small numbers), it is common to work with the logarithm of the likelihood (called the log-likelihood, which we denote by L). In this case, of inference about p given k and N, the log-likelihood is

$$L(p|k,N) = \log \binom{N}{k} + k\log(p) + (N-k)\log(1-p) \qquad (3.31)$$

Now, if we think of this as a function of p, the first term on the right hand side is a constant – it depends upon the data but it does not depend upon p. We can use the log-likelihood in inference to find the most likely value of p, given the data. We call this the maximum likelihood estimate (MLE) of the parameter and usually denote it by $\hat{p}$. To find the MLE for p, we take the derivative of $L(p|k, N)$ with respect to p, set the derivative equal to 0 and solve the resulting equation for p.

Exercise 3.7 (E)

Show that the MLE for p is $\hat{p} = k/N$. Does this accord with your intuition?

Since the likelihood is a function of p, we ask about its shape. In Figure 3.4, I show $L(p|k, N)$, without the constant term (the first term on the right hand side of Eq. (3.31) for $k=4$ and $N=10$ or $k=40$ and $N=100$. These curves are peaked at $p=0.4$, as the MLE tells us they should be, and are symmetric around that value. Note that although the ordinates both have the same range (10 likelihood units), the magnitudes differ considerably. This makes sense: both p and $1-p$ are less than 1, with logarithms less than 0, so for the case of 100 trials we are multiplying negative numbers by a factor of 10 more than for the case of 10 trials.

The most impressive thing about the two curves is the way that they move downward from the MLE. When $N=10$, the curve around the MLE is very broad, while for $N=100$ it is much sharper. Now, we could think of each value of p as a hypothesis. The log-likelihood curve is then telling us something about the relative likelihood of a particular value of p. Indeed, the mathematical geneticist A. W. F. Edwards (Edwards 1992) calls the log-likelihood function the "support for different values of p, given the data" for this very reason (Bayesian methods show how to use the support to combine prior and observed information).

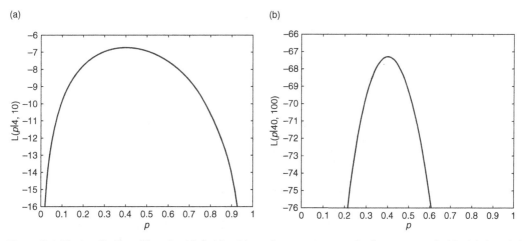

Figure 3.4. The log-likelihood function $L(p|k, N)$, without the constant term, for four successes in 10 trials (panel a) or 40 successes in 100 trials (panel b).

Of course, we never know the true value of the probability of success and in elementary statistics learn that it is helpful to construct confidence intervals for unknown parameters. In a remarkable paper, Hudson (1971) shows that an approximate 95% confidence interval can be constructed for a single peaked likelihood function by drawing a horizontal line at 2 units less than the maximum value of the log-likelihood and seeing where the line intersects the log-likelihood function. Formally, we solve the equation

$$L(p|k, N) = L(\hat{p}|k, N) - 2 \qquad (3.32)$$

for p and this will allow us to determine the confidence interval. If the book you are reading is yours (rather than a library copy), I encourage you to mark up Figure 3.4 and see the difference in the confidence intervals between 10 and 100 trials, thus emphasizing the virtues of sample size. We cannot go into the explanation of why Eq. (3.32) works just now, because we need to first have some experience with the normal distribution, but we will come back to it.

The binomial probability distribution depends upon two parameters, p and N. So, we might ask about inference concerning N when we know p and have data $K = k$ (the case of both p and N unknown will close this section, so be patient). The likelihood is now $\tilde{L}(N|k, p)$, but we can't go about blithely differentiating it and setting derivatives to 0 because N is an integer. We take a hint, however, from Eq. (3.29). If the ratio $\tilde{L}(N + 1|k, p)/\tilde{L}(N|k, p)$ is bigger than 1, then $N + 1$ is more likely than N. So, we will set that ratio equal to 1 and solve for N, as in the next exercise.

Exercise 3.8 (E)

Show that setting $\tilde{L}(N + 1|k, p)/\tilde{L}(N|k, p) = 1$ leads to the equation

$$(N + 1)(1 - p)/(N + 1 - k) = 1$$

Solve this equation for N to obtain $\hat{N} = (k/p) - 1$. Does this accord with your intuition?

Now, if $\hat{N} = (k/p) - 1$ turns out to be an integer, we are just plain lucky and we have found the maximum likelihood estimate for N. But if not, there will be integers on either side of $(k/p) - 1$ and one of them must be the maximum likelihood estimate of N. Jay Beder and I (Mangel and Beder 1985) used this method in one of the earliest applications of Bayesian analysis to fish stock assessment.

Suppose we know neither p nor N and wanted to make inferences about them from the data $K = k$. We immediately run into problems with maximum likelihood estimation, because the likelihood is maximized if we set $N = k$ and $p = 1$! Most of us would consider this a nonsensical result. But this is an important problem for a wide variety of applications: in fisheries we often know neither how many schools of fish are in the ocean nor the probability of catching them; in computer programming we know neither how many bugs are left in a program nor the chance of detecting a bug; in aerial surveys of Steller sea lions in Alaska in the summer, pups can be counted with accuracy because they are on the beach but some of the adults are out foraging at the time of the surveys, so we are confident that there are more non-pups than counted, but uncertain as to how many. William Feller (Feller 1971) wrote that problems are not solved by ignoring them, so ignore this we won't. But again, we have to wait until later in this chapter, after you know about the beta density, to deal with this issue.

The multinomial distribution: more than one kind of success

The multinomial distribution is an extension of the binomial distribution to the case of more than two (we shall assume n) kinds of outcomes, in which a single trial has probability p_i of ending in category i. In a total of N trials, we assume that k_i of the outcomes end in category i. If we let p denote the vector of the different probabilities of outcome and k denote the vector of the data, the probability distribution is then an extension of the binomial distribution

$$\Pr\{k|N,p\} = \frac{N!}{\prod\limits_{i=1}^{n} k_i!} \prod\limits_{i=1}^{n} p_i^{k_i}$$

The Poisson distribution: continuous trials and discrete outcomes

Although the Poisson distribution is used a lot in fishery science, it is named after Poisson the French mathematician who developed the mathematics underlying this distribution and not fish. The Poisson distribution applies to situations in which the trials are measured continuously, as in time or area, but the outcomes are discrete (as in number of prey encountered). In fact, the Poisson distribution that we discuss here can be considered the predator's perspective of random search and survival that we discussed in Chapter 2 from the perspective of the prey. Recall from there that the probability that the prey survives from time 0 to t is $\exp(-mt)$, where m is the rate of predation.

We consider a long interval of time [0, t] in which we count "events" that are characterized by a rate parameter λ and assume that in a small interval of time dt,

$$\text{Pr\{no event in the next } dt\} = 1 - \lambda dt + o(dt)$$
$$\text{Pr\{1 event in the next } dt\} = \lambda dt + o(dt) \qquad (3.33)$$
$$\text{Pr\{more than one event in the next } dt\} = o(dt)$$

so that in a small interval of time, either nothing happens or one event happens. However, in the large interval of time, many more than one event may occur, so that we focus on

$$p_k(t) = \text{Pr\{}k \text{ events in 0 to } t\} \qquad (3.34)$$

We will now proceed to derive a series of differential equations for these probabilities. We begin with $k=0$ and ask: how could we have no events up to time $t + dt$? There must be no events up to time t and then no events in t to $t + dt$. If we assume that history does not matter, then it is also reasonable to assume that these are independent events; this is an underlying assumption of the Poisson process. Making the assumption of independence, we conclude

$$p_0(t + dt) = p_0(t)(1 - \lambda dt - o(dt)) \qquad (3.35)$$

Note that I could have just as easily written $+o(dt)$ instead of $-o(dt)$. Why is this so (an easy exercise if you remember the definition of $o(dt)$)? Since the tradition is to write $+o(dt)$, I will use that in what follows.

We now multiply through the right hand side, subtract $p_0(t)$ from both sides, divide by dt and let $dt \to 0$ (our now standard approach) to obtain the differential equation

$$\frac{dp_0}{dt} = -\lambda p_0 \qquad (3.36)$$

where I have suppressed the time dependence of $p_0(t)$. This equation requires an initial condition. Common sense tells us that there should be no events between time 0 and time 0 (i.e. there are no events in no time), so that $p_0(0) = 1$ and $p_k(0) = 0$ for $k > 0$. The solution of Eq. (3.36) is an exponential: $p_0(t) = \exp(-\lambda t)$, which is identical to the random search result from Chapter 2. And it well should be: from the perspective of the predator, the probability of no prey found time 0 to t is exactly the same as the prey's perspective of surviving from 0 to t. As an aside, I might mention that the zero term of the Poisson distribution plays a key role in analysis suggesting (Estes *et al.* 1998) that sea otter declines in the north Pacific ocean might be due to killer whale predation.

Let us do one more together, the case of $k = 1$. There are precisely two ways to have 1 event in 0 to $t + dt$: either we had no event in 0 to t and one event in t to $t + dt$ or we had one event in 0 to t and no event in t to $t + dt$. Since these are mutually exclusive events, we have

$$p_1(t + dt) = p_0(t)[\lambda dt + o(dt)] + p_1(t)[1 - \lambda dt + o(dt)] \qquad (3.37)$$

from which we will obtain the differential equation $dp_1/dt = \lambda p_0 - \lambda p_1$, solved subject to the initial condition that $p_1(0) = 0$. Note the nice interpretation of the dynamics of $p_1(t)$: probability "flows" into the situation of 1 event from the situation of 0 events and flows out of 1 event (towards 2 events) at rate λ. This equation can be solved by the method of an integrating factor, which we discussed in the context of von Bertalanffy growth. The solution is $p_1(t) = \lambda t e^{-\lambda t}$. We could continue with $k = 2$, etc., but it is better for you to do this yourself, as in Exercise 3.9.

Exercise 3.9 (M)

First derive the general equation that $p_k(t)$ satisfies, using the same argument that we used to get to Eq. (3.37). Second, show that the solution of this equation is

$$p_k(t) = \frac{(\lambda t)^k}{k!} e^{-\lambda t} \qquad (3.38)$$

Equation (3.38) is called the Poisson distribution. We can do with it all of the things that we did with the binomial distribution. First, we note that between 0 and t something must happen, so that $\sum_{k=0}^{\infty} p_k(t) = 1$ (because the upper limit is infinite, I am going to stop writing it). If we substitute Eq. (3.38) into this condition and factor out the exponential term, which does not depend upon k, we obtain

$$e^{-\lambda t} \sum_{k=0} (\lambda t)^k / k! = 1$$

or, by multiplying through by the exponential we have $\sum_{k=0} (\lambda t)^k / k! = e^{\lambda t}$. But this is not news: the left hand side is the Taylor expansion of the exponential $e^{\lambda t}$, which we have encountered already in Chapter 2.

We can also readily derive an iterative rule for computing the terms of the Poisson distribution. We begin by noting that

$$\Pr\{\text{no event in } 0 \text{ to } t\} = p_0(t) = e^{-\lambda t} \qquad (3.39)$$

and before going on, I ask that you compare this equation with the first line of Eq. (3.33). Are these two descriptions inconsistent with each other? The answer is no. From Eq. (3.39) the probability of no event in 0 to dt is $e^{-\lambda dt}$, but if we Taylor expand the exponential, we obtain the first line in Eq. (3.33). This is more than a pedantic point, however. When one simulates the Poisson process, the appropriate formula to use is Eq. (3.39), which is always correct, rather than Eq. (3.33), which is only an approximation, valid for "small dt." The problem is that in computer simulations we have to pick a value of dt and it is possible that the value of the rate parameter could make Eq. (3.33) pure nonsense (i.e. that the first line is less than 0 or the second greater than 1).

Once we have $p_0(t)$ we can obtain successive terms by noting that

$$p_k(t) = e^{-\lambda t}\frac{(\lambda t)^k}{k!} = \frac{\lambda t}{k}\left(e^{-\lambda t}\frac{(\lambda t)^{k-1}}{(k-1)!}\right) = \frac{\lambda t}{k}p_{k-1}(t) \qquad (3.40)$$

and we use Eq. (3.40) in an iterative manner to compute the terms of the Poisson distribution, without having to compute factorials.

We will now find the mean and second moments (and thus the variance) of the Poisson distribution, showing many details because it is a good thing to see them once. The mean of the Poisson random variable K is

$$E\{K\} = \sum_{k=0}^{\infty} k\frac{e^{-\lambda t}(\lambda t)^k}{k!} = e^{-\lambda t}\left[(\lambda t) + \frac{2(\lambda t)^2}{2!} + \frac{3(\lambda t)^3}{3!} + \frac{4(\lambda t)^4}{4!} + \cdots\right]$$

and we now factor (λt) from the right hand side, simplify the fractions, and recognize the Taylor expansion of the exponential distribution

$$E\{K\} = e^{-\lambda t}(\lambda t)\left[1 + (\lambda t) + \frac{(\lambda t)^2}{2!} + \frac{(\lambda t)^3}{3!} + \cdots\right] = e^{-\lambda t}(\lambda t)e^{\lambda t} = \lambda t \quad (3.41)$$

Finding the second moment involves a bit of a trick, which I will identify when we use it. We begin with

$$E\{K^2\} = \sum_{k=0}^{\infty} k^2\frac{e^{-\lambda t}(\lambda t)^k}{k!} = e^{-\lambda t}\sum_{k=0}^{\infty} k\frac{(\lambda t)^k}{(k-1)!}$$

and as before we write out the last summation explicitly

$$E\{K^2\} = e^{-\lambda t}\left[(\lambda t) + \frac{2(\lambda t)^2}{1!} + \frac{3(\lambda t)^3}{2!} + \frac{4(\lambda t)^4}{3!} + \cdots\right]$$

$$= e^{-\lambda t}(\lambda t)\left[1 + 2(\lambda t) + \frac{3(\lambda t)^2}{2!} + \frac{4(\lambda t)^3}{3!} + \cdots\right]$$

$$= e^{-\lambda t}(\lambda t)\left[\frac{d}{d(\lambda t)}(\lambda t) + \frac{d}{d(\lambda t)}(\lambda t)^2 + \frac{d}{d(\lambda t)}\frac{(\lambda t)^3}{2!} + \frac{d}{d(\lambda t)}\frac{(\lambda t)^4}{3!} + \cdots\right]$$

$$= e^{-\lambda t}(\lambda t)\left[\frac{d}{d(\lambda t)}\left\{\lambda t\left(1 + \lambda t + \frac{(\lambda t)^2}{2!} + \frac{(\lambda t)^3}{3!} + \cdots\right)\right\}\right] \qquad (3.42)$$

and we now recognize, once again, the Taylor expansion of the exponential in the very last expression so that we have

$$E\{K^2\} = e^{-\lambda t}(\lambda t)\frac{d}{d(\lambda t)}(\lambda t e^{\lambda t}) = e^{-\lambda t}(\lambda t)[e^{\lambda t} + \lambda t e^{\lambda t}] = \lambda t + (\lambda t)^2 \quad (3.43)$$

and we thus find that $\text{Var}\{K\} = \lambda t$, concluding that for the Poisson process both the mean and variance are λt. The trick in this derivation comes in the third line of Eq. (3.42), when we recognize that the sum could be represented as the derivative of a different sum. This is a handy trick to know and to practice.

We can next ask about the shape of the Poisson distribution. As with the binomial distribution, we compare terms at $k-1$ and k. That is, we consider the ratio $p_k(t)/p_{k-1}(t)$ and ask when this ratio is increasing by requiring that it be bigger than 1.

Exercise 3.10 (E)

Show that $p_k(t)/p_{k-1}(t) > 1$ implies that $\lambda t > k$. From this we conclude that the Poisson probabilities are increasing until k is bigger than λt and decreasing after that.

The Poisson process has only one parameter that would be a candidate for inference: λ. That is, we consider the time interval to be part of the data, which consist of k events in time t. The likelihood for λ is $\tilde{L}(\lambda|k,t) = e^{-\lambda t}(\lambda t)^k/k!$ so that the log-likelihood is

$$L(\lambda|k,t) = -\lambda t + k\log(\lambda t) - \log(k!) \qquad (3.44)$$

and as before we can find the maximum likelihood estimate by setting the derivative of the log-likelihood with respect to λ equal to 0 and solving for λ.

Exercise 3.11 (E)

Show that the maximum likelihood estimate is $\hat{\lambda} = k/t$. Does this accord with your intuition?

As before, it is also very instructive to plot the log-likelihood function and examine its shape with different data. For example, we might imagine animals emerging from dens after the winter, or from pupal stages in the spring. I suggest that you plot the log-likelihood curve for $t = 5$, 10, 20, and $k = 4$, 8, 16; in each case the maximum likelihood estimate is the same, but the shapes will be different. What conclusions might you draw about the support for different hypotheses?

We might also approach this question from the more classical perspective of a hypothesis test in which we compute "p-values" associated with the data (see Connections for a brief discussion and entry into the literature). That is, we construct a function $P(\lambda|k,t)$ which is defined as the probability of obtaining the observed or more extreme data, when the true value of the parameter is λ. Until now, we have written the probability of exactly k events in time interval 0 to t as $p_k(t)$, understanding that λ was given and fixed. To be even more explicit, we could write $p_k(t|\lambda)$. With this notation, the probability of the observed or more extreme data when the true value of the parameter λ is now $P(\lambda|k, t) = \sum_{j=k}^{\infty} p_j(t|\lambda)$ where $p_j(t|\lambda)$ is the probability of observing j events, given that the value of the parameter is λ. Classical confidence intervals can be constructed, for example, by drawing horizontal lines at the value of λ for which $P(\lambda|k, t) = 0.05$ and $P(\lambda|k, t) = 0.95$.

I want to close this section with a discussion of the connection between the binomial and Poisson distributions that is often called the Poisson limit of the binomial. That is, let us imagine a binomial distribution in which N is very large (formally, $N \to \infty$) and p is very small (formally, $p \to 0$) but in a manner that their product is constant (formally, $Np = \lambda$; we will thus implicitly set $t = 1$). Since $p = \lambda / N$, the binomial probability of k successes is

$$\Pr\{k \text{ successes}\} = \frac{N!}{k!(N-k)!} \left(\frac{\lambda}{N}\right)^k \left(1 - \frac{\lambda}{N}\right)^{N-k}$$

and now let us simplify the factorials and the fraction to write

$$\Pr\{k \text{ successes}\} = \frac{N(N-1)(N-2)\dots(N-k+1)}{k!} \frac{\lambda^k}{N^k} \left(1 - \frac{\lambda}{N}\right)^{N-k}$$

which we now rearrange in the following way

$$\Pr\{k \text{ successes}\} = \frac{N(N-1)(N-2)\dots(N-K+1)}{N^k} \frac{\lambda^k \left(1 - \frac{\lambda}{N}\right)^N}{k! \left(1 - \frac{\lambda}{N}\right)^k} \tag{3.45}$$

and now we will analyze each of the terms on the right hand side. First, $N(N-1)(N-2)\dots(N-k+1)$, were we to expand it out would be a polynomial in N, that is it would take the form $N^k + c_1 N^{k-1} + \dots$, so that the first fraction on the right hand side approaches 1 as N increases. The second fraction is independent of N. As N increases, the denominator of the third fraction approaches 1, and the numerator, as you recall from Chapter 2, the limit as $N \to \infty$ of $[1 - (\lambda/N)]^N$ is $\exp(-\lambda)$. We thus conclude that in the limit of large N, small p with their product constant, the binomial distribution is approximated by the Poisson with parameter $\lambda = Np$ (for which we set $t = 1$ implicitly).

Random search with depletion

In many situations in ecology and evolutionary biology, we deal with random search for items that are then removed and not replaced (an obvious example is a forager depleting a patch of food items, or of mating pairs seeking breeding sites). That is, we have random search but the search parameter itself depends upon the number of successes and decreases with each success. There are a number of different ways of characterizing this case, but the one that I like goes as follows (Mangel and Beder 1985). We now allow λ to represent the maximum rate at which successes occur and ε to represent the decrement in the rate parameter with each success. We then introduce the following assumptions:

$$\Pr\{\text{no success in next } dt | k \text{ successes thus far}\} = 1 - (\lambda - \varepsilon k)dt + o(dt)$$

$$\Pr\{\text{exactly one success in next } dt | k \text{ successes thus far}\} = (\lambda - \varepsilon k)dt + o(dt)$$

$$\Pr\{\text{more than one success in the next } dt | k \text{ events thus far}\} = o(dt)$$

$$\text{(3.46)}$$

which can be compared with Eq. (3.33), so that we see the Poisson-like assumption and the depletion of the rate parameter, measured by ε.

From Eq. (3.46), we see that the rate parameter drops to zero when $k = \lambda / \varepsilon$, which means that the maximum number of events that can occur is λ / ε. This has the feeling of a binomial distribution, and that feeling is correct. Over an interval of length t, the probability of k successes is binomially distributed with parameters λ / ε and $1 - e^{-\varepsilon t}$. This result can be demonstrated in the same way that we derived the equations for the Poisson process. The conclusion is that

$$\Pr\{k \text{ events in}(0, t)\} = \binom{\frac{\lambda}{\varepsilon}}{k}(1 - e^{-\varepsilon t})^k (e^{-\varepsilon t})^{N-k} \qquad (3.47)$$

which is a handy result to know. Mangel and Beder (1985) show how to use this distribution in Bayesian stock assessment analysis for fishery management.

In this chapter, we have thus far discussed the binomial distribution, the multinomial distribution, the Poisson distribution, and random search with depletion. None will apply in every situation; rather one must understand the nature of the data being analyzed or modeled and use the appropriate probability model. And this leads us to the first secret of statistics (almost always unstated): there is always an under-lying statistical model that connects the source of data to the observed data through a sampling mechanism. Freedman *et al.* (1998) describe this process as a "box model" (Figure 3.5). In this view, the world consists of a source of data that we never observe but from which we sample. Each potential data point is represented by a box in this source population. Our sample, either by experiment or observation, takes boxes from the source into our data. The probability or statistical model is a mathematical representation of the sampling process. Unless you know the probability model, you do not fully understand your data. Be certain that you fully understand the nature of the trials and the nature of the outcomes.

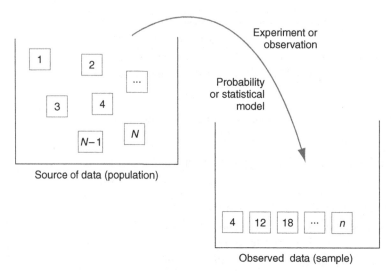

Figure 3.5. The box model of Freedman *et al.* (1998) is a useful means for thinking about probability and statistical models and the first secret of statistics. Here I have a drawn a picture in which we select a sample of size *n* from a population of size *N* (sometimes so large as to be considered infinite) using some kind of experiment or observation; each box in the population represents a potential data point in the sample, but not all are chosen. If you don't know the model that will connect the source of your data and the observed data, you probably are not ready to collect data.

The negative binomial, 1: waiting for success

In the next three sections, we will discuss the negative binomial distribution, which is perhaps one of the most versatile probability distributions used in ecology and evolutionary biology. There are two quite different derivations of the negative binomial distribution. The first, which we will do in this section, is relatively simple. The second, which requires an entire section of preparation, is more complicated, but we will do that one too.

Imagine that we are conducting a series of Bernoulli trials in which the probability of a success is p. Rather than specifying the number of trials, we ask the question: how long do we have to wait before the kth success occurs? That is, we define a random variable N according to

$$\Pr\{N = n|k,p\} = \text{Probability that the } k\text{th success occurs on trial } n \quad (3.48)$$

Now, for the kth success to occur on trial n, we must have $k - 1$ successes in the first $n - 1$ trials and a success on the nth trial. The probability of $k - 1$ successes in $n - 1$ trials has a binomial distribution with parameters $n - 1$ and p and the probability of success on the nth trial has probability p and these are independent of each other. We thus conclude

$$\Pr\{N = n|k,p\} = \binom{n-1}{k-1} p^{k-1}(1-p)^{n-k} p = \binom{n-1}{k-1} p^{k}(1-p)^{n-k}$$

$$(3.49)$$

This is the first form of the negative binomial distribution.

The negative binomial distribution, 2: a Poisson process with varying rate parameter and the gamma density

We begin with a simple enough situation: imagine a Poisson process in which the parameter itself has a probability distribution. For example, we might set up an experiment to monitor the emergence of *Drosophila* from patches of rotting fruit or vegetables in which we have controlled the number of eggs laid in the patch. Emergence from an individual patch could be modeled as a Poisson process but because individual patch characteristics vary, the rate parameter might be different for different patches. In that case, we reinterpret Eq. (3.38) as

$$\Pr\{k \text{ events in } [0, t]|\lambda\} = \frac{(\lambda t)^{k}}{k!} e^{-\lambda t} \qquad (3.50)$$

and we understand that λ has a probability distribution. Since λ is a naturally continuous variable, we assume that it has a probability density $f(\lambda)$. The product $\Pr\{k \text{ events}|\lambda\} f(\lambda)d\lambda$ is the probability that the rate parameter falls in the range λ to $\lambda + d\lambda$ and we observe k events. The probability of observing k events will be the integral of this product over all possible values of the rate parameter. Since it only makes sense to think about a positive value for the rate parameter, we conclude that

$$\Pr\{k \text{ events in } [0, t]\} = \int_{0}^{\infty} \frac{(\lambda t)^{k}}{k!} e^{-\lambda t} f(\lambda)d\lambda \qquad (3.51)$$

Equation (3.51) is often referred to as a mixture of Poisson processes. To actually compute the integral on the right hand side, we need to make further decisions. We might decide, for example, to replace the continuous probability density by an approximation involving a discrete number of choices of λ.

One classical, and very helpful, choice is that $f(\lambda)$ is a gamma probability density function. And before we go any further with the negative binomial distribution, we need to understand the gamma probability density for the rate parameter. There will be some detail, and perhaps some of it will be mysterious (why I make certain choices), but all becomes clear by the end of this section.

A gamma probability density for the rate parameter has two parameters, which we will denote by α and ν and has the mathematical form

$$f(\lambda) = \frac{\alpha^{\nu}}{\Gamma(\nu)} e^{-\alpha\lambda} \lambda^{\nu-1} \tag{3.52}$$

Since λ is a rate, we conclude that α must be a time-like variable for their product to be dimensionless (the precise meaning of α will be determined below). Similarly, ν must be dimensionless. In this equation, $\Gamma(\nu)$ is read "the gamma function of nu". Thus, before going on, we need to discuss the gamma function.

The gamma function

The gamma function is one of the classical functions of applied mathematics; here I will provide a bare bones introduction to it (see Connections for places to go learn more). You should think of it in the same way that you think about sin, cos, exp, and log. First, these functions have a specific mathematical definition. Second, there are known rules that relate functions with different arguments (such as the rule for computing $\sin(a+b)$) and there are computational means for obtaining their values. Third, these functions are tabulated (in the old days, in tables of books, and in the modern days in many software packages or on the web). The same applies to the gamma function, which is defined for $z > 0$ by

$$\Gamma(z) = \int_0^{\infty} s^{z-1} e^{-s} ds \tag{3.53}$$

In this expression, z can take any positive value, but let us start with the integers. In fact, let us start with $z = 1$, so that we consider $\Gamma(1) = \int_0^{\infty} e^{-s} ds = 1$. What about $z = 2$? In that case $\Gamma(2) = \int_0^{\infty} s e^{-s} ds$, which can be integrated by parts and we find $\Gamma(2) = 1$. We shall do one more, before the general case: $\Gamma(3) = \int_0^{\infty} s^2 e^{-s} ds$, which can be integrated by parts once again and from which we will see that $\Gamma(3) = 2$. If you do a few more, you should get a sense of the pattern: for integer values of z, $\Gamma(z) = (z-1)!$. Note, then, that we could write the binomial coefficient in Eq. (3.49) as

$$\binom{n-1}{k-1} = \frac{(n-1)!}{(k-1)!(n-k)!} = \frac{\Gamma(n)}{\Gamma(k)\Gamma(n-k+1)}$$

For non-integer values of z, the same kind of integration by parts approach works and leads us to an iterative equation for the gamma function, which is

$$\Gamma(z+1) = z\Gamma(z) \tag{3.54}$$

Finally, since $f(\lambda)$ is a probability density, its integral must be equal to 1 so that we can think of the gamma function as a normalization constant, as in $\int_0^\infty f(\lambda)d\lambda = 1$ from which we conclude

$$\int_0^\infty \frac{\alpha^\nu}{\Gamma(\nu)}e^{-\alpha\lambda}\lambda^{\nu-1}d\lambda = \frac{\alpha^\nu}{\Gamma(\nu)}\int_0^\infty e^{-\alpha\lambda}\lambda^{\nu-1}d\lambda = 1 \qquad (3.55)$$

Thus, the right hand integral in Eq. (3.55) allows us to see that

$$\int_0^\infty e^{-\alpha\lambda}\lambda^{\nu-1}d\lambda = \Gamma(\nu)/\alpha^\nu$$

which will be very handy when we find the mean and variance of the encounter rate. Note that we have just taken advantage of the information that $f(\lambda)$ is a probability density to do what appears to be a very difficult integral in our heads! Richard Feynman claimed that this trick was very effective at helping him impress young women in the 1940s (Feynman 1985).

Back to the gamma density

Now that we are more familiar with the gamma function, let us return to the gamma density given by Eq. (3.52). As with the gamma function, I will be as brief as possible, so that we can get back to the negative binomial distribution. In particular, we will examine the shape of the gamma density and find the mean and variance.

First, let us think about the shape of the gamma density (Figure 3.6). When $\nu = 1$, the algebraic term disappears and the gamma density is the same as the exponential distribution. When $\nu > 1$, the term $\lambda^{\nu-1}$ pins $f(0) = 0$ so that the gamma density will rise and then fall. Finally, when $\nu < 1$, $f(\lambda) \to \infty$ as $\lambda \to 0$. We thus see that the gamma density has a wide variety of shapes.

If we let Λ denote the random variable that is the rate of the Poisson process, then

$$E\{\Lambda\} = \int_0^\infty \lambda\frac{\alpha^\nu}{\Gamma(\nu)}e^{-\alpha\lambda}\lambda^{\nu-1}d\lambda = \frac{\alpha^\nu}{\Gamma(\nu)}\int_0^\infty e^{-\alpha\lambda}\lambda^\nu d\lambda = \frac{\alpha^\nu}{\Gamma(\nu)}\frac{\Gamma(\nu+1)}{\alpha^{\nu+1}} = \frac{\nu}{\alpha}$$

$$(3.56)$$

Be certain that you understand every step in this derivation (refer to Eq. (3.55) and to the equation just below it if you are uncertain).

Exercise 3.12 (E/M)

Use the same procedure to show that $E\{\Lambda^2\} = \nu(\nu+1)/\alpha^2$ and consequently that $Var\{\Lambda\} = \nu/\alpha^2 = (1/\nu)E\{\Lambda\}^2$.

The result derived in Exercise 3.12 has two important implications. The first is a diagnostic tool for the gamma density. That is

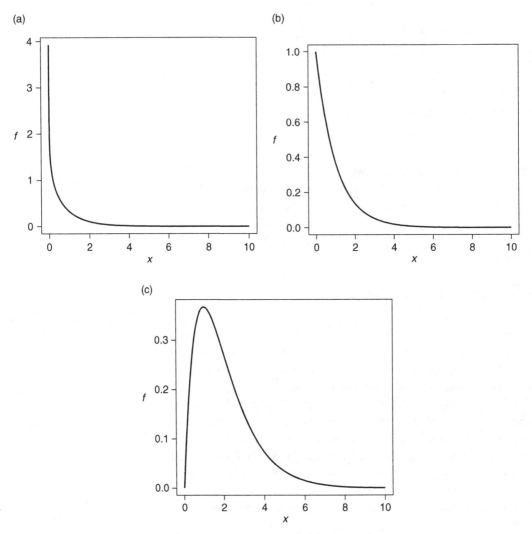

Figure 3.6. The gamma density has a wide variety of shapes depending upon the parameter ν. Here I have shown three densities, with $\alpha = 1$ and $\nu = 0.8$ (panel a), $\nu = 1$ (panel b) or $\nu = 2$ (panel c).

$$\log(\mathrm{Var}(\Lambda)) = -\log(\nu) + 2\log(E\{\Lambda\}) \qquad (3.57)$$

so that a logarithmic plot of variance versus mean will have a slope of 2 when the underlying random variable has a gamma density (but the converse is not true – see Dick (2004)). This is an example of a mean-variance power relationship; more details are provided in Connections.

Second, the coefficient of variation of Λ is directly found from Eq. (3.56) and Exercise 3.12:

$$\mathrm{CV}\{\Lambda\} = \frac{\sqrt{\mathrm{Var}\{\Lambda\}}}{E\{\Lambda\}} = \frac{\sqrt{\frac{\nu}{\alpha^2}}}{\frac{\nu}{\alpha}} = \frac{1}{\sqrt{\nu}} \qquad (3.58)$$

so that larger values of ν imply less relative variation in the distribution of the encounter rate. Let us thus conduct a thought experiment in which we hold the mean $\bar{\lambda} = \nu/\alpha$ constant (fixed from some other rule, for example), but allow $\nu \to \infty$ (obviously, then α must increase too). What happens to the probability density $f(\lambda)$? The density is becoming more and more peaked around the mean, because the coefficient of variation is getting smaller and smaller. In other words, $f(\lambda)$ will approach a delta-function centered at the mean $\bar{\lambda}$.

It has been a long, but worthwhile, detour.

Return to the negative binomial distribution

We are now ready to compute the probability of k events, as called for in Eq. (3.51). Using the gamma density we have

$$\Pr\{k \text{ events}\} = \int_0^\infty \frac{(\lambda t)^k}{k!} e^{-\lambda t} \frac{\alpha^\nu}{\Gamma(\nu)} e^{-\alpha\lambda} \lambda^{\nu-1} d\lambda = \frac{t^k \alpha^\nu}{k! \Gamma(\nu)} \int_0^\infty e^{-\lambda(\alpha+t)} \lambda^{k+\nu-1} d\lambda$$

(3.59)

and now we recognize that

$$\int_0^\infty e^{-\lambda(\alpha+t)} \lambda^{k+\nu-1} d\lambda = \Gamma(\nu+k)/(\alpha+t)^{\nu+k}$$

(why: because it is just like the integral following Eq. (3.55) except that ν and α are replaced by $\nu+k$ and $\alpha+t$). Consequently, we conclude

$$\Pr\{K = k\} = \frac{\alpha^\nu}{(\alpha+t)^{\nu+k}} \frac{t^k}{k!} \frac{\Gamma(\nu+k)}{\Gamma(\nu)}$$

(3.60)

Equation (3.60) is the second form of the negative binomial distribution.

The mean and variance of K are found using Eq. (3.60) in the same way that we found the mean and variance for the binomial or Poisson distributions. For the negative binomial distribution they are

$$E\{K\} = \frac{\nu}{\alpha} t \equiv m(t)$$

$$Var\{K\} = m(t) + \frac{m(t)^2}{\nu}$$

(3.61)

Note that I have introduced the mean $m(t)$. Clearly only two of the three of ν, α, or $m(t)$ are independent. For a variety of reasons, both operational and historical, it is good to work with $m(t)$, sometimes written just as m, and ν.

The mean of the negative binomial distribution is always bigger than the variance (for the Poisson distribution, recall, they are equal). We say then that the data are overdispersed and that ν is a measure of the overdispersion and hence is called the overdispersion parameter (eschewing the recommendation of Strunk and White about noun adjectives). Earlier, we concluded that if ν/α is held fixed, but that $\nu \to \infty$ then the probability density for λ will converge to a delta function centered on the mean $\bar{\lambda}$. Equation (3.61) is telling us the same information: that the overdispersion parameter increases, the mixture of Poisson distributions becomes more and more concentrated at a single value of the rate parameter and so the mean of the negative binomial distribution approaches the mean of the appropriate Poisson process. Indeed, as an optional (H) exercise, some readers may wish to show that in the limit of $\nu \to \infty$, Eq. (3.60) becomes the Poisson distribution.

Although I like to use the parameters ν and α, there are other forms commonly used in the ecological literature. Perhaps the most common is the "m, k" form that gained considerable popularity through the seminal book of Sir Richard Southwood (Southwood 1978). For reasons that will become clear momentarily, let us start using the random variable N for the number of events and rewrite Eq. (3.60) as

$$\Pr\{N = n\} = \left(\frac{\alpha}{\alpha + t}\right)^{\nu} \left(\frac{t}{\alpha + t}\right)^{n} \frac{1}{n!} \frac{\Gamma(\nu + n)}{\Gamma(\nu)} \tag{3.62}$$

and now introduce the mean $m = (\nu/\alpha)t$ by dividing numerator and denominator of the first two fractions on the right hand side of Eq. (3.62) by α and then multiplying by ν. We obtain

$$\Pr\{N = n\} = \frac{\Gamma(\nu + n)}{n!\Gamma(\nu)} \left(\frac{\nu}{\nu + m}\right)^{\nu} \left(\frac{m}{\nu + m}\right)^{n}$$

$$= \frac{\Gamma(\nu + n)}{n!\Gamma(\nu)} \left(\frac{\nu + m}{\nu}\right)^{-\nu} \left(\frac{m}{\nu + m}\right)^{n} \tag{3.63}$$

In the ecological literature, the overdispersion parameter is often represented by the symbol k in which case Eq. (3.63) becomes

$$\Pr\{N = n\} \equiv p_n(m, k) = \frac{\Gamma(k + n)}{n!\Gamma(k)} \left(\frac{k}{k + m}\right)^{k} \left(\frac{m}{k + m}\right)^{n} \tag{3.64}$$

and we will use this form of the negative binomial distribution in our study of host–parasitoid dynamics and of disease.

From Eq. (3.64) the probability of no events is obtained by setting $n = 0$

$$p_0(m, k) = \left(\frac{k}{k + m}\right)^{k} \tag{3.65}$$

This is a remarkable, if apparently innocuous, formula.

Exercise 3.13 (E)

Construct three plots of $p_0(m, k)$ (y-axis) vs. m (x-axis) as m runs from 10 to 500 for $k = 10$, 2, and 1. Interpret your results.

Next, we use Eqs. (3.64) and (3.65) to obtain an iterative equation relating subsequent terms, as we did for the Poisson and binomial distributions

$$p_j(m, k) = \left(\frac{j + k - 1}{j}\right)\left(\frac{m}{k + m}\right)p_{j-1}(m, k) \qquad (3.66)$$

Figure 3.7 is a comparison of the Poisson and negative binomial distributions. Here, I have set the mean equal to 10. The Poisson distribution is thus peaked around 10 and relatively symmetrical. The negative binomial distribution, with the same mean, becomes more and more skewed as the overdispersion parameter decreases from 5 to 0.5 (panels (b–d) in Figure 3.7). For $k = 0.5$ (Figure 3.6d), there is more than a 20% chance of 0 events, even though the mean is 10! Consequently, the probability of a large number of events (say 20–30 or even more) is considerable. Figure 3.8, in which I have plotted the cumulative distribution as if it were a continuous one, is another way of representing the idea. Notice that the cumulative values of the negative binomial are much higher than the cumulative values of the Poisson distribution for small values of the number of events and that they rise much more slowly than the Poisson for larger number of events. For example, at 20 events, the Poisson cumulative (with mean 10) is essentially 1, but the two negative binomial distributions that I have shown have nearly 20% of the probability still to be accounted.

We will close this section with an all too brief discussion of some aspects of inference involving the negative binomial. Let's begin with Eq. (3.61), for which the data would be the mean $\bar{K}$ and sample variance S_K^2 of a collection of random variables with a negative binomial distribution, for which we would want to estimate the parameters m and ν. If we replace the mean and variance in Eq. (3.61) by the sample average and sample variance and then solve Eq. (3.61) for the parameters, we obtain the method of moments estimates of the parameters, which are $\hat{m} = \bar{K}$ and $\nu = (\bar{K})^2/(S_K^2 - \bar{K})$. These estimates are simple, but not very accurate. More accurate estimates can be obtained by using maximum likelihood procedures with Eq. (3.60), but they are somewhat beyond the scope of what I want to do here. One good place to read about maximum likelihood for the negative binomial is Kendall and Stuart (1979) (which is also a generally good book). Dick (2004) discusses some modern methods for estimating the parameters.

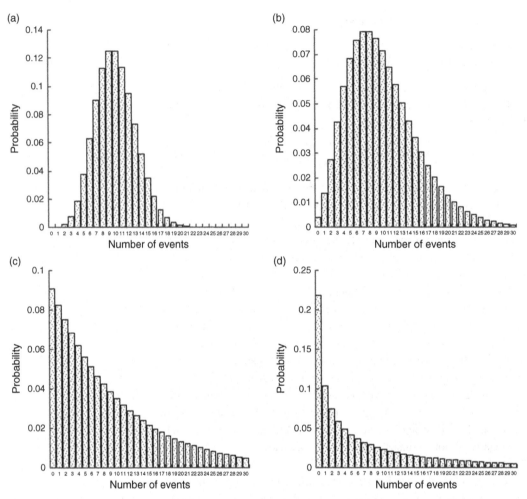

Figure 3.7. Comparison of Poisson and negative binomial frequency distributions. Panel (a) shows the Poisson distribution with mean $= 10$. Panels (b), (c), and (d) show negative binomial distributions with the same mean but with overdispersion parameter 5, 1, and 0.5 respectively. Note that although the abscissas are all the same, the ordinates differ.

There is one additional kind of inference that we should discuss, however. For this, let us return to Eq. (3.60) and recall its origins: we assumed a Poisson process, conditioned on the value of the rate parameter and assumed that the rate parameter had a gamma distribution. So, we can ask the question: given that we have observed k events, what does this tell us about the rate parameter? Formally, we want to find $\Pr\{\lambda \leq \Lambda \leq \lambda + \mathrm{d}\lambda | k$ events in 0 to $t\}$; we call this the posterior distribution of the rate parameter, given the data, and denote it by the symbol $f_{\mathrm{p}}(\lambda \,|(k,\,t))\mathrm{d}\lambda$. We apply the definitions of conditional probability and Bayes's theorem:

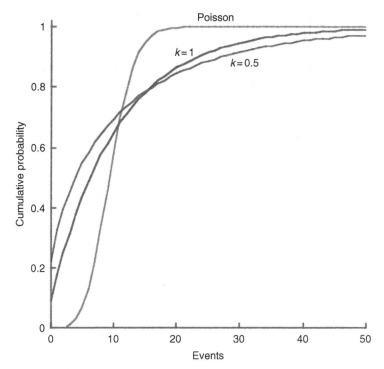

Figure 3.8. The cumulative distribution function for the Poisson (mean $= 10$) and negative binomial (mean $= 10$, two different values of the overdispersion parameter) plotted, for convenience, as continuous functions.

$$
\begin{aligned}
f_p(\lambda|(k,t))\mathrm{d}\lambda &= \frac{\Pr\{\lambda \le \Lambda \le \lambda + \mathrm{d}\lambda, k \text{ events in } 0 \text{ to } t\}}{\Pr\{k \text{ events in } 0 \text{ to } t\}} \\[6pt]
&= \frac{\Pr\{k \text{ events in } 0 \text{ to } t | \lambda\}f(\lambda)\mathrm{d}\lambda}{\Pr\{k \text{ events in } 0 \text{ to } t\}}
\end{aligned}
\tag{3.67}
$$

The key here is the transition from the numerator in the upper expression to the one in the lower expression, which relies on the definition of conditional probability, $\Pr\{A, B\} = \Pr\{A|B\}\Pr\{B\}$, and recalling that $f(\lambda)\mathrm{d}\lambda = \Pr\{\lambda \le \Lambda \le \lambda + \mathrm{d}\lambda\}$. Now, we know each of the probabilities that are called for on the right hand side of Eq. (3.67): the probability of k events given λ is Poisson, the probability density for λ is gamma, and the probability of k events is negative binomial. Substituting the appropriate distributions, we have

$$
f_p(\lambda|k) = \frac{\dfrac{(\lambda t)^k}{k!}e^{-\lambda t}\left(\dfrac{\alpha^\nu}{\Gamma(\nu)}e^{-\alpha\lambda}\lambda^{\nu-1}\right)}{\dfrac{\alpha^\nu}{(\alpha+t)^{\nu+k}}\dfrac{t^k}{k!}\dfrac{\Gamma(\nu+k)}{\Gamma(\nu)}}
\tag{3.68}
$$

Although this equation looks to be a bit of a mess, it actually simplifies very nicely and easily to become

$$
f_p(\lambda|k) = \frac{(\alpha+t)^{\nu+k}}{\Gamma(\nu+k)}e^{-(\alpha+t)\lambda}\lambda^{\nu+k-1}
\tag{3.69}
$$

We recognize this as another gamma density, with changed parameters: we started with parameters α and ν, collected data of k events in 0 to t, and update the parameters to $\alpha + t$ and $\nu + k$, while keeping the same distribution for the encounter rate. In the Bayesian literature, we say that the gamma density is the conjugate prior for the Poisson process (see Connections).

The normal (Gaussian) distribution: the standard for error distributions

We now turn to the normal or Gaussian distribution, which most readers will have encountered previously – both in other sources and in our discussion of the physical process of diffusion in Chapter 2. For that reason, I will not belabor matters and repeat much of what you already know, but will quickly move on to what I hope are new matters. However, some introduction is required.

The density function for a random variable X that is normally distributed with mean μ and variance σ^2 is

$$f(x) = \frac{1}{\sqrt{2\pi\sigma^2}} \exp\left(-\frac{(x-\mu)^2}{2\sigma^2}\right) \tag{3.70}$$

Note that I could have taken the square root of the variance, but chose to leave it within the square root. A particularly common and useful version is the normal distribution with mean 0 and variance 1; we denote this by $N(0, 1)$ and write $X \sim N(0, 1)$ to indicate that the random variable X is normally distributed with mean 0 and variance 1. In that case, the probability density function becomes $f(x) = 1/\sqrt{2\pi} \exp(-x^2/2)$. Indeed, it is easy to see that if a random variable Y is normally distributed with mean μ and variance σ^2 then the transformed variables $X = (Y - \mu)/\sigma$ will be $N(0, 1)$; we can make a normal random variable Y with specified mean and variance from a $X \sim N(0, 1)$ by setting $Y = \mu + \sigma X$.

Exercise 3.14 (E)

Demonstrate the validity of the previous sentence.

We already know that $f(x)$ given by Eq. (3.70) will approach a Dirac delta function centered at μ as $\sigma \to 0$. Recall that in Chapter 2 (Exercise 2.14), you showed where the normalization factor $1/\sqrt{2\pi}$ comes from by considering the product

$$I = \left[\int_{-\infty}^{\infty} e^{-\frac{x^2}{2}} dx\right]\left[\int_{-\infty}^{\infty} e^{-\frac{y^2}{2}} dy\right]$$

and then converting to polar coordinates in which $r^2 = x^2 + y^2$, $dxdy = rdrd\theta$, and r ranges from 0 to ∞ and θ ranges from 0 to 2π.

Abramowitz and Stegun (1974) give a variety of computational approximations for the normal probability density function in terms of

$$Z(x) = \frac{1}{\sqrt{2\pi}} \exp\left(-\frac{x^2}{2}\right) \qquad P(x) = \frac{1}{\sqrt{2\pi}} \int_{-\infty}^{x} \exp\left(-\frac{s^2}{2}\right) ds$$

$$Q(x) = \frac{1}{\sqrt{2\pi}} \int_{x}^{\infty} \exp\left(-\frac{s^2}{2}\right) ds \tag{3.71}$$

It should be apparent that $P(x) + Q(x) = 1$. In general, most of the computational formulae are not particularly transparent and, I suspect, were developed as much by trial and error as by formal analysis. There is one formula, however, which is easily understood and important; this is the behavior of $Q(x)$ when x is large. Recall from introductory statistics that hypothesis testing involves asking for the probability of obtaining the observed or more extreme data, given a certain hypothesis (whether this is a sensible question or not is, to some extent, one of the central disputes between frequentist and Bayesian statistics; see Connections for more details).

To be very specific, if somewhat trivial, let us suppose that we observe a single realization, x, of the random variable X and want to test the hypothesis that $X \sim N(0, 1)$. Our data consist of the observation x, which we will assume is positive, and the hypothesis is tested by computing the probability of obtaining a value of x or more extreme. That is, we need to evaluate $Q(x)$. The key to the computation lies in recognizing that

$$\exp\left(-\frac{s^2}{2}\right) = -\frac{1}{s}\frac{d}{ds}\left[\exp\left(-\frac{s^2}{2}\right)\right]$$

so that we can write the integral in $Q(x)$ as

$$\int_{x}^{\infty} \exp\left(-\frac{s^2}{2}\right) ds = \int_{x}^{\infty} \left(-\frac{1}{s}\frac{d}{ds}\left[\exp\left(-\frac{s^2}{2}\right)\right]\right) ds$$

We now integrate the right hand side by parts ($\int u dv = uv - \int v du$) with $u = -1/s$ and $v = \exp(-s^2/2)$ to obtain

$$Q(x) = \frac{1}{\sqrt{2\pi}}\left[\frac{1}{x}\exp\left(-\frac{x^2}{2}\right) + \int_{x}^{\infty}\frac{1}{s}\exp\left(-\frac{s^2}{2}\right) ds\right] \tag{3.72}$$

Now when x is big, the integrand on the right hand side of Eq. (3.72) is surely smaller than the original integrand. To deal with this integral, we integrate by parts again, which makes the resulting integrand even smaller. Repeated application of integration by parts will give us what is called an asymptotic expansion (see Connections for more about asymptotic expansions) of $Q(x)$ when x is large, which can be written as

$$Q(x) = \frac{Z(x)}{x} \left\{ 1 - \frac{1}{x^2} + \frac{3}{x^4} - \frac{(3)(5)}{x^6} + \cdots \right\}$$ (3.73)

and we note that the terms inside the brackets will rapidly decrease when x is even just moderately large (compute them, say for $x = 4$ or 5 and convince yourself). We will use this kind of asymptotic expansion in our study of stochastic population theory.

Gauss popularized the use of the normal distribution as an error distribution when we make measurements (he spent a lot of time observing the motion of the planets and stars). The simplest such model might go as follows. Imagine that we take n measurements of a constant but unknown quantity M, which we want to estimate from these measurements, denoted by Y_i, $i = 1, 2, \ldots n$. As a start on the estimation procedure, we could pick values of M, denoted by m, and ask how well a particular value matches the data. We thus need a means to characterize the error between the observations and our choice m. Gauss recognized that the characterization should be positive, so that errors of one sign do not counter errors of the other sign. One choice for the error between the ith observation and m would then be $|Y_i - m|$, but the absolute value has some mathematical properties that make it hard to work with. The next simplest choice is $(Y_i - m)^2$, and this is what we settle upon. It is the squared error between a single observation and our estimate of the unknown parameter. The combined squared errors are then a function $SSQ(m)$ of the estimate of the unknown parameter, given the data:

$$SSQ(m) = \sum_{i=1}^{n} (Y_i - m)^2$$ (3.74)

and it is sensible to conclude that the best estimate for M is the value of m that minimizes the sum of squared deviations.

Exercise 3.15 (E)

Show that $SSQ(m)$ is minimized when m is the sample average $1/n \sum_{i=1}^{n} Y_i$.

Suppose that our first five data points are 6.4694, 5.096, 6.0359, 5.3725, 6.5354. The curve marked $n = 5$ in Figure 3.9 shows $SSQ(m)$ using these data. Now imagine that we collect another five data points: 6.5529, 5.7963, 3.945, 6.1326, 7.5929. The curve marked $n = 10$ in Figure 3.9 shows $SSQ(m)$ using all 10 points. Both curves have a minimum around 6 (which is the true value of M used to simulate the data). The curve marked $n = 10$ is uniformly higher than that marked

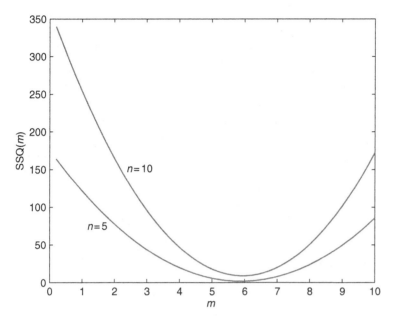

Figure 3.9. The sum of squared deviations for estimate of an unknown value M by the data 6.4694, 5.096, 6.0359, 5.3725, 6.5354, 6.5529, 5.7963, 3.945, 6.1326, 7.5929. The curve marked $n=5$ uses only the first 5 values; that marked $n=10$ uses all 10 data points.

$n = 5$ and this is to be expected since there are twice as many data points in the former case. But the curve marked $n = 10$ is also steeper than that marked $n = 5$: it rises from our best estimate of approximately 6 in a sharper manner than the curve marked $n = 5$. Based on our experience with likelihood, and Hudson's formula (Eq. (3.32)), we would expect that the steepness should tell us something about the likelihood of different values of m.

To answer that question, that is to be able to assess the likelihood of different values of m rather than just find the best estimate, we need to introduce a statistical model. And what would the simplest model be? How about this one:

$$Y_i = m + X_i \tag{3.75}$$

where X_i is $N(0, 1)$. If we accept Eq. (3.75) as the statistical model, then the sum of squared deviations is the same as $\sum_{i=1}^{n} (X_i)^2$.

The sum of the squares of n normally distributed random variables is a new statistical quantity for us. It is called the chi-square distribution with n degrees of freedom and has probability distribution function

$$P(z|n) = \Pr\{\chi^2 \le z|n\} = \frac{1}{[2^{\frac{n}{2}}\Gamma(\frac{n}{2})]} \int_0^z t^{\frac{n-2}{2}} e^{-\frac{1}{2}t} dt \tag{3.76}$$

Although Eq. (3.76) is complicated, there is nothing in it new to us (except wondering how this distribution is derived, which is beyond the scope of the book – and I am sorry that this is an exception to the rule of self-containment). We can use Eq. (3.76) to associate a probability with any observed value of SSQ(m). As with the normal distribution, we are often interested in $Q(z|n) = 1 - P(z|n)$ since this will give us the probability of observing a value of χ^2 greater than z. Many software programs provide built-in routines for computation of $Q(z|n)$.

Before leaving this simple example, let us consider it from an explicit likelihood-based perspective. That is, we wish to compute the likelihood of the data, given a particular value of m. We rely again on the notion that $X_i = Y_i - m$ is normally distributed with mean 0 and variance 1, so that the likelihood is

$$\tilde{L}(Y_1, Y_2, Y_3, \ldots Y_n | m) = \prod_{i=1}^{n} \frac{1}{\sqrt{2\pi}} \exp\left(-\frac{(Y_i - m)^2}{2}\right) \tag{3.77}$$

and the log-likelihood is

$$L(Y_1, Y_2, Y_3, \ldots Y_n | m) = -\sum_{i=1}^{n} \frac{1}{2}\log(2\pi) + \frac{(Y_i - m)^2}{2} \tag{3.78}$$

and from these equations, we see that the likelihood is maximized by making the sum of squared deviations as small as possible. That is, if the error distribution is normally distributed and the variance is known, then estimating the mean by maximum likelihood or by minimizing the sum of squared deviations is exactly the same.

Linear regression, least squares and total least squares: measurement errors in both x and y

Before moving on, I want to briefly discuss linear regression and least squares (which most readers are probably familiar with) and total least squares (probably not). The set up is that we have a set of data consisting of a set of observations of a putative causative variable X and response variable Y; the data are thus pairs $\{X_i, Y_i\}$ for $i = 1, \ldots, n$. The question is this: how do we characterize the relationship between X and Y?

The first possible answer is that there is no relationship between the two. If we were to write a formal statistical model, a natural choice is

$$Y_i = a + Z_i \tag{3.79}$$

where the parameter a is to be determined and we assume that Z_i is normally distributed with mean 0 and variance σ. We already know that the maximum likelihood estimate of a is the same as the value that minimizes the sum of squared deviations $\sum_{i=1}^{n} (Y_i - a)^2$ and that this estimate is the same as the sample average. Let us denote that estimate by $\hat{a}$. We can then compute the residual sum of squared errors according to $\sum_{i=1}^{n}(Y_i - \hat{a})^2$. This quantity can be thought of as the amount of variability in the data that is not explained by the model from Eq. (3.79).

Second, we might assume that there is a linear relationship between the causative and response variables, so that the formal statistical model is

$$Y_i = a + bX_i + Z_i \qquad (3.80)$$

where we now must determine the parameter b – and this is crucial – we assume that the causative variable X_i is measured with certainty. Proceeding as before, we can compute either the sum or squared deviations SSQ(a, b) or the log-likelihood L(a, b). They are

$$\mathrm{SSQ}(a, b) = \sum_{i=1}^{n}(Y_i - a - bX_i)^2$$

$$\mathrm{L}(a, b) = -n\left[\log(\sigma) + \frac{1}{2}\log(2\pi)\right] - \frac{1}{2\sigma^2}\sum_{i=1}^{n}(Y_i - a - bX_i)^2 \qquad (3.81)$$

and we see that once again maximizing the likelihood is the same as minimizing the sum of squared deviations. To do that, we take the derivatives of SSQ(a, b) first with respect to a, then with respect to b, and set them equal to 0, in order to obtain equations for the maximum likelihood estimates for a and b.

Exercise 3.16 (E/M)

Show that the maximum likelihood estimates for a and b are the solution of the equations

$$\sum_{i=1}^{n} Y_i = n\hat{a} + \hat{b}\sum_{i=1}^{n} X_i$$

$$\sum_{i=1}^{n} X_i Y_i = \hat{a}\sum_{i=1}^{n} X_i + \hat{b}\sum_{i=1}^{n} X_i^2 \qquad (3.82)$$

and then solve these equations for the maximum likelihood values of the parameters.

Once we have computed these parameters, the remaining variation, which is unexplained by the linear model, is $\sum_{i=1}^{n}(Y_i - \hat{a} - \hat{b}X_i)^2$. We might then ask if the linear model is an improvement over no model at all. One way of assessing this is to ask how much the variation is explained by the linear model, relative to the constant model. The common way to do that is with the ratio

$$r^2 = 1 - \frac{\sum_{i=1}^{n}(Y_i - \hat{a} - \hat{b}X_i)^2}{\sum_{i=1}^{n}(Y_i - \hat{a})^2} \qquad (3.83)$$

Figure 3.10. In ordinary least squares (OLS) we minimize the vertical distance between the regression line $y = a + bx$ and the data points. In total least squares (TLS), we minimize the actual distance between the data points and the regression line.

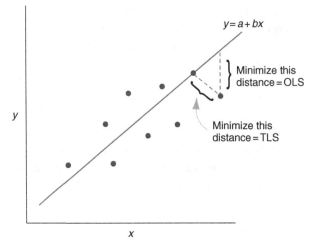

Figure 3.10. In ordinary least squares (OLS) we minimize the vertical distance between the regression line $y = a + bx$ and the data points. In total least squares (TLS), we minimize the actual distance between the data points and the regression line.

where we understand that the $\hat{a}$s in the fraction are different – coming from minimizing the sum of squares in the linear model (numerator) or the constant model (denominator). This ratio is the fraction of the variance explained by the linear model relative to the constant model and is a handy tool for measuring how much understanding we have gained from the increase in complexity of the model.

The least squares procedure that we have been discussing is called ordinary least squares (OLS) and works under the presumption that the causative variables are measured without error. But often that condition cannot be guaranteed – we simply cannot measure the X_i accurately. The general formulation of this problem, in terms of a statistical model, is very complicated (see Connections) and is called the errors in variables problem. However, we can discuss an extension of least squares, called total least squares, to this case. In ordinary least squares (Figure 3.10) we minimize the vertical distance between the regression line $y = a + bx$ and the data points (that is, we assume no error in the measurement of X). In total least squares, we minimize the actual distance between the data points and the regression line. That is, we let $\{x_{\hat{c}i}, y_{\hat{c}i}\}$ denote the point on the regression line closest to the data point $\{X_i, Y_i\}$, and find it by choosing $\sum_{i=1}^{n} (X_i - x_{\hat{c}i})^2 + (Y_i - y_{\hat{c}i})^2$ to be a minimum. To operationalize this idea, we need to find the closest point. The line has slope b, so that the line segment perpendicular to the regression line will have slope $-1/b$ and the equation of the line joining the data and the closest point is $(Y_i - y_{\hat{c}i})/(X_i - x_{\hat{c}i}) = -(1/b)$. Since we know that the closest point is on the regression line $y = a + bx$ we conclude that

$$\frac{Y_i - (a + bx_{\hat{c}i})}{X_i - x_{\hat{c}i}} = -\frac{1}{b} \tag{3.84}$$

and view this as an equation for $x_{\hat{c}i}$. We solve Eq. 3.84 for $x_{\hat{c}i}$ and obtain

$$x_{\hat{c}i} = \frac{1}{(1+b^2)}[X_i + b(Y_i - a)] \tag{3.85}$$

and then $y_{\hat{c}i} = a + bx_{\hat{c}i}$. We thus know the description of the closest points in terms of the data and the parameters, and can now evaluate the sum of squares and minimize it over the choice of parameters. For more details about this, see Connections.

The *t*-distribution and the second secret of statistics

We now come to the *t*-distribution, presumably known to most readers because they have encountered the *t*-test in an introductory statistics course. The apocrypha associated with this distribution is fascinating (Freedman *et al.* 1998): all agree that W. S. Gossett (1876–1936) worked for Guiness Brewery, developed the ideas, and published under the name of "Student." Whether he did this to keep industrial secrets (Yates 1951) or because his employers did not want him to be doing such work so that he tried to keep it secret from them, is part of the legend. According to Yates (1951), Gossett set out to find the exact distribution of the sample standard deviation, of the ratio of the mean to the standard deviation, and of the correlation coefficient. He was trained as a chemist, not a mathematician, and ended up using experiment and curve fitting to obtain the answers (which R. A. Fisher later proved to be correct).

There are three ways, of increasing complexity, of thinking about the *t*-distribution. The first is a simple empirical observation: very often – especially with ecological data – the normal distribution does not give a good fit to the data because the tails of the data are "too high." That is, there are too many data points with large deviations for the data to be likely from a normal distribution.

The second is this: whenever we take measurements with error (i.e. almost always when we take measurements), we need to estimate the standard deviation of the normal distribution assumed to characterize the errors. But with a limited number of measurements, it is hard to estimate the standard deviation accurately. And this is the second secret of statistics: we almost never know the standard deviation of the error distribution.

The third approach is a more formal, mathematical one. To begin, we note that if Y is normally distributed with mean 0 and variance σ^2, then $X = Y/\sigma$ will be normally distributed with mean 0 and variance 1. Now, when we take a series of measurements and compute the squared deviations, we will end up with a chi-square random variable. If we let χ_n^2 denote a chi-square random variable with n degrees of freedom and X denote a $N(0, 1)$ random variable, then the ratio $T = X/\chi_n^2$ is

said to be a Student's t-random variable with n degrees of freedom. It
has probability density function

$$f(t) = c\left(1 + \frac{t^2}{n}\right)^{-\frac{n+1}{2}}$$

(3.86)

where c is a normalization constant, chosen so that $\int_{-\infty}^{\infty} f(t)\,dt = 1$.
I have not explicitly written it out because c involves the beta function,
which we have not encountered yet, but will soon.

Since c is a constant, we can learn a bit about $f(t)$ by examining
Eq. (3.86) as a function of t. For example, note that $f(t)$ is symmetrical
because t appears only as a square; thus we conclude that $f(-t) = f(t)$
and from that $E\{T\} = 0$. Second, recalling the definition of the exponential function and writing $f(t)$ as $f(t) = c[1 + (t^2/n)]^{-1/2}[1 + (t^2/n)]^{-n/2}$
we conclude that as $n \to \infty$, $f(t) \to ce^{-t^2/2}$, which is the normal probability density function. Finally, and this we cannot see from Eq. (3.86)
so you have to take my word for it (and that of Abramowitz and Stegun
(1974), p. 948), $\mathrm{Var}\{T\} = n/(n-2)$, which goes to 1 as n goes to infinity,
but is larger than 1. Clearly, we need $n > 2$ for the variance to be
defined; the t-distribution with one degree of freedom is also known
as the Cauchy distribution. In Figure 3.11 I show the t-distribution for

Figure 3.11. The t-distribution with $n = 4$ or $n = 10$ degrees of freedom. Note that the shape is normal-like, but that the tails are "fatter" when $n = 4$.

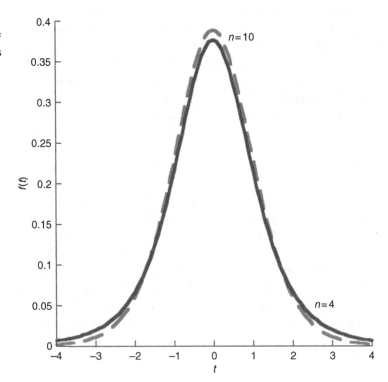

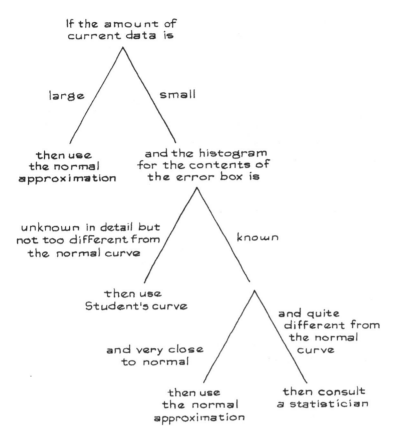

Figure 3.12. The decision tree recommended by Freedman *et al.* (1998) for deciding when to use the normal distribution, the *t*-distribution, another distribution, or go for help. Here a "large" amount of current data means upwards of 25 measurements or so. Reprinted with permission.

$n = 4$ and $n = 10$. The key features to see from this graph are the normal-like shape and the difference in the tails of the distribution.

Freedman *et al.* (1998) offer a decision tree (reproduced in Figure 3.12) that can guide one's thinking about when to characterize the data using the normal distribution, the *t*-distribution, some other probability distribution, or going for professional counseling.

The log-normal distribution and non-negative measurements

We are still not done with the normal distribution and its variants, because very often we take measurements that can only have positive values. Thus our very first, and simplest model, in Eq. (3.75), $Y_i = m + X_i$ where m is fixed but unknown and X_i are $N(0, 1)$, will fail when the observed values are required by biological or physical law to be positive (e.g. masses, lengths, or gene frequencies).

The way to avoid breaking natural law is to use the log-normal distribution, which we construct as follows. First, note that if

$X \sim N(0, 1)$, then σX will be normally distributed with mean 0 and variance σ^2. We then define a new random variable Y by

$$Y = Ae^{\sigma X} \tag{3.87}$$

so that $\log(Y)$ is normally distributed with mean $\log(A)$ and variance σ^2. Now, although X takes values from $-\infty$ to ∞, the exponential will only take values from 0 upwards. Thus, although $E\{X\} = 0$, we conclude that $E\{Y\} > A$. As it happens, we can compute all of the moments for Y in one calculation, which is pretty snazzy. That is, let us consider

$$E\{Y^n\} = E(A^n e^{n\sigma X}) = A^n E(e^{n\sigma X}) \tag{3.88}$$

so that we have to compute

$$E(e^{n\sigma X}) = \frac{1}{\sqrt{2\pi}} \int_{-\infty}^{\infty} \exp(n\sigma x) \exp\left(-\frac{1}{2}x^2\right) dx$$

$$\tag{3.89}$$

$$= \frac{1}{\sqrt{2\pi}} \int_{-\infty}^{\infty} \exp\left(-\frac{1}{2}x^2 + n\sigma x\right) dx$$

We now use a technique from high school algebra, completing the square, by recognizing that

$$\frac{1}{2}x^2 - n\sigma x = \frac{1}{2}[x^2 - 2n\sigma x] = \frac{1}{2}[(x - n\sigma)^2 - n^2\sigma^2] \tag{3.90}$$

and using Eq. (3.90) in Eq. (3.89), we have

$$E(e^{n\sigma X}) = \frac{1}{\sqrt{2\pi}} \int_{-\infty}^{\infty} \exp\left(-\frac{1}{2}(x - n\sigma)^2 + \frac{n^2\sigma^2}{2}\right) dx = \exp\left(\frac{n^2\sigma^2}{2}\right) \tag{3.91}$$

and we have thus shown that if Y is defined by Eq. (3.87), then $E\{Y^n\} = A^n\exp(n^2\sigma^2/2)$, from which we can compute the mean and the variance.

This calculation also suggests that if we want to create a log-normally distributed random variable with a specified mean A, rather than using Eq. (3.87), we should use the definition

$$Y = A\exp\left(\sigma X - \frac{1}{2}\sigma^2\right) \tag{3.92}$$

Exercise 3.17 (E)

Show that if Y is defined by Eq. (3.92) then $E\{Y\} = A$ and $Var\{A\} = A^2(e^{\sigma^2} - 1)$. Note that these two relationships will be sufficient to find A and σ so that we can construct a log-normally distributed random variable with any mean and variance desired.

The beta density and patch leaving

Recall the end of our discussion about the binomial distribution: that if we had data of only k successes, then the maximum likelihood estimates for N and p are $N = k$ and $p = 1$, but these are just plain silly. We will now show how to get around this problem by using the beta probability density. This probability density will be defined below, but first let's examine a slightly different motivation for the beta density, based on a question in foraging theory.

Imagine a forager moving in a patchy environment, seeking food that comes in discrete units, such as seeds. This same example also applies to a foraging parasitoid, seeking hosts in which to place its eggs; we will discuss parasitoids in great detail in the next chapter. When the forager enters a new patch, the probability of finding food, P, will be unknown, although there might be a prior distribution for it, perhaps described by the environmental average or the forager's history until now.

In the current patch, the forager collects data that consist of having found $S = s$ items in $A = a$ attempts at finding food. Clearly, the natural estimate of P is s/a. If $P = p$ were known, the probability of the data (s, a) is given by the binomial distribution

$$\Pr\{(s, a)|p\} = \binom{a}{s} p^s (1 - p)^{a-s} \tag{3.93}$$

We now ask: "given the data (s, a), what can be said about P?". That is, we want to know the probability that P falls in the interval $[p, p + dp]$ given the data (s, a) of s successes in a attempts at finding food. We followed a similar line of reasoning when discussing the Poisson process in which the rate parameter was unknown. Recall that the idea is we begin with a prior probability density for the unknown parameter and use the data and Bayes's Theorem to update the prior and construct the posterior probability density for the parameter, given the data.

If we denote the prior by $f_0(p)$ with the interpretation that $f_0(p)dp = \Pr\{p \leq P \leq p + dp\}$, the posterior distribution is computed from Bayes's Theorem

$$f(p|\text{data})dp = \frac{\Pr\{p \leq P \leq p + dp, \text{data}\}}{\Pr\{\text{data}\}} = \frac{\Pr\{\text{data}|p\}f_0(p)dp}{\Pr\{\text{data}\}} \tag{3.94}$$

Using Eq. (3.93) and dividing by dp gives

$$f(p|(s, a)) = \frac{\binom{a}{s} p^s (1 - p)^{a-s} f_0(p)}{\int_0^1 \binom{a}{s} p^s (1 - p)^{a-s} f_0(p) dp} \tag{3.95}$$

As before, we need to pick a choice for the prior before we can go any further. The classical answer to this choice is to use the beta density with parameters α and β, defined according to

$$f_0(p) = \frac{\Gamma(\alpha + \beta)}{\Gamma(\alpha)\Gamma(\beta)} p^{\alpha-1}(1-p)^{\beta-1} \tag{3.96}$$

The reason for this choice will become clear momentarily. Just a few details about it. First, the beta function $B(\alpha, \beta)$ is defined by $B(\alpha, \beta) = \int_0^1 p^{\alpha-1}(1-p)^{\beta-1} dp$, so that

$$f_0(P) = [1/B(\alpha, \beta)]p^{\alpha-1}(1-p)^{\beta-1}$$

(which also tells us how to relate the beta and gamma functions). Second, the mean and variance of P are given by

$$E\{P\} = \frac{\alpha}{\alpha + \beta} \quad \text{Var}\{P\} = \frac{\alpha\beta}{(\alpha+\beta)^2(\alpha+\beta+1)} \tag{3.97}$$

Third, the shape of the beta density varies according to the choices of the parameters (Figure 3.13). Fourth (Abramowitz and Stegun 1974, p. 944), if χ_1^2 and χ_2^2 are chi-squared random variables with ν_1 and ν_2 degrees of freedom respectively, then the ratio $\chi_1^2/(\chi_1^2 + \chi_2^2)$ has a beta probability density with parameters $\alpha = \nu_1/2$ and $\beta = \nu_1/2$.

Why pick the beta probability density? The following exercise should answer the question.

Exercise 3.18 (E/M)

Show that if $f_0(p)$ is a beta probability density with parameters α and β and that the data are s successful searches in a attempts, then the posterior probability density for P obtained from Eq. (3.96) is also a beta density, with updated parameters $\alpha + s$ and $\beta + a - s$. Clark and Mangel (2000) provide further elaborations of how this updating result can be used in the prediction of patch leaving rules.

We close with one more consideration of that pesky problem of estimation of N and p for the binomial distribution, given k successes. If we assume that N is uniformly distributed (see Raftery (1988) for the case where N has a Poisson prior), that p has a beta density with parameters α and β and that the probability of k events is binomially distributed given p and N, then the posterior distribution for N and p, found in a manner analogous to Eq. (3.97) is proportional to $\binom{n}{k}p^{k+\alpha-1}(1-p)^{N-k+\beta-1}$, which is nice and well behaved (e.g. we can find sensible maximum posterior estimates for both N and p, but we won't do that here – on to parasitoids!).

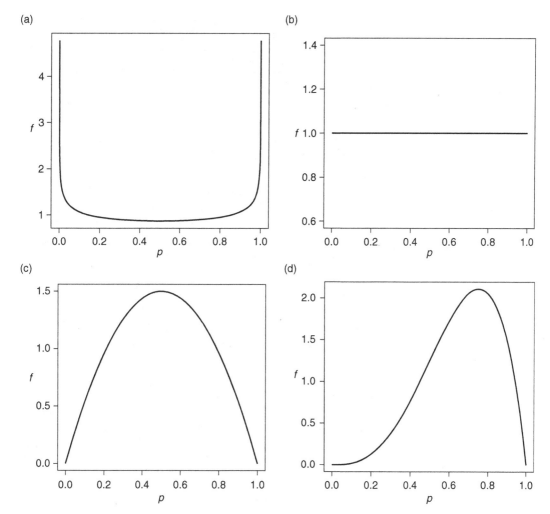

Figure 3.13. The beta probability density with mean 1/2 can be U-shaped ($\alpha = \beta = 0.8$, panel a), flat ($\alpha = \beta = 1$, panel b), bell-shaped ($\alpha = \beta = 2$, panel c) or peaked and asymmetrical ($\alpha = 4$, $\beta = 2$, panel d).

Connections

Thomas Bayes and the frequentist–Bayesian debate in statistics

The two main approaches to statistical analysis are called frequentist (usually associated with R. A. Fisher and Jerzy Neyman) and Bayesian (associated with the Reverend Thomas Bayes). Fisher is well known to biologists because of his work on genetics (Hotelling (1951) gives an interesting perspective of 50 years ago, which was then 25 years after the publication of Fisher's book on statistical methods for scientists);

Bayes is less well known to biologists, but is becoming more so. Bayes was a British minister born in 1702. His essay on "solving a problem in the doctrine of chances" was published posthumously and is generally hard to get, so it was republished in 1958 by *Biometrika*, with a nice historical introduction (Barnard and Bayes 1958). Frequentist and Bayesian statistics differ in both operational (how one does certain kinds of calculations) and philosophical (what exactly one is trying to accomplish) aspects. It is impossible to review the issues here in any comprehensive manner; the articles by Suter (1996), Ludwig *et al.* (2001) and Ellison (1996, 2004) and the book by Taper and Lele (2004) point out the variety of issues and towards the primary literature. Here, we shall briefly focus on just one question that allows us to see the difference. When dealing with any kind of statistics, we have both data (*D*) and hypotheses (*H*). The approach of classical, frequentist statistics, is to ask questions about the probability of the data, given a hypothesis. That is, formally we compute $\Pr\{D|H\}$. For example, a standard hypothesis test (with a 5% significance value) asks: what is the probability of obtaining these or more extreme data, given that the hypothesis is true? If this probability is less than 5%, then the hypothesis is rejected (the choice of 5% is arbitrary, but now more or less accepted). The alternative approach is to ask what kind of support the data provide for the hypothesis. Formally, we compute $\Pr\{H|D\}$. The big problem is that in general $\Pr\{D|H\} \neq \Pr\{H|D\}$ so that testing a statistical hypothesis often does not give us what we need for scientific understanding. This difficulty has been recognized for a long time (Yates 1951, Royall 1997) but it is only with modern computing that application of the Bayesian approach in general has become practicable. (Another way to think about the difference is that frequentists believe that unknown parameters are fixed and real and that the data are drawn from a distribution of possible observations while Bayesians believe that the data are real and that the unknown parameters are drawn from a distribution.) The second difference between frequentist and Bayesian statistics is how we deal with prior information. In many problems arising in ecology or evolutionary biology, we have such prior information. For example, when managing a fish stock, we may know life history information for similar stocks or the same species elsewhere; when computing an evolutionary tree, we may know something about the relationships between different species in the tree. Bayesian statistics provides a consistent means for dealing with this prior information, while frequentist statistics does not. A general introduction to the Bayesian approach is the book by DeGroot (1970); while old, is still a great read and one of the classic texts in the field. Efron (2005) offers a view for the twentyfirst century. Bayesian approaches are now used extensively in phylogeny and evolutionary biology (Huelsenbeck and Ronquist 2001, Huelsenbeck *et al.* 2001) and in fishery management (McAllister *et al.* 1994, McAllister and Kirkwood 1998a, b, 1999, McAllister *et al.* 2001).

More about likelihood

Likelihood underlies both frequentist and Bayesian approaches to statistics. The books by Edwards (1992) and Royall (1997) are key sources that belong on one's shelf. Here, I want to make one general connection to what we have done already. Suppose that we have data X_i, $i = 1, 2, \ldots n$, from a probability density function $f(x, p)$ where p is a parameter to be estimated from the data. The likelihood of the parameter given the data is then $\tilde{L}(p|\{X_i\}) = \prod_{i=1}^{n} f(X_i|p)$ and the log-likelihood is $L(p|\{X_i\}) = \sum_{i=1}^{n} \log(f(X_i|p))$. Suppose that we find the maximum likelihood estimate of the parameter in the usual way by setting the derivative of the log-likelihood with respect to p equal to 0 and solving for p. We then obtain a maximum likelihood estimate for the parameter that depends upon the data. If one Taylor expands around the maximum likelihood estimate, keeping only the first two terms, the result is a quadratic – reminding us of the sum of squared deviations and the Gaussian likelihood. This is the reason that "asymptotic normal theory" is such a powerful statistical tool – for most cases when there is a considerable amount of data a normal approximation can prevail because of the Taylor expansion.

The gamma and beta functions

The gamma function is one of the classical functions used in applied mathematics. If you don't yet own it, Abramowitz and Stegun (1974) is a very good investment (it is also coming to the web someday soon); it will stand by you for many years. In Chapter 6 there is all about the gamma function and its close relatives. These special functions often arise in the solution of different kinds of ordinary or partial differential equations. One should learn to think about them in the same way that one thinks about the simpler functions such as log, exp, or the trigonometric functions.

Mean-variance power laws

In the ecological literature, the mean-variance power laws are most often associated with the name L. R. Taylor, who popularized them in

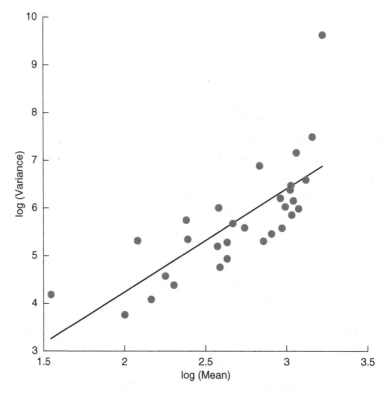

Figure 3.14. The diagnostic for a mean-variance power law is a log–log plot of variance versus mean, as shown here for some groundfish survey data from the West Coast of North America (with thanks to E. J. Dick). When the slope is 2, the diagnostic suggests gamma/negative binomial models may be appropriate.

the study of the spatial distributions of insect populations. His 1984 review (Taylor 1984) is still well worth reading, as are the series of papers published in the late 1970s and early 1980s (Taylor *et al.* 1978, 1979, 1980). The general ideas have a rich and long history in applied mathematical statistics (Greenwood and Yule 1920, Anscombe 1950). A generally used diagnostic for a mean variance power law is to use a log–log plot (variance of the data versus the mean of the data). In Figure 3.14 I show an example of such a diagnostic plot.

Conjugate priors; non-informative priors

The gamma density as a prior for the Poisson parameter or the beta density as the prior for the binomial parameter are called conjugate priors. The meaning is this: we begin with a density (say the gamma), collect data (from a Poisson process), and update using Bayes's theorem to end with a posterior that is also the same type of density, but with changed parameters. We say that the density is "closed" during updating. When computation was very difficult to do, conjugate priors played a key and important role in Bayesian analysis because they allowed operational implementation of the Bayesian approach. Modern computational methods allow us to operationalize Bayesian approaches without resort to conjugate priors. Another common prior used in Bayesian analysis is called the non-informative prior. The rough idea with these is that one chooses the prior so that the location but not the shape of the posterior is changed by the data; this is not the same as choosing a uniform prior. A simple illustration of these differences is provided by Mangel and Beder (1985). For more general details, I suggest any of Martz and Waller (1982), Leonard and Hsu (1999), or Congdon (2001).

Asymptotic expansions

The calculation that we did for normal cumulative distribution function when x was large is an example of an asymptotic expansion, in which one exploits the largeness (or smallness, since then the reciprocal will be large) of a variable or parameter to obtain approximations to the solution of an equation or evaluation of an integral. This is a standard tool in classical applied mathematics (see, for example, de Bruijn (1981), Bleistein and Handelsman (1975), Bender and Orszag (1978), or Lin and Segel (1988 (1974))). As shown with our simple example, such expansions can be very powerful and give intuition about much more complicated quantities. We will use asymptotic expansions in our study of stochastic population theory in Chapter 8.

Testing your methods with simulated data and then some of the real data

The analysis of models is pretty much a science, but the development of models is an art. There has grown up a large literature concerning "model validation," which generally intends that one tests or validates the model by comparison with the data. Here, I offer three suggestions about testing a model that you have developed (also see Mangel *et al.* (2001)). First, always try to test the assumptions that go into the model independent of the predictions of the model. Second, always test your model or method with simulated data in which you know exactly what is happening. If the method does not work on simulated data, it is almost surely not going to work on real data. Third, set aside some of your data at the beginning of an analysis. Estimate parameters with the remainder and then use the set-aside data as a means of testing the predictions of the method. Hilborn and Mangel (1997) discuss model size and model validation in more detail.

Total least squares and errors in variables

Schnute (1993) explains how one can construct likelihood models when both the independent and dependent variable are measured with error. The key to these approaches is to assume that one knows the ratio of the variance of X to the variance in Y. Ludwig (1999) applies this approach to questions of extinction times (a topic that we shall visit in Chapter 8).

Using the t-distribution in ecological models and Bayesian updating of the parameter

Our simplest statistical model involving a normal random variable was $Y = m + X$, where m is unknown and to be estimated and X was a normally distributed random variable. A more complicated version, in which Y is constrained to be positive, is $\log(Y) = \log(m) + X$. However, ecological or evolutionary data are often non-normal, with tails that are fatter than normal. This suggests that we might work with the model $Y = m + T_\nu$, where T_ν is a random variable following a t-distribution with ν degrees of freedom. (Alternatively, of course, we might work with $\log(Y) = \log(m) + T_\nu$.) Carpenter *et al.* (1999) used such an approach in development of a model for the eutrophication of a lake. Furthermore, we know that as $\nu \to \infty$, $T_\nu \to X$, so that by applying Bayesian methods, we can allow the data to tell us the kind of error distribution to use (see Gelman *et al.* (1995) pp. 350–361).

Model selection via likelihood ratio, AIC and BIC

In ecology and evolutionary biology, we often do not know the precisely correct form of the model to use for a situation. (This differs, for example, from classical mechanics, which often gives rise to physics envy in biologists, but only slightly – see the Epilogue in Mangel and Clark (1988)). It is thus worthwhile to have a variety of models and allow the data to arbitrate among them.

As an example, suppose that we are interested in a species accumulation curve that relates the number of species to the number of individuals in a sample. We might construct a wide variety of models (Flather 1996, Burnham and Anderson 1998) all of the fundamental form $S_j = f(I_j) + X_j$, where S, I, and X are respectively the number of species in a sample, the number of individuals in a sample and a normally distributed random variable with mean 0 and unknown variance (which is one parameter that we need to estimate). In this case, the model is $f(I)$; some candidates with the number of parameters (which is 1, for the

variance, plus the number of parameters in the functional form) are
shown in the following table.

Form of $f(I)$	Number of parameters
aI^b	3
$a + b\log(I)$	3
$aI/(b+I)$	3
$a(1 - \exp(-bI))$	3
$a - bc^I$	4
$(a + bI)/(1 + cI)$	4
$a(1 - \exp(-bI))^c$	4
$a(1 - [1 + cI^d]^{-b})$	5

It is silly to think that one of these models is "right" or "true" –
they are all approximations to nature. It is not silly, however, to expect
that some of these models will be better descriptions of nature than
others. Also that some of these models are nested, in the sense that one
can obtain one of the models from another by setting a parameter equal
to 0. Other of these models are not nested at all because there is no way
to travel between them by eliminating parameters. When models are
nested, the appropriate way to compare them is to use the likelihood
ratio test (Kendall and Stuart 1979). When the models are not nested,
the Aikaike Information Criterion or one of its extensions (Burnham
and Anderson 1998) is the appropriate tool to use. The AIC, and its
various extensions, is built from the maximized log-likelihood, given
the data, and the number of parameters in the model. The basic AIC is
given by $AIC = -2\log\{L(\hat{p}|X)\} + 2K$, where $\hat{p}$ is the MLE estimate of
the parameters given the data X, and K is the number of parameters. One
minimizes the AIC across the choice of models. The choice of minimizing AIC, rather than maximizing $-AIC$, and the use of 2 in the
definition are both the results of history. Burnham and Anderson (1998)
recommend the use of AIC differences defined by $\Delta_i = AIC_i - \min AIC$,
where AIC_i is the AIC for model i and $\min AIC$ is the minimum AIC,
over the different models. They suggest that models for which the
difference is less than about 2 have substantial support from the data,
those with differences in the range 4–7 have considerably less support,
and those with differences greater than 10 have essentially no support
and can be omitted from future consideration. Furthermore, it is possible to compute AIC weights from the differences of AIC values according to the formula $w_i = \exp(-\Delta_i)/\sum_{j-1}^m \exp(-\Delta_j)$, where m is the
number of models. Note that if we ignored the number of parameters,

the weight would simply be a measure of the relative likelihoods, as if we were doing a Bayesian calculation in which each model had the same prior probability. There is, in fact, a Bayesian viewpoint for the information criterion, called the Bayesian Information Criterion (BIC) in which one assumes equal prior probability for the different models and very broad prior distributions for the parameters (Schwarz 1978, Burnham and Anderson 1998, p. 68). This information criterion is $BIC = -2 \log\{L(\hat{p}|X)\} + K \log(N)$, where N is the number of data points. There is also a correction for the AIC when the number of parameters is comparable to the number of data points (also see Burnham and Anderson). Burnham and Anderson (1998) is a volume well worth owning. The use of AIC or its extension is becoming popular in the ecological literature as a means of selecting the best of disparate models (Morris $et\ al.$ 1995, Klein $et\ al.$ 2003).

Chapter 4

The evolutionary ecology of parasitoids

Insect parasitoids – those insects that deposit their eggs on or in the eggs, larvae or adults of other insects and whose offspring use the resources of those hosts to fuel development – provide a rich area of study for theoretical and mathematical biology. They also provide a broad collection of examples of how the tools developed in the previous chapters can be used (and they are some of my personally favorite study species; the pictures shown in Figure 4.1 should help you see why).

There is also a rich body of experimental and theoretical work on parasitoids, some of which I will point you towards as we discuss different questions. The excellent books by Godfray (1994), Hassell (2000a), and Hochberg and Ives (2000) contain elaborations of some of the material that we consider. These are well worth owning. Hassell (2000b), which is available at *JSTOR*, should also be in everyone's library.

It is helpful to think about a dichotomous classification scheme for parasitoids using population, behavioral, and physiological criteria (Figure 4.2). First, parasitoids may have one generation (univoltine) or more than one generation (multivoltine) per calendar year. Second, females may lay one egg (solitary) or more than one egg (gregarious) in hosts. Third, females may be born with essentially all of their eggs (pro-ovigenic) or may mature eggs (synovigenic) throughout their lives (Flanders 1950, Heimpel and Rosenheim 1998, Jervis *et al.* 2001). Each dichotomous choice leads to a different kind of life history.

Figure 4.1. Some insect parasitoids and insects that have life histories that are similar to parasitoids. (a) *Halticoptera rosae*, parasitoid of the rose hip fly *Rhagoletis basiola,* (b) *Aphytis lingannensis*, parasitoid of scale insects, and (c) *Leptopilinia heterotoma*, parasitoid of *Drosophila subobscura*. (d, e) Tephritid (true) fruit flies have life styles that are parasitoid-like: adults are free living, but lay their eggs in healthy fruit. The larvae use the resources of the fruit for development, then drill a hole out of the fruit and burrow into the ground for pupation. Here I show a female rose hip fly *R. basiola* (d) ovipositing, and two males of the walnut husk fly *R. compleata* (e) fighting for an oviposition site (the successful male will then try to mate with females when they come to use the oviposition site). The black trail under the skin of the walnut is the result of a larva crawling about and creating damage between the husk and the shell as it uses the resource of the fruit.

(a) Generations per year:

Multivoltine

Univoltine

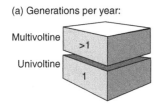

(b) Eggs per host:

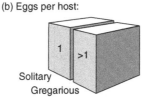

Solitary

Gregarious

Figure 4.2. A method of classifying parasitoid life histories according to population, behavioral and physiological criteria.

(c) Egg production after emergence:

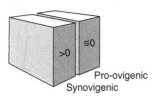

Pro-ovigenic

Synovigenic

(d) Combining the characteristics:

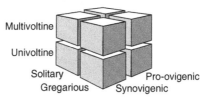

Multivoltine

Univoltine

Solitary

Gregarious

Pro-ovigenic

Synovigenic

The Nicholson–Bailey model and its generalizations

The starting point for our (and most other) analysis of host–parasitoid dynamics is the Nicholson–Bailey model (Nicholson 1933, Nicholson and Bailey 1935) for a solitary univoltine parasitoid. We envision that hosts are also univoltine, in a season of unit length, in which time is measured discretely and in which $H(t)$ and $P(t)$ denote the host and parasitoid populations at the start of season t. Each host that survives to the end of the season produces R hosts next year. The parasitoids search randomly for hosts, with search parameter a, so that the probability that a single host escapes parasitism from a single parasitoid is e^{-a}. Thus, the probability that a host escapes parasitism when there are $P(t)$ parasitoids present at the start of the season is $e^{-aP(t)}$. These absolutely sensible assumptions lead to the dynamical system

$$H(t+1) = RH(t)e^{-aP(t)}$$
$$P(t+1) = H(t)(1 - e^{-aP(t)})$$
(4.1)

Note that in this case the only regulation of the host population is by the parasitoid. Hassell (2000a, Table 2.1) gives a list of 11 other sensible assumptions that lead to different formulations of the dynamics.

The first question we might ask concerns the steady state of Eq. (4.1), obtained by assuming that $H(t+1)=H(t)$ and $P(t+1)=P(t)$. These are easy to find.

Exercise 4.1 (E)

Show that the steady states of Eqs. (4.1) are

$$\bar{P} = \frac{1}{a}\log(R) \qquad \bar{H} = \frac{R}{a(R-1)}\log(R)$$
(4.2)

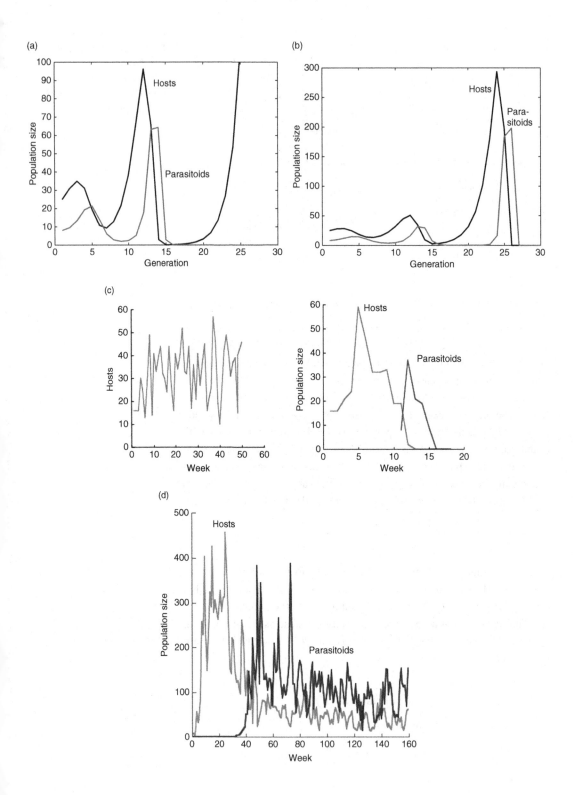

which shows that $R > 1$ is required for a steady state (as it must be) and that higher values of the search effectiveness reduce both host and parasitoid steady state values.

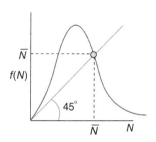

The sad fact, however, is that this perfectly sensible model gives perfectly nonsensical predictions when the equations are iterated forward (Figure 4.3): regardless of parameters, the model predicts increasingly wild oscillations of population size until either the parasitoid becomes extinct, after which the host population is not regulated, or both host and parasitoid become extinct. To be sure, this sometimes happens in nature, usually this is not the situation. Instead, hosts and parasitoids coexist with either relative stable cycles or a stable equilibrium.

Figure 4.4. The stability of the steady state of the one dimensional dynamical system $N(t+1) = f(N(t))$ is determined by the derivative of $f(N)$ evaluated at the steady state $\bar{N}$.

In a situation such as this one, one can either give up on the theory or try to fix it. My grade 7 PE teacher, Coach Melvin Edwards, taught us that "quitters never win and winners never quit," so we are not going to give up on the theory, but we are going to fix it. The plan is this: for the rest of this section, we shall explore the origins of the problem. In the next section, we shall fix it.

As a warm-up, let us consider a discrete-time dynamical system of the form

$$N(t+1) = f(N(t)) \tag{4.3}$$

where $f(N)$ is assumed to be shaped as in Figure 4.4, so that there is a steady state $\bar{N}$ defined by the condition $\bar{N} = f(\bar{N})$. To study the stability of this steady state, we write $N(t) = \bar{N} + n(t)$ where $n(t)$, the perturbation from the steady state, is assumed to start off small, so that $|n(0)| \ll \bar{N}$. We then evaluate the dynamics of $n(t)$ from Eq. (4.3) by Taylor expansion of the right hand side keeping only the linear term

$$\bar{N} + n(t+1) = f(\bar{N} + n(t)) \approx f(\bar{N}) + \left.\frac{\mathrm{d}f}{\mathrm{d}N}\right|_{\bar{N}} n(t) \tag{4.4}$$

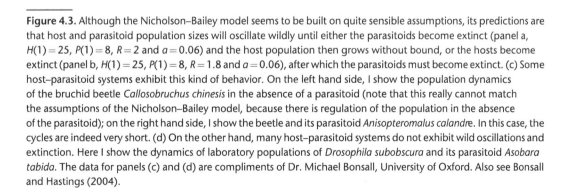

Figure 4.3. Although the Nicholson–Bailey model seems to be built on quite sensible assumptions, its predictions are that host and parasitoid population sizes will oscillate wildly until either the parasitoids become extinct (panel a, $H(1) = 25$, $P(1) = 8$, $R = 2$ and $a = 0.06$) and the host population then grows without bound, or the hosts become extinct (panel b, $H(1) = 25$, $P(1) = 8$, $R = 1.8$ and $a = 0.06$), after which the parasitoids must become extinct. (c) Some host–parasitoid systems exhibit this kind of behavior. On the left hand side, I show the population dynamics of the bruchid beetle *Callosobruchus chinesis* in the absence of a parasitoid (note that this really cannot match the assumptions of the Nicholson–Bailey model, because there is regulation of the population in the absence of the parasitoid); on the right hand side, I show the beetle and its parasitoid *Anisopteromalus calandre*. In this case, the cycles are indeed very short. (d) On the other hand, many host–parasitoid systems do not exhibit wild oscillations and extinction. Here I show the dynamics of laboratory populations of *Drosophila subobscura* and its parasitoid *Asobara tabida*. The data for panels (c) and (d) are compliments of Dr. Michael Bonsall, University of Oxford. Also see Bonsall and Hastings (2004).

Since $\bar{N} = f(\bar{N})$ and setting $f_N = df/dN|_{\bar{N}}$ we conclude that $n(t)$ approximately satisfies

$$n(t+1) = f_N n(t) \tag{4.5}$$

and we conclude that the steady state will be stable, in the sense that perturbations from it decay, if $|f_N(\bar{N})| < 1$.

Exercise 4.2 (E)

For more practice determining when a steady state is stable, do the computation for the discrete Ricker map

$$N(t+1) = N(t)\exp\left[r\left(1 - \frac{N(t)}{K}\right)\right]$$

and show that the condition is $|1 - r| < 1$, or $0 < r < 2$.

But we have a two dimensional dynamical system. Since what follows is going to be a lot of work, we will do the analysis for the more general host–parasitoid dynamics. Basically, we do for the steady state of a two dimensional discrete dynamical system the same kind of analysis that we did for the two dimensional system of ordinary differential equations in Chapter 2. Because the procedure is similar, I will move along slightly faster (that is, skip a few more steps) than we did in Chapter 2. Our starting point is

$$\begin{align} H(t+1) &= RH(t)f(H(t), P(t)) \\ P(t+1) &= H(t)(1 - f(H(t), P(t))) \end{align} \tag{4.6}$$

which we assume has a steady state $(\bar{H}, \bar{P})$. We now assume that $H(t) = \bar{H} + h(t)$ and $P(t) = \bar{P} + p(t)$, substitute back into Eq. (4.6), Taylor expand keeping only linear terms and use $o(h(t), p(t))$ to represent terms that are higher order in $h(t), p(t)$, or their product to obtain

$$\begin{align} \bar{H} + h(t+1) &= R(\bar{H} + h(t))[f(\bar{H}, \bar{P}) + f_H h(t) + f_P p(t)] + o(h(t), p(t)) \\ \bar{P} + p(t+1) &= (\bar{H} + h(t))[1 - f(\bar{H}, \bar{P}) - f_H h(t) - f_P p(t)] + o(h(t), p(t)) \end{align} \tag{4.7}$$

where $f_H = (\partial/\partial H)f(H, P)|_{(\bar{H}, \bar{P})}$ and f_P is defined analogously. Now, from the definition of the steady states we know that $\bar{H} = R\bar{H}f(\bar{H}, \bar{P})$, which also means that $Rf(\bar{H}, \bar{P}) = 1$, and that $\bar{P} = \bar{H}(1 - f(\bar{H}, \bar{P}))$. We now use these last observations concerning the steady state as we multiply through, collect terms, and simplify to obtain

$$h(t+1) = h(t)(1 + R\bar{H}f_H) + R\bar{H}f_P p(t) + o(h(t), p(t))$$

$$p(t+1) = h(t)(1 - (1/R) - \bar{H}f_H) - \bar{H}f_P p(t) + o(h(t), p(t))$$

(4.8)

Unless you are really smart (probably too smart to find this book of any use to you), these equations should not be immediately obvious. On the other hand, you should be able to derive them from Eqs. (4.7), with the intermediate clues about properties of the steady states in about 3–4 lines of analysis for each line in Eqs. (4.8). If we ignore all but the linear terms in Eqs. (4.8) we have the linear system

$$h(t+1) = ah(t) + bp(t)$$

$$p(t+1) = ch(t) + dp(t)$$

(4.9)

with the coefficients a, b, c, and d suitably defined; as before, we can show that this is the same as the single equation

$$h(t+2) = (a+d)h(t+1) + (bc - ad)h(t)$$

(4.10)

by writing $h(t+2) = ah(t+1) + bp(t+1)$, $p(t+1) = ch(t) + dp(t) = ch(t) + (d/b)(h(t+1) - ah(t))$ and simplifying. (Once again you should not necessarily see how to do this in your head, but writing it out should make things obvious quickly.) If we now assume that $h(t) \sim \lambda^t$ (there is actually a constant in front of the right hand side, as in Chapter 2, but also as before it cancels), we obtain a quadratic equation for λ:

$$\lambda^2 - (a+d)\lambda + ad - bc = 0$$

(4.11)

which I am going to write as $\lambda^2 - \beta\lambda + \gamma = 0$ with the obvious identification of the coefficients. Also as before, Eq. (4.11) will have two roots, which we will denote by $\lambda_1 = (\beta/2) + (\sqrt{\beta^2 - 4\gamma}/2)$ and $\lambda_2 = (\beta/2) - (\sqrt{\beta^2 - 4\gamma}/2)$. The steady state will be stable if perturbations from the steady state become smaller in time, this requires that $|\lambda_{1,2}| < 1$. We will now find conditions on the coefficients that makes this true. The analysis which we do follows Edelstein-Keshet (1988), who attributes it to May (1974). We will do the analysis for the case in which the eigenvalues are real (i.e. for which $\beta^2 \geq 4\gamma$); this is our first condition. Figure 4.5 will be helpful in this analysis. The parabola $\lambda^2 - \beta\lambda + \gamma$ has a minimum at $\beta/2$, and because we require $-1 < \lambda_2 < \beta/2 < \lambda_1 < 1$ we know that one condition for stability is that $|\beta/2| < 1$, so that $|\beta| < 2$. The parabola is symmetric around the minimum. Now, if the roots lie between -1 and 1, the distance between the minimum and either root, which I have called D_1, must be smaller than the distance between the minimum and -1 or 1, depending upon

Figure 4.5. The construction
needed to determined when
the solutions of the equation
$\lambda^2 - \beta\lambda + \gamma = 0$ have absolute
values less than 1, so that the
linearized system in Eq. (4.9)
has a stable steady state.

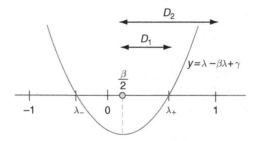

Figure 4.5. The construction
needed to determined when
the solutions of the equation
$\lambda^2 - \beta\lambda + \gamma = 0$ have absolute
values less than 1, so that the
linearized system in Eq. (4.9)
has a stable steady state.

whichever is closer. Thus, for example, for the situation in Figure 4.5 we must have $1 - |\beta/2| > \sqrt{\beta^2 - 4\gamma}/2$; if we square both sides of this expression the condition becomes $1 - |\beta| + (\beta^2/4) > (\beta^2/4) - \gamma$ and this simplifies to $1 + \gamma > |\beta|$. Our first condition was $\beta^2 \geq 4\gamma$ and we have agreed that $|\beta| < 2$ so that $\beta^2 < 4$. Therefore, $4 > \beta^2 \geq 4\gamma$, so that $1 > \gamma$ or $2 > 1 + \gamma$. When we combine the two conditions, we obtain the criterion for stability that (Edelstein-Keshet 1988)

$$2 > 1 + \gamma > |\beta| \tag{4.12}$$

Hassell (2000a) gives (his Eqs. (2.2), (2.3)) the application of this condition to the general Eqs. (4.6).

In the case of Nicholson–Bailey dynamics, $f(H, P) = \exp(-aP)$, so that $f_H = 0$ and $f_P = -a\exp(-aP)$; these need to be evaluated at the steady states and the coefficients a, b, c, and d in Eq. (4.9) evaluated so that we can then determine β and γ.

Exercise 4.3 (M/H)

For Nicholson–Bailey dynamics show that $\beta = 1 + [\log(R)/(R-1)]$ and that $\gamma = R\log(R)/(R-1)$. Then show that since $R > 1$, $1 + \gamma > \beta$. However, also show that $1 + \gamma > 2$ by showing that $\gamma > 1$ (to do this, consider the function $g(R) = R\log(R) - R + 1$ for which $g(1) = 0$ and show that $g'(R) > 0$ for $R > 1$) thus violating the condition in Eq. (4.12), and thus conclude that the Nicholson–Bailey dynamics are always unstable.

What biological intuition underlies the instability of the Nicholson–Bailey model? There are two answers. First, the per capita search rate of the parasitoids is independent of population size of parasitoids (which are likely to experience interference when population is high). Second, there is no refuge for hosts at low density – the fraction of hosts killed depends only upon the parasitoids and is independent of the number of hosts. We now explore ways of stabilizing the Nicholson–Bailey model.

Stabilization of the Nicholson–Bailey model

I now describe two methods that are used to stabilize Nicholson–Bailey population dynamics, in the sense that the unbounded oscillations disappear. Note that we implicitly define that a system that oscillates but stays within bounds is stable (Murdoch, 1994). The methods of stabilization rely on variation and refuges.

Variation in attack rate

The classic (Anderson and May 1978) means of stabilizing the Nicholson–Bailey model is to recognize that not all hosts are equally susceptible to attack, for one reason or another. To account for this variability, we replace the attack rate a by a random variable A, with $E\{A\} = a$, so that the fraction of hosts escaping attack is $\exp(-AP)$. However, to maintain a deterministic model, we average over the distribution of A; formally Eq. (4.1) becomes

$$H(t+1) = RH(t)E_A\{e^{-AP(t)}\}$$
$$P(t+1) = H(t)(1 - E_A\{e^{-AP(t)}\}) \tag{4.13}$$

where $E_A\{\ \}$ denotes the average over the distribution of A. For the distribution of A, we choose a gamma density with parameters α and k. We then know from Chapter 3 that the resulting average of $\exp(-AP(t))$ will be the zero term of a negative binomial distribution, so that

$$E_A\{e^{-AP(t)}\} = \left(\frac{\alpha}{\alpha + P}\right)^k \tag{4.14}$$

Since the mean of a gamma density with parameters α and k is k/α, it would be sensible for this to be the average value of the attack rate so that $a = k/\alpha$; we choose $\alpha = k/a$. We then multiply top and bottom of the right hand side of Eq. (4.14) by k/α to obtain

$$H(t+1) = RH(t)\left[\frac{k}{k + aP(t)}\right]^k$$
$$P(t+1) = H(t)\left(1 - \left[\frac{k}{k+aP(t)}\right]^k\right) \tag{4.15}$$

This modification of the Nicholson–Bailey model is sufficient to stabilize the population dynamics (Figure 4.6). To help understand the intuition that lies behind this stabilization, I note the following remarkable feature (Pacala *et al.* 1990): the stabilization occurs as long as the overdispersion parameter $k < 1$. I have illustrated this point in

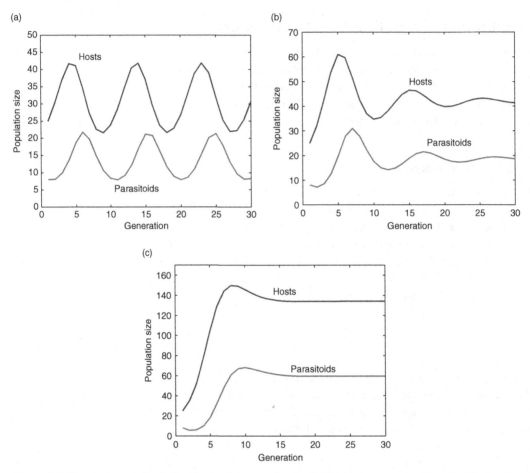

Figure 4.6. Allowing for variation in the attack rate by assuming that it follows a gamma density, so that the fraction of hosts escaping parasitism is the zero term of the negative binomial distribution, stabilizes the Nicholson–Bailey dynamics. All parameters are as in Figure 4.3b, except that the mean attack rate is now $a = 0.06$ and the overdispersion parameter k is 0.99 (panel a), 0.5 (panel b), or 0.2 (panel c).

Figure 4.7, showing that if $k = 0.99$ the dynamics are stable (the oscillations have decreasing amplitude), but if $k > 1$ they are not (the oscillations have increasing amplitude).

Recall that the coefficient of variation of the gamma density with parameters α and k is $1/\sqrt{k}$, so that $k < 1$ is equivalent to the rule that the coefficient of variation is greater than 1. Pacala *et al.* (1990) call this the $CV^2 > 1$ rule (but also see Taylor (1993) who notices that the specific properties of the dynamics will depend not only upon k but also upon R). Also recall that when $k < 1$, the probability density for the attack rate is large when the attack rate is small 0. This means that arbitrarily small values of the attack rate have substantial probability associated with

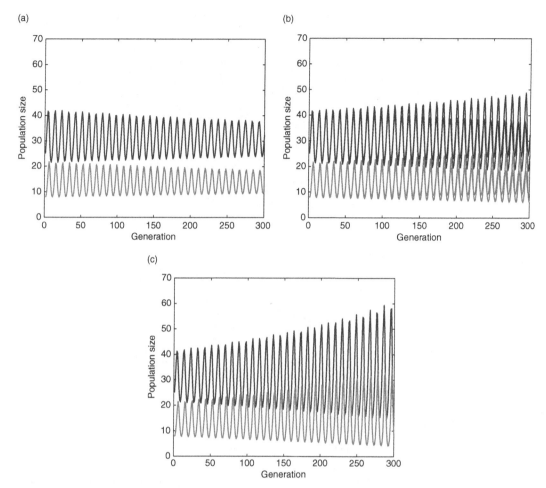

Figure 4.7. The dynamics determined by Eq. (4.14) when $k = 0.99$ (panel a), 1.01 (panel b), or 1.02 (panel c) showing that the dynamics are unstable when $k > 1$. All other parameters as in Figure 4.6. In each case, the hosts are the upper curve, the parasitoids the lower curve.

them, even though the mean attack rate is held constant. But very small attack rates mean that some hosts are essentially invulnerable to attack or that a refuge from attack exists. A host refuge is clearly one way to stabilize the dynamics. For example, the stable dynamics shown in Figure 4.3d involve a 30% refuge for the host.

Multiple attacks may provide a different kind of refuge

Solitary parasitoids lay only a single egg in a host, but often they do not perfectly discriminate when laying eggs (Figure 4.8). When that happens, there will be larval competition with the host (Taylor 1988a, b,

Figure 4.8. The parasitoid
Nasionia vitrepennis is solitary
and attacks a variety of hosts
(shown here are pupae of
Phormia regina). However,
sometimes more than one egg
is laid in a host, in which case
larval competition of the
parasitoids occurs.
Photographs compliments of
Robert Lalonde, University of
British Columbia, Okanagan
Campus.

1993) and this competition may have profound effects on the dynamics
of the parasitoids, with associated effects on the dynamics of the host.
Taylor (1988a, b, 1993) provides a general treatment of the effects of
within-host competition; here we will consider a simplification that Bob
Lalonde (University of British Columbia, Okanagan Campus) taught me.

Let us suppose that a host that is attacked and receives only one
parasitoid egg produces a parasitoid in the next generation with certainty, but that hosts that receive more than one egg fail to produce a
parasitoid because of competition between the parasitoid larvae within
the host (that is, they fight each other to the point of being unable to
complete development but kill the host too). Now the standard
Nicholson–Bailey dynamics correspond to random search, so that the
probability that a host receives exactly one egg is a $aP(t) \exp(-aP(t))$.
Thus, the original Nicholson–Bailey dynamics become

$$H(t+1) = RH(t)e^{-aP(t)}$$
$$P(t+1) = H(t)aP(t)e^{-aP(t)} \tag{4.16}$$

The first line in Eq. (4.16) corresponds to hosts that escape parasitism
entirely (the zero term of the Poisson distribution); the second line

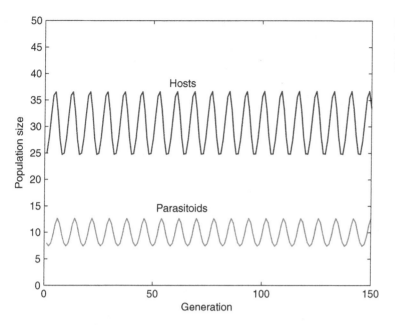

Figure 4.9. If multiple attacks on a host lead to no emergent parasitoids, the Nicholson–Bailey dynamics are stabilized.

corresponds to hosts that receive exactly one parasitoid egg. These dynamics stabilize the Nicholson–Bailey distribution (Figure 4.9) because a new kind of refuge is provided through regulation of the parasitoid population.

Exercise 4.4 (M)

Show that if a fraction ϕ of multiple attacks on hosts lead to the emergence of a parasitoid, then Eqs. (4.16) are replaced by

$$
\begin{aligned}
H(t+1) &= RH(t)\mathrm{e}^{-aP(t)} \\
P(t+1) &= H(t)aP(t)\mathrm{e}^{-aP(t)} + H(t)(1 - \mathrm{e}^{-aP(t)} - aP(t)\mathrm{e}^{-aP(t)})\phi
\end{aligned}
\tag{4.17}
$$

then explore the dynamical properties of Eqs. (4.17) by iterating them forward. As a hint: be certain to use sufficiently long time horizons that allow you to see the full range of effects.

There are other means of stabilizing the Nicholson–Bailey dynamics; these include various kinds of density dependence (Hassell 2000a, b) and spatial models (see Connections).

More advanced models for population dynamics

In many biological systems generations overlap so that a population of hosts and parasitoids simultaneously consists of eggs, larvae, pupae and adults. In that case, a more appropriate formulation of the models

The evolutionary ecology of parasitoids

Figure 4.10. A diagrammatic formulation of the life history of hosts (horizontal) and parasitoids (vertical) with overlapping generations, useful for the continuous time model. Here T_E and T_L are the development times of host eggs and larvae; the development time of the parasitoid from egg to adult consists of some time as an egg and development time T_J as a juvenile.

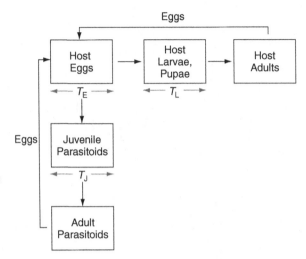

involves differential, rather than difference, equations and delays to account for development in the different stages (Murdoch *et al.* 1987, MacDonald 1989, Briggs 1993). There have been literally volumes written about these approaches; in this section I give a flavor of how the models are formulated and analyzed. In Connections, I point towards more of the literature.

Our goal is to capture the dynamics of hosts and parasitoids in continuous time with overlapping generations. Figure 4.10 should be helpful. After a host egg is laid, there is a development time T_E, during which the egg may be attacked by an adult parasitoid. Surviving eggs become larvae and then pupae (both of which are not attacked by the parasitoid) with a development time T_L, after which they emerge as adults with average lifetime T_A. Parasitoids are characterized in a similar way. It is customary to use different notation to capture the various stages of the host life history, so we now introduce the following variables

$$
\begin{aligned}
E(t) &= \text{number of host eggs at time } t \\
L(t) &= \text{number of host larvae at time } t \\
A(t) &= \text{number of adult hosts at time } t \\
P(t) &= \text{number of adult parasitoids at time } t
\end{aligned}
\tag{4.18}
$$

We will derive equations for each of these variables. The rate of change of eggs, dE/dt, is the balance between the rate at which eggs are produced (assumed to be proportional to the adult population size, with no density dependent effects) and the rate at which eggs are lost. Eggs are lost in three ways: due to parasitism (assumed to be proportional to both the number of eggs and the number of parasitoids), due to other sources of mortality, not related to the parasitoid, and due to survival through development and movement into the larval class, which we denote by $M_E(t)$, for maturation of eggs at time t. Combining these different rates, we write

$$\frac{dE(t)}{dt} = rA(t) - aP(t)E(t) - \mu_E E(t) - M_E(t) \qquad (4.19)$$

The new parameters in this equation, r, a, and μ_E have clear interpretations as the per capita rate at which adults lay eggs, the per capita attack rate by parasitoids and the non-parasitoid related mortality rate. Note that I have written the argument of these equations explicitly on both sides of Eq. (4.19). The reason becomes clear when we explicitly write the maturation function. Eggs that mature into larva at time t had to be laid at time $t - T_E$ and survived from then until time t. The rate of egg laying at that earlier time was $rA(t - T_E)$ and if we assume random search by parasitoids and other sources of mortality, the probability of survival from the earlier time to time t is $\exp[-\int_{(t-t_E)}^{t} (aP(s) + \mu_E)ds]$. Combining these, we conclude that

$$M_E(t) = rA(t - T_E) \exp\left[- \int_{(t-t_E)}^{t} (aP(s) + \mu_E)ds \right] \qquad (4.20)$$

The same kind of logic applies to the larval stage, for which the rate of change of larval numbers is a balance between maturation of eggs into the larval stage, maturation of larvae/pupae into the adult stage, and natural mortality. Hence, we obtain

$$\frac{dL(t)}{dt} = M_E(t) - M_L(t) - \mu_L L(t) \qquad (4.21)$$

where, in analogy with Eq. (4.20), we have

$$M_L(t) = M_E(t - T_L) \exp(-\mu_L T_L) \qquad (4.22)$$

Adult hosts are produced by maturing larvae and lost due to natural mortality, so that

$$\frac{dA(t)}{dt} = M_L(t) - \mu_A A(t) \qquad (4.23)$$

and this completes the description of the host population dynamics.

The reasoning is similar for the parasitoids. Adult parasitoids emerge from eggs that were laid at a time T_P before the current time and that survive to produce a parasitoid (assumed to occur with probability σ and disappear due to natural mortality), so that we have

$$\frac{dP(t)}{dt} = a\sigma E(t - T_P)P(t - T_P) - \mu_P P(t) \qquad (4.24)$$

Equations (4.19)–(4.24) constitute the description of a host parasitoid system with overlapping generations and potentially different developmental periods. They are called differential-difference equations, for the obvious reason that both derivatives and time differences are involved. What can we say about these kinds of equations in general? Three things. First, finding the steady states of these equations is easy. Second, the numerical solution of these equations is harder than the numerical solution of corresponding solutions without delays (although some software packages might do this automatically for you). Third, the analysis of the stability of these kinds of equations is much, much harder than the work we did in Chapter 2 or in this chapter until now. However, these are important tools so that we now consider a simple version of such an equation and in Connections, I point you towards literature with more details.

Our analysis will focus on the logistic equation with a delay and will follow the treatment given by Murray (2002). We consider the single equation

$$\frac{dN}{dt} = rN(t)\left[1 - \frac{N(t - \tau)}{K}\right] \tag{4.25}$$

for which the delay τ is fixed and for which K is a steady state. As we did in the past, we write $N(t) = K + n(t)$, and assume that $n(0)$ is small. Substituting this $N(t) = K + n(t)$ into Eq. (4.25) leads to

$$\frac{dn}{dt} = r(K + n(t))\left[1 - \frac{K + n(t - \tau)}{K}\right] = r(K + n(t))\left[-\frac{n(t - \tau)}{K}\right] \tag{4.26}$$

and if we keep only the linear term, we need to understand the dynamics of

$$\frac{dn}{dt} = -rn(t - \tau) \tag{4.27}$$

Note that in order to reach Eq. (4.27) we have assumed that both $n(t)$ and $n(t - \tau)$ are small. This could be a big assumption, but our goal in this analysis is – to some extent – to determine when it fails, which occurs when the deviation $n(t)$ grows in time. As before, we start with the guess $n(t) = ce^{\lambda t}$ and substitute this into Eq. (4.27).

Exercise 4.5 (E/M)

Show that λ has to satisfy the equation $\lambda = -re^{-\lambda t}$ and then explain why if τ is sufficiently small there is a solution of this equation corresponding to decay towards the steady state (it may be easiest to sketch a graph with λ on the x-axis and $y = \lambda$ or $y = -re^{-\lambda \tau}$ on the y-axis and look for their intersections and note that if $\tau = 0$ then $\lambda = -r$).

To further increase our intuition, note the following. Suppose we knew that $n(t) = A\cos(\pi t/2\tau)$, so that $dn/dt = -(A\pi/2\tau)\sin(\pi t/2\tau)$. Furthermore, $n(t - \tau) = A\cos[(\pi t/2\tau) - (\pi/2)]$ and if we recall the angle addition formula from trigonometry, $\cos(a + b) = \cos(a)\cos(b) + \sin(a)\sin(b)$, we conclude that $n(t - \tau) = -A\sin(\pi t/2\tau)$ from which we conclude that in this specific case $dn/dt = (\pi/2\tau)n(t - \tau)$. Thus, if we start with an oscillatory solution, we know that we can derive a differential equation similar to Eq. (4.27). This suggests that we might seek oscillatory solutions for the more general delay-differential equations.

An oscillatory solution would mean that we assume $\lambda = \mu + i\omega$, where μ is the amplitude of the oscillations and ω is the frequency of the oscillations. We take this and use it in Eq. (4.27) to obtain

$$\mu + i\omega = -re^{-(\mu+i\omega)\tau} = -re^{-\mu\tau}e^{-i\omega\tau} = -re^{-\mu\tau}[\cos(\omega\tau) - i\sin(\omega\tau)] \quad (4.28)$$

We now equate that the real and imaginary parts to obtain equations for μ and ω:

$$\mu = -re^{-\mu\tau}\cos(\omega\tau) \qquad \omega = re^{-\mu\tau}\sin(\omega\tau) \quad (4.29)$$

We want to understand the conditions for which $\mu < 0$, which will mean that the dynamics are stable. From the first equation in (4.29), we conclude that one condition for $\mu < 0$ is that $\omega\tau < \pi/2$. Furthermore, we know that when $\tau = 0$, we have the solution $\mu = -r$, $\omega = 0$, so that perturbations decay without oscillation. The classic result is that the steady state $N(t) = K$ of Eq. (4.25) is stable if $0 < r\tau < \pi/2$ (see Murray 2002, p. 19; I have not been able to find a better way to explain the derivation, so simply send you there). If the condition is violated, then perturbations from the steady state will exhibit oscillatory behavior. Although the logistic equation with a delay seems to be highly simplistic, it both provides insight for us and, in some cases, leads to good fits between theory and data. In Figure 4.11, I show the fit obtained by May (1974) to the data of Nicholson (1954) on the Australian sheep-blowfly

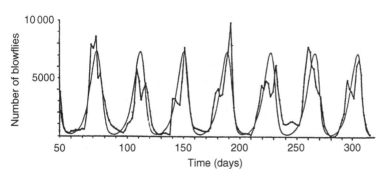

Number of blowflies
10 000
5000
50 100 150 200 250 300
Time (days)

Figure 4.11. Figure based on data from Nicholson (1954) on blowfly population dynamics that were fit by May (1974) to the logistic equation with a delay. Reprinted with permission.

Lucilia cuprina. In this case, the theory with $r\tau = 2.1$ provides a remarkably good fit to the data, given the simplicity of the model, and since the stability condition is violated, the steady state is unstable and we expect oscillations around it.

Evolution of host choice in parasitoids

Hosts that are attacked by parasitoids will come in a range of varieties. For example, hosts will vary in size and larger hosts will often provide more hemolymph for developing parasitoids than smaller hosts. When the hosts are pupae (and thus do not move around), hosts that are in the sun may provide quicker development times for the parasitoids than those that are in the shade (and if they are not hidden may also provide higher risks of mortality). For solitary parasitoids, then, an issue is in which kind of host to lay eggs.

Gregarious parasitoids have an additional problem of how many eggs to lay in a host. More eggs in a host may imply more daughters, but they could be smaller, and since size is tied to fecundity, the overall representation of genes in future generations may be reduced. For example, Rosenheim and Rosen (1992) studied clutch size in *Aphytis lingnaensis*, which attacks scale insects that are pests of citrus. They found that the average size of a daughter emerging from a clutch of size c laid in a host was $S(c) = \max\{0.2673{-}0.0223c, 0\}$ and that the number of eggs a female can lay depends upon her size according to $E(S) = \max\{181.8S{-}26.7, 0\}$, where $\max\{A, B\}$ means take the larger of A or B. Thus, for example, there is a minimum size below which a female does not have any eggs. A simple measure of fitness for an ovipositing female is the number of grand-offspring produced from a clutch of size c and this is $cE(S(c))$. This computation (Figure 4.12) shows that for a gregarious parasitoid there may be an optimal number of eggs to lay in a single host.

In the early 1980s, Eric Charnov and Sam Skinner recognized that many of the ideas from foraging theory could be applied to understand the evolution of host choice in parasitoids (Charnov and Skinner 1984, 1985, 1988). I subsequently wrote on similar topics; the relevant chapters in Mangel and Clark (1988) and Clark and Mangel (2000) are entry points to the broader literature. In this section, we will consider a relatively simple version of these kinds of models, which will set up the next two sections as well as introduce new methods.

We begin with a pro-ovigenic, univoltine, solitary parasitoid in which the season length is T and the parasitoid has two different kinds of hosts to attack. Oviposition in host type i leads to an offspring in the next generation with probability f_i and we assume that host type 1 is

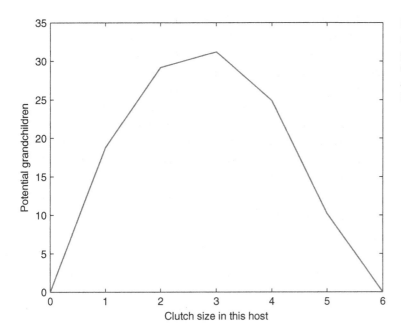

Figure 4.12. The potential number of grandoffspring produced from the daughters of a clutch of size c, for the *Aphytis* data of Rosenheim and Rosen (1992).

better, in the sense that $f_1 > f_2$, and that hosts are encountered at different rates. Because there is one generation per year, we choose as a measure of fitness the expected lifetime reproductive success of the parasitoid and ask for the pattern of host acceptance that maximizes expected reproductive success. The question is interesting because both time within the season and current egg complement may affect the oviposition behavior. For example, early in the season, we might anticipate that females will be more choosy, for the same egg complement, than late in the season. We also might predict that females with many eggs will be less choosy than those with fewer eggs. The question is then how are we able to more formally characterize these predictions and use them in guiding our thinking about additional theory, experiments and field work.

The method we use is called stochastic dynamic programming (SDP, see Connections for more details; in a previous draft I apologized – perhaps to Strunk and White – for using a noun adjective. But Nick Wolf, upon reading this apology, wrote "I don't see the problem: 'programming' is a noun (a gerund, actually), 'dynamic' is an adjective modifying 'programming', and 'stochastic' is an adjective acting as an adverb because it modifies another adjective. No problem!") and is very straightforward in this particular case. To begin, we define a fitness function $F(x, t)$ as the maximum expected reproductive success from the current time t until T (when the season ends) given that the egg complement at time t is $X(t) = x$. Here "maximum" refers to the choices over alternative oviposition behaviors at a particular time and egg complement, and "expected" refers to the mathematical average over possible host encounters and natural mortality. Thus, expected lifetime reproductive success at emergence is $F(x, 1)$.

We will work in discrete time, and characterize the probability of encountering a host of type i in an unit of time by λ_i, so that the sum $\lambda_1 + \lambda_2$, which must be less than or equal to 1, is a measure of the richness of the environment. We assume that the probability of surviving a single period of time is e^{-m}, where m is the rate of mortality.

Since the season ends at time T, there is no gain in fitness thereafter and any eggs that the parasitoid holds are wasted. This end of season condition is represented mathematically by

$$F(x, T) = 0 \qquad (4.30)$$

For previous times, we need to think about the balance between current and future fitness. At any previous time t, three mutually exclusive events may occur: no host is encountered, host type 1 is encountered or host type 2 is encountered. If no host is encountered (with probability $1 - \lambda_1 - \lambda_2$) and the parasitoid survives, she will start the next time interval with the same number of eggs. If a host of type 1 (the better host) is encountered we assume that she oviposits in it; if she survives to the next period she begins that period with one less egg. If a host type 2 is encountered, the parasitoid may reject the host (thus beginning the next period with the same number of eggs, but having gained no fitness from the encounter) or she may oviposit in (thus beginning the next period with one fewer egg but having gained fitness from the encounter). We assume that the order of the processes is oviposition, then survival (so that she always gets credit for an oviposition, even if she does not survive to the next period). These three possibilities and their consequences lead to a relationship between fitness at time t and at time $t + 1$:

$$F(x, t) = (1 - \lambda_1 - \lambda_2)e^{-m}F(x, t+1) + \lambda_1[f_1 + e^{-m}F(x-1, t+1)]$$
$$+ \lambda_2 \max\{e^{-m}F(x, t+1); f_2 + e^{-m}F(x-1, t+1)\} \qquad (4.31)$$

Equation (4.31) is called an equation of stochastic dynamic programming. We solve this equation backwards in time, since $F(x, T)$ is known. Hence, this method is called "backwards induction"; details of doing this can be found in Mangel and Clark (1988) and Clark and Mangel (2000).

Note that each term on the right hand side involves current accumulation to fitness (which may be 0 if no host is encountered) and future accumulations of fitness, discounted by the chance of mortality. The most interesting term, of course, is the third one in which the balance is complicated by the loss of an egg that can be used in the future accumulation of fitness.

The solution of this equation generally must be done by numerical methods, which means that specific parameter values must be chosen. For Eq. (4.31), these parameters are the encounter and mortality rates, the fitnesses associated with oviposition in the two kinds of hosts, the time horizon and the maximum egg complement that the parasitoid may have. Once these are specified, the solution of Eq. (4.31) comes rapidly (Figure 4.13a), especially these days: in Mangel and Clark (1988), we had to introduce a variety of means for getting around the limited computer power of then extant machines and software.

Although Figure 4.13a is interesting, we are often more interested in the behavior of the parasitoids than in their lifetime reproductive success. Happily, predictions about behavior come freely as we solve Eq. (4.31). That is, when we consider the maximization step in this equation, we also determine the predicted optimal behavior $b^*(x, t)$, which is to either accept the inferior host or reject it for oviposition at time t when $X(t) = x$. We thus are able to construct a boundary curve that separates the $x - t$ plane into regions in which the parasitoid is predicted to reject the inferior host and regions in which the parasitoid is predicted to accept the inferior host (Figure 4.13b). Studying this figure as we move horizontally (forward in time with egg complement fixed), we see a formalization of the intuition that individuals are predicted to become less choosy as time increases. Holding time constant and moving vertically upwards, we see a formalization of the intuition that individuals are predicted to become less choosy as they have higher egg complements.

How might such an idea be tested? One method is to use a photoperiod manipulation to signal to the parasitoids that it is either earlier or later in the season than it is. For example, by rearing parasitoids in a late summer photoperiod, we send the signal that t is closer to T than it actually is and the consequence would be that if the real point in the $x - t$ plane were at A in Figure 4.13b, we predict that the parasitoids will behave as if they were located at point B in the plane. That is, the photoperiod manipulation is predicted to cause parasitoids that would otherwise reject an inferior host to accept it. Roitberg et al. (1992) did exactly that manipulation, using a theory somewhat more complicated than the one here. The more elaborate theory lead to a wide range of predictions and the experimental results were in concordance with the predictions.

Our predictions will change as parameters vary. For example, in Figure 4.13c, I show the boundary curves for the previous case in which the mortality rate was 0.05 and for the case in which the mortality rate is 0.10. In the latter case, we predict that the balance between current and

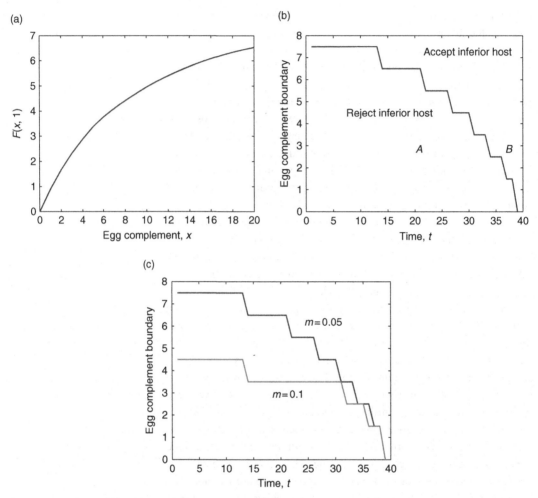

Figure 4.13. The total lifetime expected reproductive success $F(x, 1)$ (panel a) and the boundary for oviposition in the inferior host (panel b) obtained by the solution of Eq. (4.31) for the parameters $\lambda_1 = 0.3$, $\lambda_2 = 0.3$, $m = 0.05$, $T = 40$, $f_1 = 1$, $f_2 = 0.3$, and maximum egg complement 20. If egg complement exceeds the boundary value at a particular time, we predict that the parasitoid will oviposit in the inferior host; otherwise we predict that she will reject the inferior host. (c) The boundary curve changes as parameters change. When mortality increases, the balance between current and future fitness shifts towards current fitness and thus the boundary curve lowers.

future reproduction should favor current reproduction (because the chance of surviving to achieve future reproduction is lower) and that the parasitoids will accept inferior hosts both earlier and at lower egg complements. Roitberg *et al.* (1993) tested these ideas in an experiment that simulated an impending thunder storm (via a dropping barometric pressure), which has the potential of high mortality rates for small insects, and found that the predictions and experimental results were once again consistent.

Exercise 4.6 (M/H)

Write your own program to solve Eq. (4.31) and create the boundary curve. Then conduct numerical experiments to investigate how the boundary curve varies as you change parameters. It is always good to make a prediction by thinking about the result that you anticipate before you run the code. (If you need help getting the coding going, I suggest that you consult Mangel and Clark (1988) and Clark and Mangel (2000).)

In nature we do not observe boundary curves or states. Rather we observe behaviors manifested in time. One way of capturing these behavioral observations is through the simulation of a large number of individuals that follow the rules generated by the dynamic programming equation. Colin Clark and I (Mangel and Clark 1988, Clark and Mangel 2000) called such individual based models forward iterations (see Connections for more about these), to distinguish them from the backward iterations that generate the decision rules. To implement them, we envision simulating a large number, N, of individuals in which the egg complement of individual i at time t is denoted by $X_i(t)$. We then use the random number generator to connect the state of each individual at time t to time $t - 1$. If we let the state -1 correspond to death, the forward state dynamics associated with Eq. (4.31) are: $X(t + 1) = -1$ if the parasitoid does not survive from t to $t + 1$; $X(t + 1) = X(t)$ if no host is encountered or an inferior host is encountered and the parasitoid survives from t to $t + 1$; and $X(t + 1) = X(t) - 1$ if a superior host is encountered or an inferior host is encountered and accepted. By simulating forward, we are able to track variables that are measurable in the field or laboratory such as behaviors, mean egg complements, and survival. Sometimes these can even be done by purely analytical (Markov Chain) methods; see Mangel and Clark (1988) and Houston and McNamara (1999) for examples, but many times simulation is required because the analytical methods are simply too hard.

Exercise 4.7 (M/H)

Use the solution of the backward iteration from Exercise 4.6 in a forward iteration and use that forward iteration to predict population level properties that might be of interest to you in a field or laboratory setting.

Combining behavior and population dynamics

Population dynamics interest us and behavior interests us; how much more interesting then would be the combination of behavior and population dynamics? At the same time, the study of population dynamics is hard and the study of behavior is hard. How much more difficult then

Figure 4.14. In order to couple behavior and population dynamics, we need to think about annual time scales (on which Nicholson–Bailey dynamics occur) and within-season time scales (on which behavior occurs).

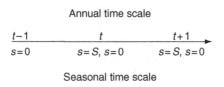

will be the combination of behavior and population dynamics? Pretty difficult, but as the example in this section shows, we can make some progress. This example draws heavily on a paper by Bernie Roitberg and me (Mangel and Roitberg 1992). Citations to other literature on this subject are found in Connections.

We return to Eq. (4.1), which connects the annual dynamics of hosts and parasitoids. In order to incorporate behavior, we need to think about both the annual time scale (as in Eq. (4.1)) and about a time scale within the season, which we shall denote by s (Figure 4.14). Thus, the season runs from $s = 0$ to $s = S$ (which we might set equal to 1 to obtain exactly the same form as Eq. (4.1)) and we now think of the populations as $H(t, s)$ and $P(t, s)$, the number of hosts and parasitoids at time s within year t. First, we observe that to obtain the between-season Nicholson–Bailey dynamics, we solve the within-season dynamics

$$\frac{dH}{ds} = -aP(t,s)H(t,s) \qquad H(t,0) = H_0$$
$$\frac{dP}{ds} = 0 \qquad\qquad\qquad P(t,0) = P_0 \tag{4.32}$$

in the sense that the solution of these equations is Eqs. (4.1) when we couple the appropriate within- and between-season dynamics by linking $H(t, S)$ with $H(t + 1, 0)$ and $P(t, S)$ with $P(t + 1, 0)$. Our first insight thus comes virtually for free: the Nicholson–Bailey dynamics assume a constant number of parasitoids throughout the season. How might this occur? We could assume, for example, that parasitoids are emerging from last season's hosts at the same rate at which they are dying during this season. This assumption means that we can think of the number of parasitoids as a function only of the season and not worry about parasitoid numbers within the season.

In order to incorporate behavior, we require some kind of variation in the hosts. Following the lead of the previous section, let us assume that hosts come in two phenotypes in which the first phenotype is preferred, for whatever biological reasons. We thus consider the dynamics of $H_1(t, s)$ and $H_2(t, s)$, the numbers of superior and inferior hosts at time s within year t. To capture parasitoid behavior, we assume that there is a time s^* (which we will find) before which only the superior hosts are attacked and after which both hosts are attacked. The within-season dynamics in Eq. (4.32) must now be expanded to account for these two cases. We will assume that parasitoids search randomly and that when both hosts are attacked, they are attacked in proportion to their initial abundance (so that the parasitoid cannot distinguish between previously attacked hosts and unattacked hosts). The dynamics for the two host types thus become

$$\frac{dH_1}{ds} = \begin{array}{ll} -aP(t)H_1(t,s) & \text{for } s < s^* \\ -a\left[\dfrac{H_1(t,0)}{H_1(t,0)+H_2(t,0)}\right]P(t)H_1(t,s) & \text{for } s \geq s^* \end{array} \tag{4.33a}$$

and

$$\frac{dH_2}{ds} = \begin{array}{ll} 0 & \text{for } s < s^* \\ -a\left[\dfrac{H_2(t,0)}{H_1(t,0)+H_2(t,0)}\right]P(t)H_2(t,s) & \text{for } s \geq s^* \end{array} \tag{4.33b}$$

and solution of these equations will tell us the within-season population dynamics of hosts, given the behavior of the parasitoids. Equations (4.33) are linear equations and are very easy to solve (so much so that I do not even make finding the solution an exercise). We conclude that at the end of season t

$$H_1(t,S) = H_1(t,0)\exp(-aP(t)s^*)\exp\left(-\frac{H_1(t,0)}{H_1(t,0)+H_2(t,0)}aP(t)(S-s^*)\right)$$

$$H_2(t,S) = H_2(t,0)\exp\left(-\frac{H_2(t,0)}{H_1(t,0)+H_2(t,0)}aP(t)(S-s^*)\right) \tag{4.34}$$

and Eqs. (4.34) tell us the whole story about the within-season effects of behavior.

Next, we must construct the between-season population dynamics of hosts and parasitoids. If we assume that each surviving host produces R offspring, then the total number of hosts at the start of the next season will be $H_T(t+1, 0) = R\{H_1(t, S) + H_2(t, S)\}$, which needs to be distributed across superior and inferior phenotypes. In general, we might imagine that these are functions of the total host population, so that $H_i(t+1, 0) = f_i(H_T(t+1, 0))$. The simplest function is a constant proportion and Hassell (1978) called this the "proportional refuge model" because the inferior hosts provide a refuge from attack by parasitoids. An alternative, which captures density-dependent effects, is

$$f_2(H_T(t+1,0)) = w_1(1 - \exp(-w_2 H_T(t+1,0))) \tag{4.35}$$

where ω_1 and ω_2 are parameters. As total host population declines, the number that are inferior declines as well.

Next season's parasitoids emerge from hosts that have been attacked. If we assume that the parasitoid is solitary, then one offspring will emerge from a superior host, and one from an inferior host, but only with probability β. In that case, parasitoid numbers at the start of the next season are given by

$$P(t + 1) = H_1(t, 0) - H_1(t, S) + \beta[H_2(t, 0) - H_2(t, S)] \qquad (4.36)$$

in which the first two terms on the right hand side of Eq. (4.36) represent the number of superior hosts attacked and the second two terms represent the number of inferior hosts attacked.

We still need to find s^*. If $s^* = 0$, then we have the standard Nicholson–Bailey model and if $s^* = S$, then we have a version of Taylor's stabilization of the Nicholson–Bailey dynamics because of a host refuge. To find s^* in the more general situation, we will use an especially simple version of stochastic dynamic programming by assuming that the parasitoids never run out of eggs (a topic discussed in the next section). Given that there are a total of H_T hosts for the parasitoid to attack, and the random search assumption of the Nicholson–Bailey model, we have

$$Pr\{\text{encounter a host in a unit interval of time}|H_T \text{ hosts present}\}$$
$$= 1 - \exp(-aH_T) \qquad (4.37)$$

so that if $\lambda_i(t)$ is the probability of encountering host type i within one unit of time in season t, we have

$$\lambda_i(t) = \frac{H_i(t, 0)}{H_T(t, 0)}(1 - \exp(-aH_T(t, 0))) \qquad (4.38)$$

and we assume that hosts are sufficiently plentiful that the parasitoid never re-encounters a previously attacked host (see Exercise 4.8 below).

Now let $F(s|t)$ denote the maximum expected accumulated reproduction between within-season time s and S, given host parameters of season t. We then have $F(S|t) = 0$ and if m_s and m_o are the probabilities of mortality during search and oviposition, the dynamic programming equation is

$$F(s|t) = (1 - \lambda_1(t) - \lambda_2(t))(1 - m_s)F(s + 1|t) + \lambda_1(t)\{1 + (1 - m_o)F(s + 1|t)\}$$
$$+ \lambda_2(t)\max\{\beta + (1 - m_o)F(s + 1|t); (1 - m_s)F(s + 1|t)\}$$
$$(4.39)$$

which we solve backwards, as before.

Exercise 4.8 (M/H)

Equation (4.39) does not include a physiological state of the parasitoid. How would it be modified to include egg complement? The equation also does not include the possibility that a host which is encountered has previously been attacked. How would it be modified to account for that?

Further exploration of these ideas requires numerical solution; the parameters that Mangel and Roitberg (1992) fix are 50 parasitoids, 500 hosts at the start of the first season, a season of length 20, host per capita reproduction $R = 2$, quality of inferior host $\beta = 0.1$, search parameter $a = 0.0001$, and mortality probabilities during search and oviposition of 0.001 and 0.2 respectively.

There are at least two ways that we can conceive of viewing the results. The first is through the host–parasitoid phase plane in which steady states are represented by single points and oscillatory solutions by closed orbits (limit cycles). The second is the distribution of s^* across different years. These results are shown in Figure 4.15. We conclude from Figures 4.15a–c that the dynamics of the interaction between host and parasitoid can be very rich when behavior and population dynamics are coupled, including strange attractors such as Figure 4.15c (Mangel and Roitberg (1992) show some even stranger cases). Perhaps more importantly, the result of Figure 4.15d, which shows the distribution of s^* across generations and years, tells us that we should expect variation in behavior of parasitoids in the field. Nature is indeed complicated and variable, but much of that complication and variation can be captured and understood.

In both this model, Eq. (4.39), and the previous model, Eq. (4.31), a shift in host preference occurs during the season. However, the mechanisms are very different. Here, the shift in host preference occurs because there is so little time left in the season that β exceeds the loss in lifetime fitness that occurs when the parasitoid chooses an inferior host, i.e. $\beta > (m_o - m_s)F(s + 1|t)$. On the other hand, in Eq. (4.31) the shift in host preference occurs because there is too little time left in the season for the ovipositing female to find good hosts for all of her eggs. Thus, Eq. (4.31) involves a mixture of time limitation and egg limitation, whereas Eq. (4.39) is purely time limitation (via survival). It would be interesting to design experiments to separate these; in the next section we consider some additional theory.

Are parasitoids egg- or time-limited?

Parasitoids live in uncertain and risky worlds and natural selection acts on patterns of development and behavior through expected reproductive success. One topic of considerable interest in the late 1990s was the

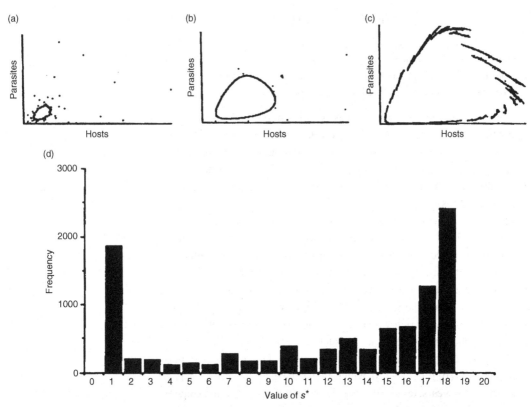

Figure 4.15. Linking behavior and population dynamics together leads to a rich panoply of dynamical results. In panels (a–c), the superior and inferior hosts are determined in a manner that is free of density dependence. For panel (a), the fraction of superior hosts is 0.5 and the behavior of the parasitoids is determined by the SDP model Eq. (4.39). For panel (b), the fraction is 0.75 and inferior hosts are never attacked (the proportional refuge model). For panel (c), the fraction is 0.75 and the behavior of parasitoids is determined by the SDP model. In panel (d), I show the frequency distribution of s^* over 10 000 generations for the case in which the fraction is 0.75. Reproduced with permission from Mangel and Roitberg (1992).

nature of factors that limit parasitoid reproductive success (a set of citations to these questions can be found in Driessen and Hemerik (1992), Heimpel and Collier (1996), Rosenheim (1996), Heimpel and Rosenheim (1998), Heimpel *et al.* (1998), Sevenster *et al.* (1998), Rosenheim (1999), Rosenheim and Heimpel (2000), and van Baalen (2000)). Factors limiting parasitoid reproductive success include mortality risk while searching and foraging, egg complement, and time available for searching for oviposition sites and ovipositing. Natural selection could act on the latter two through development, either by increasing the capacity for eggs or the rate of egg maturation or by selecting for individuals who are more efficient at handling hosts.

We then come to the question: is reproductive success of parasitoids limited by time or by eggs (that is, are they likely to die with some eggs still in their bodies, unlaid and thus "wasted," or are they likely to spend some of the end of their lives without eggs, but with opportunities for laying them, thus being reproductively senescent).

One example of how this problem might be attacked is provided by van Baalen (2000). If at the start of the season there are $H(t)$ hosts and $P(t)$ parasitoids, then the number of hosts per parasitoid is on average $h(t) = H(t)/P(t)$. When the parasitoid searches randomly – as assumed in the Nicholson–Bailey model – the number of hosts attacked by an individual parasitoid will follow a Poisson distribution with parameter a. However, the number of attacks will be limited by either the number of hosts per parasitoid or the egg complement, e, of the parasitoid. Thus, the expected number of hosts attacked per parasitoid

$$E\{\text{number of hosts attacked per parasitoid}\} =$$

$$\sum_{k=0}^{min(h(t),\, e)-1} k\frac{e^{-a}a^k}{k!} + \sum_{k=min(h(t),\, e)}^{\infty} min(h(t), e)\frac{e^{-a}a^k}{k!} \qquad (4.40)$$

and using this gives a sense of the limitation due either to eggs or to hosts by comparison with the average egg complement of individual parasitoids (and provides another way of stabilizing the Nicholson–Bailey dynamics).

Here, I take a slightly different approach than either van Baalen or my own work in collaboration with George Heimpel and Jay Rosenheim (Heimpel *et al.* 1996, 1998; Rosenheim and Heimpel 2000). We will use the method of stochastic dynamic programming and begin by considering a pro-ovigenic, univoltine parasitoid. The expected lifetime reproductive success of such a parasitoid will be characterized by her maximum egg complement, x_{max}, and the handling time per host, h. We can formally write this as $F(x_{max}, 1|h, T)$ to emphasize egg complement, lifetime expected reproductive success at the beginning of a season of length T when handling time is h.

Now we need to define what is meant by egg limitation and time limitation. I propose the following: a parasitoid is egg-limited rather than time-limited if an increase in maximum egg complement of one egg will increase her lifetime expected reproductive success more than a decrease in handling time of one unit (appropriately defined) does. This suggests that there will be a boundary in egg-complement/handling-time space separating regions in which parasitoids are egg-limited from those in which they are time-limited, and we are going to find the boundary, defined by the condition

$$F(x_{max}, 1|h - 1, T) = F(x_{max} + 1, 1|h, T) \qquad (4.41)$$

because the left hand side of Eq. (4.41) is expected lifetime fitness with current maximum egg complement but a decreased handling time and the right hand side is expected lifetime fitness if handling time were the same but maximum egg complement were increased by one egg. The solution of Eq. (4.41) is the sought-after boundary curve $h(x_{max})$.

To begin, consider a solitary parasitoid searching for hosts that, as before, are of two kinds (clearly many generalizations are possible). We fully characterize the problem by describing the encounter rate and fitness accrued from oviposition in host type i, λ_i and f_i respectively, and the mortality rate while searching or ovipositing, m. The dynamic programming equation is then similar to Eq. (4.31):

$$F(x, t|h, T) = (1 - \lambda_1 - \lambda_2)e^{-m}F(x, t+1|h, T) + \lambda_1[f_1 + e^{-m}F(x-1, t+h|h, T)]$$
$$+ \max \lambda_2\{e^{-m}F(x, t+1|h, T); f_2 + e^{-m}F(x-1, t+h|h, T)\}$$
$$(4.42)$$

Also as before, we solve this equation backwards in time, starting with the condition that $F(x, T|h, T) = 0$. One additional decision needs to be made, however, and that concerns how we treat oviposition for situations in which $t + h > T$ (so that the oviposition being considered is the last). The decision that I made for the results presented here is to credit the parasitoid with this last oviposition, even if she dies immediately thereafter.

The numerical results that I show correspond to $T = 100$, $f_1 = 1$, $f_2 = 0.1$, $\lambda_1 = 0.4$, $\lambda_2 = 0.4$, and $m = 0.01$ or 0.001. (I have also limited handling time to a maximum of 15 time units.) The boundary curve (Figure 4.16) separates regions in which another egg would increase fitness more than a decrease in handling time. The general shape of the boundary should make sense; for example, parasitoids with lots of eggs are more likely to be time-limited than egg-limited. In addition, the dependence of the boundary on mortality rate should also make sense in that when mortality rates are higher the tendency will be towards a greater likelihood of time limitation than egg limitation.

Exercise 4.9 (M/H)

Instead of considering a solitary parasitoid, consider a gregarious parasitoid which encounters only one kind of host, with probability λ per unit time, and which lays clutches in that host. Let $f(c)$ denote the increment of expected lifetime reproductive success when she lays a clutch of size c in a host. Explain why an appropriate dynamic programming equation is

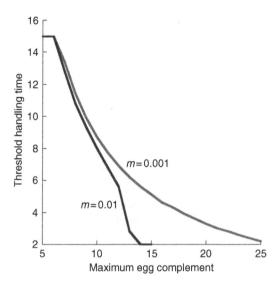

Figure 4.16. The boundary curve implied by Eq. (4.41), obtained by the numerical solution of Eq. (4.42). For values of maximum egg complement and handling time that fall below the curve, the parasitoid is egg-limited, in the sense that her lifetime fitness would be incremented more by an increase in egg complement than by a decrease in handling time. For values above the curve, the parasitoid is time-limited, in the sense that lifetime fitness would be incremented more by a decrease in handling time than by an increase in maximum egg complement.

$$F(x, t|h, T) = (1 - \lambda)e^{-m}F(x, t + 1|h, T)$$
$$+ \lambda \, \max_c\{f(c) + e^{-mch}F(x - c, t + ch|h, T)\} \qquad (4.43)$$

and identify the other assumptions that are implicitly included in this equation. (As you might imagine, the results obtained using Eq. (4.43) are similar, but not identical, to those obtained using Eq. (4.42).)

Let us close by considering how all this might be modified for a synovigenic parasitoid. In such a case, we require two state variables to describe the parasitoid: mature eggs (denoted by x) and oocytes (denoted by y) and also need to specify the egg maturation time τ, in the sense that oocytes mature into eggs at rate $1/\tau$. A very simple example is one in which in a single period of time the parasitoid either encounters food (understood to be both protein and carbohydrate) or a host, with probabilities λ_f and λ_h respectively. If she encounters food, her oocyte reserve is incremented by an amount δ and if she encounters a host, her lifetime fitness is incremented by 1 offspring and eggs are decremented by 1 egg. In such a case, the dynamic programming equation is (with the dependence on h and T suppressed)

$$F(x, y, t) = (1 - \lambda_h - \lambda_f)e^{-m}F\left(x + \frac{y}{\tau}, y - \frac{y}{\tau}, t + 1\right)$$
$$+ \lambda_f e^{-m}F\left(x + \frac{y}{\tau}, y - \frac{y}{\tau} + \delta, t + 1\right)$$
$$+ \lambda_h\left\{1 + e^{-m}F\left(x + \frac{y}{\tau} - 1, y - \frac{y}{\tau}, t + 1\right)\right\} \qquad (4.44)$$

and we can now think about time limitation in two contexts: handling time and egg maturation time. That would be a pretty interesting project.

Connections

Space can be stabilizing, sometimes destabilizing, and always interesting

Another means of stabilizing the Nicholson–Bailey model is through the explicit introduction of space. There is a large literature on such spatial models; entry points include Hassell and May (1988), Taylor (1988a, b), Walde and Murdoch (1988), Comins *et al.* (1992), Adler (1993), Ruxton and Rohani (1996), and Briggs and Hoopes (2004). Spatial aspects of host–parasitoid interaction are always interesting.

Delay differential models for host–parasitoid dynamics

The recent volume of Murdoch *et al.* (2003) is a wonderful general reference on the topics considered in this chapter. Gordon *et al.* (1991) use a delay differential model to understand the synchrony of host and parasitoid development. Bonsall *et al.* (2002) use such models to investigate the roles of ecological trade-offs, resource partitioning and coexistence in host–parasitoid assemblages. Delays can enter into the discrete time models too, for example through age structure (Bellows and Hassell 1988).

Delay differential equations in general

The simplest such equation is

$$\frac{\mathrm{d}}{\mathrm{d}t}N(t) = rN(t - \tau) - mN(t)$$

in which we understand r to be the per-capita reproduction of adults who require a developmental period of length τ, and m is natural mortality. The still classic text in this area is Bellman and Cooke (1963). Nisbet and Gurney and their colleagues (see, for example, Gurney *et al.* (1983), Nisbet and Gurney (1983), Blythe *et al.* (1984), Nisbet *et al.* (1985)) have developed relatively complete methods for formulating population models with delays (see also May *et al.* (1974)). MacDonald (1989) shows how a single differential-difference equation can sometimes be converted to a system of linear equations without delays.

Evolution of host choice in parasitoids, marking pheromones, superparasitism and patch leaving

The parasitoids are a rich source of ideas in ecology and evolutionary biology; everyone should be excited by them! The examples considered in the text are literally only the tip of the iceberg. Some of my other favorite topics include the evolution of pheromones used to mark hosts after oviposition (Roitberg *et al.* 1984, Roitberg and Prokopy 1987, Roitberg and Lalonde 1991); whether parasitoids feed on a host (to make more eggs) or lay an egg in it (de Bach 1943, Edwards 1954, Bartlett 1964, Jervis and Kidd 1986, Walter 1988, Rosenheim and Rosen 1992, Heimpel *et al.* 1994, Heimpel and Rosenheim 1995, Heimpel and Collier 1996, McGregor 1997, Giron *et al.* 2002); whether parasitoids (or tephritid fruit flies) lay an egg in a host that has already been attacked (which they can tell because of the marking pheromone, for example) or, rather, ignore superparasitism of such hosts (Pritchard 1969, Hubbard and Cook 1978, Hubbard *et al.* 1987, van Alphen *et al.* 1987, van Alphen and Visser 1990, van Randen and Roitberg 1996); whether parasitoids will search for oviposition sites or food sites (to find both carbohydrate, to run the operation, and protein, to make more eggs if they are synovigenic) (Lewis and Takasu 1990, Waeckers 1994, Heimpel *et al.* 1997, Olson *et al.* 2000); what happens to parasitoids in the field (Janssen *et al.* 1988, Janssen 1989); information as a state variable (Roitberg 1990, Haccou *et al.* 1991, Hemerik *et al.* 2002); the effects of intraspecific competition between ovipositing females, making the oviposition behavior a dynamic game (Visser and Rosenheim 1998); the effects of hyperparasitism (parasitoids of parasitoids) on oviposition behavior (Mackauer and Voekl 1993); the role of sex ratio in population dynamics (Hassell *et al.* 1983) and behavior (Olson and Andow 1997); and patch-leaving behavior (Hemerik *et al.* 1993, Rosenheim and Mangel 1994, Vos *et al.* 1998, Wajnberg *et al.* 2000, van Alphen *et al.* 2003). Stochastic dynamic programming provides a natural means for making predictions about the interactions of age, egg complement and behavior. These predictions are also imminently testable (Cook and Hubbard 1977, Hubbard and Cook 1978, Marris *et al.* 1986, Rosenheim and Rosen 1992, Fletcher *et al.* 1994, Heimpel *et al.* 1996, Rosenheim and Heimpel 1998, Vos and Hemerik 2003).

Stochastic dynamic programming

The words "stochastic dynamic programming" were invented by the American mathematician Richard Bellman, who was both incredibly prolific and incredibly creative (often these do not go together).

Bellman invented the name somewhat before he had the method (Bellman 1984) and this makes sense if you think that the method is based on two principles: (1) with probability 1 something will happen; and (2) whatever happens, act optimally. The origin of the method goes back at least to the 1800s and the Irish mathematician Hamilton in his study of the motion of the planets (Courant and Hilbert 1962). The method of stochastic dynamic programming as a tool in behavioral ecology was popularized by Colin Clark, Alasdair Houston, John McNamara and me (Mangel and Clark 1988, Houston and McNamara 1999, Clark and Mangel 2000, McNamara *et al.* 2001). The parasitoid problem that we considered is basically one of investment (in this case, of eggs) in situations with uncertainty (in this case maternal survival). Such problems abound and the methods have a wide range of applications in economic settings. One of my favorite volumes is Dixit and Pindyck (1994). Owen-Smith (2002) gives some interesting applications of these methods to herbivore diet choice and ecology. More mathematical sources include Bertsekas (1976), Whittle (1983), Mangel (1985), Puterman (1994), and Bertsekas (1995).

Individual-based models

Forward iterations of the sort described in the text are examples of individual-based models (IBMs), which are becoming a common tool in ecology (DeAngelis and Gross 1992, van Winkle *et al.* 1998, Railsback 2001). Such models are used to simulate the behavior of many individuals, so that population processes emerge from individual interactions. In order to do this, one must have the rules that the individuals follow and a major question is then, from where do such rules come? One possibility, of course, is simply to make them up from one's sense of how the system works. Another is to use deterministic or stochastic dynamic programming, as in here, but use the results to predict general rules of thumb, rather than state and time dependent tables of behavior. Still another is to use the method of genetic algorithms (Haupt and Haupt 1998), in which behavior is the result of genetic predispositions that evolve through simulated genetic processes (Robertson *et al.* 1998, Huse *et al.* 1999, McGregor and Roitberg 2000, Giske *et al.* 2002, Strand *et al.* 2002). Other means of achieving individual adaptation are described in Belew and Mitchell (1996).

Combining behavior and population dynamics

Ecologists and evolutionists should be interested in behavior because of its impact on population dynamics (behaviorists, on the other hand, find behavior itself interesting). Our analysis of a simple situation shows how quickly things can become complex. Meier *et al.* (1994) provide a more formal mathematical analysis of strange attractors associated with the classical models of host–parasitoid systems. Comins and Hassell (1979) aim to bridge the gap between foraging models and population models and conclude that although foraging behavior is important for the quantitative details, the qualitative ones do not change much; but there is still much to be done.

Chapter 5
The population biology of disease

We now turn to a study of the population biology of disease. We will consider both microparasites – in which populations increase in hosts by multiplication of numbers – and macroparasites – in which populations increase in hosts by both multiplication of numbers and by growth of individual disease organisms. The age of genomics and bioinformatics makes the material in this chapter more, and not less, relevant for three reasons. First, with our increasing ability to understand type and mechanism at a molecular level, we are able to create models with a previously unprecedented accuracy. Second, although biomedical science has provided spectacular success in dealing with disease, failure of that science can often be linked to ignoring or misunderstanding aspects of evolution, ecology and behavior (Schrag and Weiner 1995, de Roode and Read 2003). Third, there are situations, as is well known for AIDS but is true even for flu (Earn *et al.* 2002), in which ecological and evolutionary time scales overlap with medical time scales for treatment (Galvani 2003).

To begin, a few comments and caveats. At a meeting of the (San Francisco) Bay Delta Modeling Forum in September 2004, my colleague John Williams read the following quotation from the famous American jurist Oliver Wendell Holmes: "I would not give a fig for simplicity this side of complexity, but I would give my life for simplicity on the other side of complexity". It could take a long time to fully deconstruct this quotation but, for our purposes, I think that it means that models should be sufficiently complicated to do the job, but no more complicated than necessary and that sometimes we have to become more complicated in order to see how to simplify. In this chapter, we will develop models of increasing complexity. The building-up feeling of the progression of sections is not intended to give the impression that more complicated models are better. Rather, the scientific question is paramount, and the simplest model that helps you answer the question is the one to aim for.

Furthermore, the mathematical study of disease is a subject with an enormous literature. As before, I will point you toward the literature in the main body of the chapter and in Connections. As you work through this material, you will develop the skills to read the appropriate literature. That said, there is a warning too: disease problems are inherently nonlinear and multidimensional. They quickly become mathematically complicated and there is a considerable literature devoted to the study of the mathematical structures themselves (very often this is described by the authors as "mathematics motivated by biology"). As a novice theoretical biologist, you might want to be chary of these papers, because they are often very difficult and more concerned with mathematics than biology.

There are two general ways of thinking about disease in a population. First, we might simply identify whether individuals are healthy or sick, with the assumption that sick individuals are able to spread infection. In such a case, we classify the population into susceptible (S), infected (I) and recovered or removed (R) individuals (more details on this follow). This classification is commonly done when we think of micro-parasites such as bacteria or viruses. An alternative is to classify individuals according to the parasite burden that they carry. This is typically done when we consider parasitic worms. We will begin with the former (classes of individuals) and move towards the latter (parasite burden).

The *SI* model

As always, it is best to begin with a simple and familiar story. Lest you think that this is too simple and familiar, it is motivated by the work of Pybus *et al.* (2001), published in *Science* in June 2001. Since this is our first example, we begin with something relatively simple.

Envision a closed population of size N and let $S(t)$ and $I(t)$ denote respectively the number of individuals who are susceptible to infection (susceptibles) and who are infected (infecteds) with the disease at time t. Since the population is closed, $S(t) + I(t) = N$, which we will exploit momentarily. New cases of the disease arise when an infected individual comes in contact with a susceptible individual. One representation of this rate of new infections is bSI, which is called the mass action formulation of transmission, and which we will discuss in more detail in the next section. Note that because the population is closed, the rate of

The population biology of disease

Figure 5.1. The solution of the *SI* model (Eq. (5.1)) is logistic growth if $bN > v$ and decline of the number of infected individuals if $bN < v$. Parameters here are $N = 500$, $v = 0.1$ and $b = 2v/N$ or $b = 0.95v/N$.

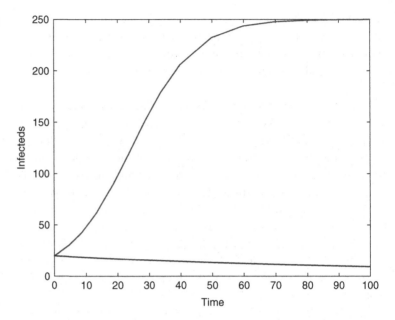

new infections is also $b(N - I)I$; this is often called the force of infection. We assume that individuals lose infectiousness at rate v, so that the rate of loss of infected individuals is vI. Combining these, we obtain an equation for the dynamics of infection:

$$\frac{dI}{dt} = bI(N - I) - vI \qquad (5.1)$$

If we combine the linear terms together we have

$$\frac{dI}{dt} = I(bN - v) - bI^2 \qquad (5.2)$$

and we see from this equation that if $bN < v$, the number of infecteds will decline from its initial value. However, if $bN > v$, then Eq. (5.2) is the logistic equation, written in a slightly different format (what would the r and K of the logistic equation be in terms of the parameters in Eq. (5.2)?). The resulting dynamics are shown in Figure 5.1. If $bN < v$, the disease will not spread in the population, but if it does spread, the growth will be logistic – an epidemic will occur, leading to a steady level of infection in the population $\bar{I} = (bN - v)/b$. Furthermore, whether the disease spreads or not can be determined by evaluating bN/v without having to evaluate the parameters individually. Pybus *et al.* (2001) fit this model to a number of different sets of data on hepatitis C virus.

Since the population is closed, we could also work with the fraction of the population that is infected, $i(t) = I(t)/N$. Setting $I(t) = Ni(t)$ in

Eq. (5.1) gives $N(di/dt) = bNi(N - Ni) - vNi$ and if we divide by N, and set $\beta = bN$ we obtain

$$\frac{di}{dt} = \beta i(1 - i) - vi \qquad (5.3)$$

as the equation for the dynamics of the infected fraction. Note that the parameter β has the units of a pure rate, whereas b has somewhat funny units: 1/time-individuals-infected, such as per-day-per-infected individual. I have more to say about this in the next section.

Now let us consider these disease dynamics from the perspective of the susceptible population. Furthermore, suppose that the initial number of infected individuals is 1. We can then ask, if the disease spreads in the population, how many new infections will occur as a result of contact with this one individual? Since the rate of new infections is bIS, the dynamics for $S(t)$ are $dS/dt = -bIS$, which we will solve with the initial condition $S(0) = N - 1$, holding $I(t) = 1$. This will allow us to ask how many cases arise, approximately, from the one infected individual (you could think about why this is approximate). The solution for the dynamics of susceptibles under these circumstances is $S(t) = (N - 1)\exp(-bt)$. Recall that the recovery rate for infected individuals is v, so that $1/v$ is roughly the time during which the one infected individual is contagious. The number of susceptible individuals remaining at this time will be $S(1/v) = (N - 1)\exp(-b/v)$, so that the number of new cases caused by the one infected individual is $S(0) - S(1/v) = (N - 1) - (N - 1)\exp(-b/v) = (N - 1)(1 - \exp(-b/v))$. If we assume that the population is large, so that $N - 1 \approx N$ and we Taylor expand the exponential, writing $\exp(-b/v) \sim 1 - (b/v)$, we conclude that the number of new infections caused by one infected individual is approximately Nb/v. This value – the number of new infections caused by one infected individual entering a population of susceptible individuals – is called the basic reproductive rate of the disease and is usually denoted by R_0. Note that $R_0 > 1$ is the condition for the spread of the disease, and it is exactly the same condition that we arrived at by studying the Eq. (5.2) for the dynamics of infection. In this case, R_0 tells us something interesting about the dynamics of the disease too, since we can rewrite Eq. (5.1) as $(1/v)(dI/dt) = (R_0 - 1)I - (b/v)I^2$; see Keeling and Grenfell (2000) for more on the basic reproductive rate.

Characterizing the transmission between susceptible and infected individuals

Before going any further, it is worthwhile to spend time thinking about how we characterize the transmission of disease between infected and susceptible individuals. This is, as one might imagine, a topic with an immense literature. Here, I provide sufficient information for our needs, but not an overall discussion – see the nice review paper of McCallum *et al.* (2001) for that.

In the previous section, we modeled the dynamics of disease transmission by bIS. This form might remind you of introductory chemistry and of chemical kinetics. In fact, we call this the mass action model for transmission. Since $dS/dt = -bIS$, and the units of the derivative are individuals per time, the units of b must be 1/(time)(individuals); even more precisely, we would write 1/(time)(*infected* individuals). Thus, b is not a rate, but a composite parameter.

The simplest alternative to the mass action model of transmission is called the frequency dependent model of transmission, in which we write $dS/dt = -b(I/N)S$. Now b becomes a pure rate, because I/N has no units. Note that we assume here that the rate at which disease transmission occurs depends upon the frequency, rather than absolute number, of infected individuals. If we were working with an open, rather than closed, population in which infected individuals are removed by death or recovery, instead of N we could use $I + S$.

A third model, which is phenomenological (that is, based on data rather than theory) is the power model of transmission, in which we write $dS/dt = -bS^pI^q$ where p and q are parameters, both between 0 and 1. In this case, the units of b could be quite unusual.

A fourth model, to which we will return in a different guise, is the negative binomial model of transmission, for which

$$\frac{dS}{dt} = -kS \log\left(1 + \frac{bI}{k}\right) \tag{5.4}$$

where k is another parameter – and is intended to be exactly the overdispersion parameter of the negative binomial distribution. This model is due to Charles Godfray (Godfray and Hassell 1989) who reasoned as follows. Over a unit interval of time, let us hold I constant and integrate Eq. (5.4) by separating variables

$$\frac{dS}{S} = -k \log\left(1 + \frac{bI}{k}\right) dt$$

$$S(1) = S(0) \exp\left(\log\left(1 + \frac{bI}{k}\right)^{-k}\right) = S(0)\left(1 + \frac{bI}{k}\right)^{-k} \tag{5.5}$$

$$= S(0)\left(\frac{k + bI}{k}\right)^{-k} = S(0)\left(\frac{k}{k + bI}\right)^{k}$$

so that we see that in one unit of time, the fraction of susceptibles escaping disease is given by the zeroth term of the negative binomial distribution.

As in Chapter 3, where you explored the negative binomial distribution, it is valuable here to understand the properties of the negative binomial transmission model.

Exercise 5.1 (M)

(a) Show that as $k \to \infty$, the negative binomial transmission model approaches the mass action transmission model. (Hint: what is the Taylor expansion of $\log(1 + x)$? Alternatively, set $k = 1/x$ and apply L'Hospital's rule.) (b) Define the relative rate of transmission by

$$R(k) = \frac{kS \log\left(1 + \frac{bI}{k}\right)}{bIS}$$

and do numerical investigations of its properties as k varies. (c) Note, too, that your answer depends only on the product bI, and not on the individual values of b or I. How do you interpret this? (d) The force of infection is now $kS\log(1 + (bI/k))$. Holding S and I constant, investigate the level curves of the force of infection in the $b - k$ plane.

In most of what follows, we will use the mass action model for disease transmission. In the literature, mass action and frequency dependent transmission models are commonly used, but rarely tested (for an exception, see Knell *et al.* 1996). Because of this, one must be careful when reading a paper to know which is the choice of the author and why.

The *SIR* model of epidemics

The mathematical study of disease was put on firm footing in the early 1930s in a series of papers by Kermack and McKendrick (1927, 1932, 1933); a discussion of these papers and their intellectual history, c. 1990, is found in R. M. Anderson (1991). When Kermack and McKendrick did their work, computing was difficult, so that good thinking (analytic ability, finding closed forms of solutions and their approximations) was even more important than now (of course, one might argue that since these days it is so easy to blindly solve a set of equations on the computer, it is even more important now to be able to think about them carefully).

We consider a closed population in which individuals are either susceptible to disease (S), infected (I) or recovered or removed by death (R). Since the population is closed, at any time t we have $S(t) + I(t) + R(t) = N$. If we assume mass action transmission of the disease and that removal occurs at rate v, the dynamics of the disease become

$$\frac{dS}{dt} = -bIS$$

$$\frac{dI}{dt} = bIS - vI \qquad (5.6)$$

$$\frac{dR}{dt} = vI$$

and in general, the initial conditions would be $S(0) = S_0$, $I(0) = I_0$ and $R(0) = N - S_0 - I_0$ (since the population may already contain individuals who have experienced and recovered from the disease).

Let us begin with the special case of $S(0) = N - 1$ and $I(0) = 1$. As in the model of hepatitis, we can ask the following question: how many new cases of the disease are caused directly by this one infected individual entering a population in which everyone else is susceptible. We proceed in very much the same way as we did with hepatitis. If we set $I = 1$ in the first line of Eq. (5.6), the solution is $S(t) = (N - 1)\exp(-bt)$. The one infected individual is infectious for a period of time approximately equal to $1/v$, at which time the number of susceptibles is $(N - 1)\exp(-b/v)$. The number of new cases caused by this one infected individual is then $N - 1 - [(N - 1)\exp(-b/v)] = (N - 1)(1 - \exp(-b/v))$ and if we Taylor expand the exponential, keeping only the linear term, and assume that the population is large so that $N - 1 \approx N$ we conclude that $R_0 \approx bN/v$, just as with the model for hepatitis C.

Now let us think about Eq. (5.6) in general. The only steady state for the number of infected individuals is $I = 0$, but there are two choices for the steady states of S: either $S = 0$ (in which case an epidemic has run through the entire population) or $S = v/b$ (in which case an epidemic has run its course, but not every individual became sick). We would like to know which is which, and how we determine that. The phase plane for Eq. (5.6) is shown in Figure 5.2, and it is an exceptionally simple phase plane. Indeed, from this phase plane we conclude the following remarkable fact: if $S(0) > v/b$ then there will be a wave of epidemic in the population in the sense that $I(t)$ will first increase and then decrease. Note that this condition, $S(0) > v/b$, is the same as the condition that

Figure 5.2. The phase plane for the *SIR* model. This is an exceptionally simple phase plane: since dS/dt is always negative, points in the phase plane can move only to the left. If $S(0) > v/b$, then $I(t)$ will increase, until the line $S = v/b$ is crossed. If $S(0) < v/b$, then $I(t)$ only declines.

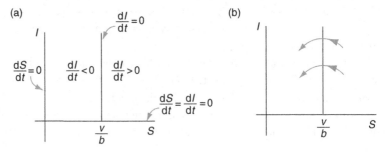

$R_0 > 1$. Thus the heuristic analysis and the phase plane analysis lead to the same conclusion. This remarkable result is called the Kermack–McKendrick epidemic theorem. Note that once again, the threshold depends upon the number of susceptible individuals, not the number of infected individuals.

We can actually do more by noting that $dI/dS = (dI/dt)/(dS/dt)$ from which we conclude

$$\frac{dI}{dS} = -1 + \frac{v}{bS} \qquad (5.7)$$

If we think of I as a function of S, then I will takes its maximum when $dI/dS = 0$; this occurs when $S = b/v$. We already know this from the phase plane, but Eq. (5.7) allows us to find an explicit representation for $I(t)$ and $S(t)$.

Exercise 5.2 (E/M)

Separate the variables in Eq. (5.7) to show that

$$I(t) + S(t) - \frac{v}{b}\log(S(t)) = I(0) + S(0) - \frac{v}{b}\log(S(0)) \qquad (5.8)$$

Note that this equation allows us to find the relationship between $I(t)$ and $S(t)$ at any time in terms of their initial values.

How about computation of trajectories? That involves the solution of Eq. (5.6.) We might work with the variables $S(t)$ and $I(t)$ themselves, which could involve dealing with relatively large numbers. For those who want to write their own iterations by treating the differential equation as a difference equation, I remind you of the warning that we had in Chapter 2 on the logistic equation. The following observation is helpful. If we set $S(t + dt) = S(t)\exp(-bI(t)dt)$, then in the limit that $dt \to 0$, we get back the first line of Eq. (5.6) (if this is unclear to you, Taylor expand the exponential, subtract $S(t)$ from both sides, divide by dt and take the limit). This reformulation also provides a handy interpretation: $\exp(-bI(t)dt) < 1$ and can be interpreted as the fraction of susceptible individuals who escape infection in the interval $(t, t + dt)$ when the number of infected individuals is $I(t)$.

However, because the population is closed and $R(t) = N - S(t) - I(t)$, we can focus on fraction of susceptible and infected individuals, rather than absolute numbers. That is, if we set $S(t) = s(t)N$, $I(t) = i(t)N$ and $\beta = bN$ as in Eq. (5.3), the first two lines of Eq. (5.6) become

$$\frac{ds}{dt} = -\beta is$$

$$\frac{di}{dt} = \beta is - vi \qquad (5.9)$$

to which we append initial conditions $s(0) = s_0$ and $i(0) = i_0$. Note that the critical susceptible fraction for the spread of the epidemic is now v/β. These equations can be solved by direct Euler iteration or by more complicated methods, or by software packages such as MATLAB.

Exercise 5.3 (M)

Solve Eqs. (5.9) for the case in which the critical susceptible fraction is 0.4, for values of $s(0)$ less than or greater than this and for $i(0) = 0.1$ or 0.2.

Kermack and McKendrick, who did not have the ability to compute easily, obtained an approximate solution of the equations characterizing the epidemic. To do this, they began by noting that since the population is closed we have $dR/dt = vI = v(N - S - R)$, which at first appears to be unhelpful. But we can find an equation for S in terms of R by noting the following

$$\frac{dS}{dR} = \left(\frac{dS}{dt}\right) \bigg/ \left(\frac{dR}{dt}\right) = -\left(\frac{b}{v}\right)S \qquad (5.10)$$

and so we see that S, as a function of R, declines exponentially with R; that is $S(R) = S(0)\exp(-(b/v)R)$. When we use this in the equation for R, we thus obtain

$$\frac{dR}{dt} = v\left[N - S(0)\exp\left(-\frac{bR}{v}\right) - R\right] \qquad (5.11)$$

to which we add the condition $R(0) = N - S_0 - I_0$ and from which we would like to find $R(t)$, after which we compute $S(t) = S(0)\exp(-(b/v)R(t))$ and from that $I(t) = N - S(t) - R(t)$. However, Eq. (5.11) cannot be solved either. In order to make progress, Kermack and McKendrick (1927) assumed that $bR \ll v$ (how do you interpret this condition?), so that the exponential could be Taylor expanded. Keeping up to terms of second order in the expansion, we obtain

$$\frac{dR}{dt} = v\left[N - S(0)\left(1 - \frac{bR}{v} + \frac{1}{2}\left(\frac{b}{v}\right)^2 R^2\right) - R\right] \qquad (5.12)$$

and this equation can be solved (Davis 1962). In Figure 5.3, I have reprinted a figure from Kermack and McKendrick's original paper, showing the general agreement between this theory and the observed data, the solution of Eq. (5.12) (although their notation is slightly different than ours), and their comments on the solution.

To close this section, and give a prelude to what will come later in the chapter, let us ask what will happen to the dynamics of the disease if individuals can either recover or die. Thus, let us suppose that the

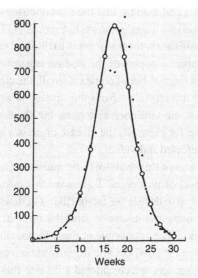

Figure 1. Deaths from plague in the island of Bombay over the period 17 December 1905 to 21 July 1906. The ordinate represents the number of deaths per week, and the abscissa denotes the time in weeks. As at least 80–90% of the cases reported terminate fatally, the ordinate may be taken as approximately representing dz/dt as a function of t. The calculated curve is drawn from the formula:

$$y = \frac{dz}{dt} = 890 \; \text{sech}^2 (0.2t - 3.4)$$

We are, in fact, assuming that plague in man is a reflection of plague in rats, and that with respect to the rat: (1) the uninfected population was uniformly susceptible; (2) that all susceptible rats in the island had an equal chance of being infected; (3) that the infectivity, recovery, and death rates were of constant value throughout the course of sickness of each rat; (4) that all cases ended fatally or became immune; (5) that the flea population was so large that the condition approximated to one of contact infection. None of these assumptions are strictly fulfilled and consequently the numerical equation can only be a very rough approximation. A close fit is not to be expected, and deductions as to the actual values of the various constants should not be drawn. It may be said, however, that the calculated curve, which implies that the rates did not vary during the period of epidemic, conforms roughly to the observed figures.

Figure 5.3. Reproduction of Figure 1 from Kermack and McKendrick (1927), showing the solution of Eq. (5.12) and a comparison with the number of deaths from the plague in Bombay. Reprinted with permission.

mortality rate for the disease is m. The dynamics of susceptible and infected individuals are now

$$\frac{dS}{dt} = -bIS$$

$$(5.13)$$

$$\frac{dI}{dt} = bIS - (v + m)I$$

and the basic reproductive rate of the disease is now $R_0 = bS_0/(v + m)$. How might the mortality from the disease, m, be connected to the rate at which the disease is transmitted, b? We will call m the virulence or the

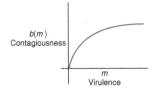

$b(m)$
Contagiousness

m
Virulence

Figure 5.4. The assumed relationship between contagion or infectiousness, $b(m)$ and virulence or infectedness, m.

infectedness and assume that the contagiousness or infectiousness is a function $b(m)$ with shape shown in Figure 5.4. The easiest way to think about a justification for this form is to think of m and $b(m)$ as a function of the number of copies of the disease organism in an infected individual. When the number of copies is small, the chance of new infection is small, and the mortality from the disease is small. As the number of copies rises, the virulence also rises, but the contagion begins to level off because, for example, the disease organism is saturating the exhaled air of an infected individual.

If we accept this trade-off, the question then becomes what is the optimal level of virulence? To answer this question, which we will do later, we need to decide the factors that will determine the optimal level, and then figure out a way to find the optimal level. For example, is making m as large as possible optimal for the disease organism? I leave this question for now, but you might want to continue to think about it.

In this section, we considered a disease that is epidemic: it enters a population, and runs its course, after which there are no infected individuals in the population. We now turn to a case in which the disease is endemic – there is a steady state number of infected individuals in the population.

The *SIRS* model of endemic diseases

We now modify the basic *SIR* model to assume that recovered individuals may lose resistance to the disease and thus become susceptible again, but continue to assume that the population is closed. Assuming that the rate at which resistance to the disease is lost is f, the dynamics of susceptible, infected, and recovered individuals becomes

$$\frac{dS}{dt} = -bIS + fR$$

$$\frac{dI}{dt} = bIS - vI \tag{5.14}$$

$$\frac{dR}{dt} = vI - fR$$

One possible steady state for this system is $I = R = 0$ and $S = N$, in which case we conclude that the disease is extirpated from the population. If this is not the case, we then set $R = N - S - I$ and work with the dynamics of susceptible and infected individuals:

$$\frac{dS}{dt} = -bIS + f(N - S - I)$$

$$\frac{dI}{dt} = bIS - vI \tag{5.15}$$

The number of infected individuals is at a steady state if $\bar{S} = v/b$. We then set $dS/dt = 0$ and solve for the steady state number of infected individuals (this is why the assumption of a closed population is such a nice one to make):

$$\bar{I} = \frac{f(N - \bar{S})}{b\bar{S} + f} \qquad (5.16)$$

and if we evaluate this at the steady number of susceptible individuals, we obtain

$$\bar{I} = \frac{f[N - (v/b)]}{v + f} \qquad (5.17)$$

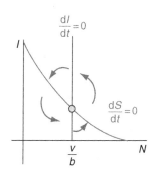

Figure 5.5. The phase plane for the *SIRS* model for the case in which the disease is predicted to be endemic.

so that we conclude the steady number of infecteds is positive if $N > v/b$ (a quantity which should now be familiar). That is, we have determined a condition for endemicity of the disease, in the sense that the steady state number of infected individuals is greater than 0.

The next question concerns the dynamics of the disease. In Figure 5.5, I show the phase plane for the case in which the disease is predicted to be endemic. The phase plane suggests that we should, in general, expect oscillations in the case of an endemic disease – that is periodic outbreaks that are not caused by anything other than the fundamental population biology of the disease.

Furthermore, from this analysis we conclude that, although whether the disease is endemic or not depends only upon the ratio v/b and the size of the population N, the level of endemicity (determined by the steady state number of infected individuals) will also depend, as Eq. (5.17) shows us, upon the ratio v/f. Through this analysis, we thus learn what critical parameters to measure in the study of an endemic disease.

A numerical example is found in the next section.

Adding demography to *SIR* or *SIRS* models

Until now, we have ignored all other biological processes that might occur concomitantly with the disease. One possibility is population growth and mortality that is independent of the disease. There are many different ways that one may add demographic processes to the *SIR* or *SIRS* models. Here, I pick an especially simple case, to illustrate how this can be done and how the conclusions of the previous sections might change.

When adding demography, we need to be careful and explicit about the assumptions. Let us assume that (1) only susceptible individuals reproduce, and do so at a density-independent rate r, (2) all individuals experience mortality μ that is independent of the disease with $r > \mu$, and (3) there is no disease-dependent mortality. In that case, the *SIR* equations (5.6) become

$$\frac{dS}{dt} = -bIS + (r - \mu)S$$

$$\frac{dI}{dt} = bIS - vI - \mu I \qquad (5.18)$$

$$\frac{dR}{dt} = vI - \mu R$$

The term representing demographic process of net reproduction is $(r - \mu)S$. Other choices are possible; for example we might assume that both susceptible and recovered individuals could reproduce, that all individuals can reproduce (still with no vertical transmission) or that birth rate is simply a constant (e.g. proportional to N). Each of these could be justified by a different biological situation and may lead to different insights than using Eqs. (5.18); μI and μR are demographic sources of mortality. If one particularly appeals to you, I encourage you to redo the analysis that follows with the assumption that you find most attractive.

We proceed to find the steady states by setting the left hand side of Eqs. (5.18) equal to 0. When we do this, we obtain (from $dS/dt = dI/dt = dR/dt = 0$ respectively)

$$\bar{I} = \frac{r - \mu}{b} \qquad \bar{S} = \frac{v + \mu}{b} \qquad \bar{R} = \frac{v\bar{I}}{\mu} = \frac{v}{\mu}\left(\frac{r - \mu}{b}\right) \qquad (5.19)$$

We learn an enormous amount just from the steady states. First, recall that for the *SIR* model without demography, the only steady state is $I = 0$. However, from Eqs. (5.19), we conclude that in the presence of demographic factors, a disease that would be epidemic becomes endemic. Second, we see that the steady state levels of susceptible, infected, and recovered individuals depends upon a mixture of demographic and disease parameters. Third, and perhaps most unexpected, note that the steady state level of susceptibles is independent of r! (You should think about the assumptions and results for a while and explain the biology that underlies it.) It is helpful to summarize the various versions of the *SIR* model in a single figure (Figure 5.6). Here I show the *SIR* model for an epidemic (panel a), the *SIRS* model for an endemic disease (which approaches the steady state in an oscillatory fashion) (panel b), and the *SIR* model with demography (panel c). Note the progression of increasing dynamic complexity (also see Connections).

Equations (5.19) beg at least two more questions: first, what is the nature of this steady state; second, what happens if there is more

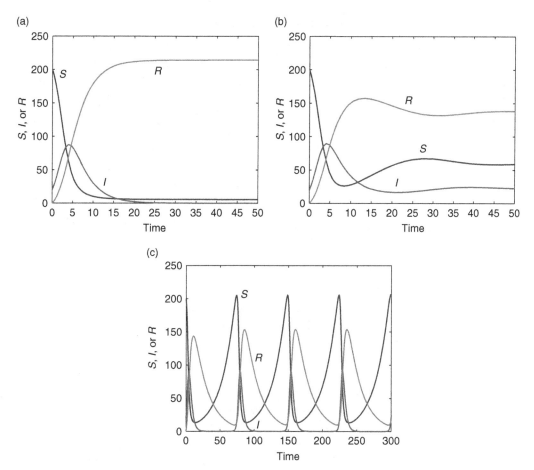

Figure 5.6. Solutions of various forms of the *SIR* model. (a) The basic *SIR* model for an epidemic ($b=0.005$, $v=0.3$; true for panels b and c); (b) the *SIRS* model for an endemic disease ($f=0.05$); and (c) the *SIR* model with demography ($f=0$, $r=0.1$, $\mu=0.05$).

complicated demography? These are good questions, but since I want to move on to other topics, I will leave them as exercises.

Exercise 5.4 (M/H)

Conduct an eigenvalue analysis of the steady state in Eqs. (5.19). Note that there will be three eigenvalues. How are they to be interpreted?

Exercise 5.5 (E/M)

How do Eqs. (5.19) change if we assume logistic growth rather than exponential growth as the demographic term. That is, what happens if we replace $(r-m)S$ by $rS(1-(S/K))$?

The evolution of virulence

In the same way that demographic processes can occur simultaneously with disease processes, evolutionary processes can occur simultaneously with ecological processes in the dynamics of a disease. Although we tend to think of population dynamics and evolution occurring on different time scales, contemporary evolution (evolution observed in less than a few hundred generations) is receiving more attention (Stockwell *et al.* 2003). One of the most impressive and well-known examples is the AIDS virus, which shows evolution of drug resistance within patients during the course of their care.

In this section, we will consider three examples, with the goal of giving you a sense of how one can think about the evolution of virulence.

The optimal level of virulence

Recall that we closed the section on the *SIR* model with a discussion of the basic reproductive rate for a disease when the disease related mortality rate is m and recovery rate is v

$$R_0(m) = \frac{b(m)S_0}{v + m} \tag{5.20}$$

where I have made explicit the dependence of the contagion on the virulence, still assumed to have the shape as in Figure 5.4. How might natural selection act on the reproductive rate of a disease? A reasonable starting point is to assume that the disease strain that spreads the fastest (i.e. has the greatest value of $R_0(m)$) will be the most prevalent. If we accept this assumption as a starting point, we then ask for the value of m that maximizes $R_0(m)$ given by Eq. (5.20).

Now you should compare Eq. (5.20) with Eq. (1.6). They are essentially the same equation: a saturating function of a variable divided by that variable plus a constant. Thus, from the marginal value construction in Chapter 1, we instantly know how to find the optimal level of virulence. First, we plot $b(m)$ versus m. Second, we draw the tangent line from $(-v, 0)$ to the curve $b(m)$. Third, we read the predicted optimal level of virulence from the intersection of the tangent line and the x-axis (Figure 5.7). Thus, the marginal value theorem, developed for foraging in patchy environments, is also useful here.

The unbeatable (ESS) level of virulence

We will now look at the problem in a slightly different manner, from the perspective of invasions. Recall that the dynamics of the infected

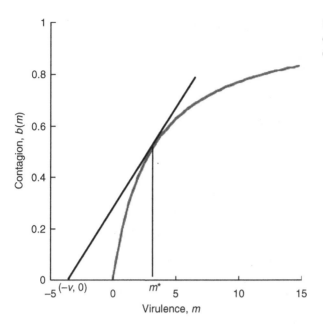

individuals are $dI/dt = bIS - (v + m)I$ from which we conclude that the steady state level of susceptibles is $\bar{S}(m) = (v + m)/b(m)$. Now let us consider an invader, which is rare and which uses an alternative level of virulence $\tilde{m}$. Because the invader is rare, we assume that it has no effect on the steady state level of the susceptible population, and we ask "when will the invader increase?". Under these assumptions, if $\tilde{I}$ denotes the number of invaders, the dynamics of the invader are

$$\frac{d\tilde{I}}{dt} = b(\tilde{m})\tilde{I}\bar{S}(m) - (v + \tilde{m})\tilde{I} \tag{5.21}$$

and we now substitute for the steady state level of susceptibles and factor out the number of infecteds to obtain

$$\frac{d\tilde{I}}{dt} = \tilde{I}\left[b(\tilde{m})\left(\frac{v + m}{b(m)}\right) - (v + \tilde{m})\right] \tag{5.22}$$

and the invader will spread if the term in brackets is greater than 0. This is true when $b(\tilde{m})((v + m)/b(m)) > (v + \tilde{m})$, which is, of course, the same as $b(\tilde{m})/(v + \tilde{m}) > b(m)/(v + m)$. We thus conclude that the strategy that maximizes $b(m)/(v + m)$ is unbeatable because it cannot be invaded. This is exactly the same condition that arises in the maximization of R_0. In other words, the strategy that optimizes the basic reproductive rate is also unbeatable and cannot be invaded. This is a very interesting result, in part because optimality and ESS analyses may

Figure 5.8. The infection process modeled by Koella and Restif (2001) in their study of the coevolution of virulence and host age at maturity. The host becomes infected by disease propagules (such as spores) independent of the density of other infected individuals.

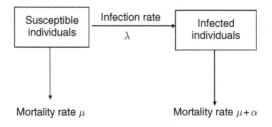

often lead to different conclusions (Charlesworth 1990, Mangel 1992) but here they do not.

The coevolution of virulence and host response

As the virulence of the parasite evolves, the host response may also change. Thus, we have a case of coevolution of parasite virulence and host response. Here, we develop, in a slightly different manner, a model due to Koella and Restif (2001) and I encourage you to seek out and read the original paper.

For the host, we assume a semelparous organism following von Bertalanffy growth with growth rate k, asymptotic size L_∞, disease independent mortality μ, and allometric parameter β connecting size at maturity and reproductive success. With these assumptions, we know from Chapter 2 that if age at maturity is t, then an appropriate measure of fitness is $F(t) \propto e^{-\mu t}(1 - e^{-kt})^\beta$ and we also know from Chapter 2 that the optimal age at maturity is $t_m^* = (1/k)\log[(\mu + \beta k)/\mu]$.

For the disease, we assume horizontal transmission between disease propagules and susceptible hosts at rate λ that is independent of the number of infected individuals (think of a disease transmitted by propagules such as spores). The virulence of the disease can be characterized by an additional level of host mortality α, so that the mortality rate of infected hosts is $\mu + \alpha$. (Figure 5.8). We then immediately predict that hosts that are infected will reproduce at a different age, given by

$$t_{m,i}^* = \frac{1}{k}\log\left(\frac{\mu + \alpha + \beta k}{\mu + \alpha}\right)$$

Exercise 5.6 (E/M)

Determine the corresponding values for size at maturity.

Our first prediction is that if there are no constraints acting on age at maturity, then infected individuals will mature at earlier age (and smaller size) than non-infected individuals. However, suppose all individuals are forced to use the same age at maturity (e.g. the physiological machinery required for maturity is slow to develop, so that the age of maturity has to be set long in advance of potential infection). We could then ask, as do Koella and Restif (2001), what is the best age at maturity, taking into account the potential effect of infection on the way to maturation.

In that case, we allow the age at maturity to be different from either of the values determined above and proceed as follows. First, we will determine the optimal level of virulence for the pathogen, given that the age of maturity is t_m. This optimal level of virulence can be denoted as $\alpha^*(t_m.)$. Given the optimal level of virulence in response to an age at maturity, we then allow the host to determine the best age at maturity. This procedure, in which the age at maturity is fixed, the pathogen's optimal response to an age at maturity is determined, and then the host's choice of optimal age at maturity is then determined is a special form of dynamic game theory called a leader–follower or Stackelberg game (Basar and Olsder 1982). The general way that these games are approached is to first find the optimal response of the follower (here the parasite), given the response of the leader (here the host), and then find the optimal response of the leader, given the optimal response of the follower. So, let's begin.

If hosts mature and reproduce at age t_m, then they may become infected at any time τ between 0 and t_m. Horizontal transmission of the disease will then be determined by transmission rate λ and the length of time that that individual is infected. To find the latter, we set

$$D(t_m, \tau) = E\{\text{length of time an individual is alive, given infection at } \tau\}$$
$$(5.23)$$

This interval is composed of two kinds of individuals: those who survive to reproduction (and thus whose remaining lifetime is $t_m - \tau$) and those who die before reproduction. We thus conclude

$$D(t_m, \tau) = (t_m - \tau)\Pr\{\text{survive to reproduction}\}$$
$$+ E\{\text{lifetime}|\text{death before } t_m, \text{infection at } \tau\}$$
$$(5.24)$$

Since the mortality rate of an infected individual is $\mu + \alpha$, the probability that an individual dies before age s is $1 - \exp(-(\mu + \alpha)s)$ and the probability density for the time of death is $(\mu + \alpha)\exp(-(\mu + \alpha)s)$. Consequently, the expected lifetime of individuals who die before t_m and who are infected at age τ is $(\mu + \alpha) \int_0^{(t_m - \tau)} t e^{-(\mu+\alpha)t} dt$. The integral in this expression can be evaluated using integration by parts (or the $1 - F(z)$ trick mentioned in Chapter 3).

Exercise 5.7 (M)

Evaluate the integral, and combine it with the term corresponding to individuals who survive to reproduction to show that

$$D(t_{\mathrm{m}}, \tau) = \frac{1}{\mu + \alpha}[1 - e^{-(\mu+\alpha)(t_{\mathrm{m}}-\tau)}] \tag{5.25}$$

Now this equation is conditioned on the time at which an individual becomes infected, so to find the average duration of the disease, we need to average over the distribution of the time of infection. Since the rate of horizontal transmission is λ, the probability that an individual is infected in the interval $(\tau, \tau + d\tau)$ is $\lambda e^{-\lambda\tau}d\tau$. Consequently, the average duration of infection, when individuals reproduce at age t_{m} is

$$D(t_{\mathrm{m}}) = \int_0^{t_{\mathrm{m}}} \lambda e^{-\lambda\tau}e^{-\mu\tau}D(t_{\mathrm{m}}, \tau)d\tau \tag{5.26}$$

To analyze the evolution of virulence, Koella and Restif separate transmission of disease propagules by contact between susceptible and infected hosts (with rate λ) and the efficiency of the transmission, which they denote by $\epsilon(\alpha)$ and which is assumed to have the same kind of form as $b(m)$ that we encountered previously: $\epsilon(\alpha) = \epsilon_{\max}\alpha/(\alpha + \alpha_0)$, where $\epsilon_{\max}$ is the maximum efficiency and α_0 is the level of virulence at which half of this efficiency is reached. We then combine Eq. (5.26) with the efficiency to obtain a measure of the success of horizontal transmission when the host matures at age t_{m} and the level of virulence is α:

$$H(t_{\mathrm{m}}, \alpha) = \varepsilon(\alpha)D(t_{\mathrm{m}}) \tag{5.27}$$

and we assume that natural selection has acted on virulence to maximize $H(t_{\mathrm{m}}, \alpha)$ with respect to the level of virulence α.

In Figure 5.9, I show the optimal level of virulence (i.e. that maximizes $H(t_{\mathrm{m}}, \alpha)$) as a function of the age at which the host reproduces. The results accord with the intuition that we have developed thus far: slowly developing hosts select for reduced virulence in parasites because there is more time for the transmission of the disease. Let us denote the curve in Figure 5.9 by $\alpha^*(t_{\mathrm{m}})$, to remind ourselves that it is the optimal level of parasite virulence when the hosts mature at age t_{m}.

We now turn to the computation of the optimal age of maturity for the hosts. Since we have assumed a semelparous host, the appropriate measure of fitness is expected lifetime reproductive success. Imagine a cohort of hosts, with initial population size N, and in which all individuals begin susceptible. At a later time, the population will consist of

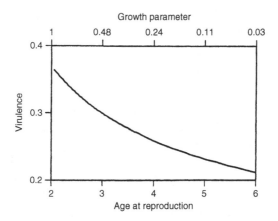

Figure 5.9. Optimal virulence of the parasites when hosts mature at age t (reprinted from Koella and Restif (2001) with permission). Parameters are $\mu = 0.15$, $\epsilon_{max} = 5$, $\alpha_0 = 0.1$, $\lambda = 0.05$.

$S(t)$ uninfected individuals and $I(t)$ infected individuals (with $S(0) = N$ and $I(0) = 0$). Recall that we assumed that hosts become infected at rate λ, independent of the density of infected individuals. Consequently, the dynamics for susceptible and infected individuals is slightly different than before:

$$\frac{dS}{dt} = - (\lambda + \mu)S$$

$$\frac{dI}{dt} = \lambda S - [\mu + \alpha^*(t_m)]I \tag{5.28}$$

We now solve these equations subject to the initial conditions. The first equation can be solved by inspection, so that $S(t) = Ne^{-(\lambda+\mu)t}$. The solution of the second equation is slightly more complicated. We separate the case in which $\lambda = \alpha^*(t_m)$ and the case in which they are not equal. In the latter case, we solve the equation for infected individuals by the use of an integrating factor and we obtain

$$I(t) = \left[\frac{\lambda}{\alpha^*(t_m) - \lambda}\right] N(0)[e^{-(\lambda+\mu)t} - e^{-(\alpha^*(t_m)-\mu)t}]$$

Exercise 5.8 (M)

For the case in which $\lambda = \alpha^*(t_m)$ show that $I(t) = \lambda t_m N e^{-(\lambda+\mu)t_m}$.

Given $S(t)$ and $I(t)$, we next compute the probability that an individual survives to age t as $p(t) = [S(t) + I(t)]/N$ and thus the expected lifetime reproductive success is $F(t_m) \propto p(t_m)(1 - e^{-kt_m})^\beta$. We may then assume that natural selection acts to maximize this expression through the choice of age at maturity, which you should now be able to find. This approach differs somewhat from that of Koella and Restif (2001) and I encourage you to read their paper, both for the approach and the discussion of the advantages and limitations of this model in the study of the evolution of virulence.

Vector-based diseases: malaria

Diseases that are transmitted from one host to another via vectors rather than direct contact are common and important. For example (Spielman and D'Antonio 2001), mosquitoes transmit malaria (*Anopheles* spp.), dengue and yellow fever (*Aedes* spp.), West Nile Virus and filariasis, the worm that causes elephantitis (*Culex* spp.) (Figure 5.10). In this section, we will focus on malaria, which continues to be a deadly disease, killing more than one million people per year and being widespread and endemic in the tropics. The history of the study of malaria is itself an interesting topic and the book by Spielman and D'Antonio (2001) is a good place to start reading the history; Bynum (2002) gives a two page summary, from the perspective of Ronald Ross. From our perspective, some of the highlights of that history include the following.

- 1600s: Quinine derived from tree bark in Peru is used to treat the malarial fever.
- 1875: Patrick Manson uses a compound microscope and discovers the organism responsible for elephantitis.
- 1880: Pasteur develops the germ theory of disease.
- 1880: Charles Levaran is the first to see the malarial parasite in the blood.
- 1893: Neocide (DDT) is invented by Paul Mueller as a moth killer.
- 1890s–1910s: A world-wide competition for understanding the malarial cycle involves Ronald Ross (UK), Amico Bignami (Italy), Giovanni Grassi (Italy), Theobald Smith (US), W. G. MacCallum (Canada). The win is usually attributed to Ross, who also develops a mathematical model for the malarial cycle. In 1911 Ross writes the second edition of *The Prevention of Malaria*.
- 1939–45: During World War II, atabrine, a synthetic quinine, is developed, as is chloroquinine; DDT is used as a delouser in prisoner of war camps.
- 1946–1960s: Attempts are made to eradicate malaria and they fail to do so; resistance to DDT develops.
- 1950s: G. MacDonald publishes his model of malaria and studies the implications of this model. In 1957 he writes *The Epidemiology and Control of Malaria*.
- 1960: The first evidence of resistance of the malaria parasite (*Plasmodium* spp.) to chloroquinine is discovered.
- 1962: Rachel Carson's *Silent Spring* is published. John McNeil (2000) has called *Silent Spring* "the most important book published by an American." If you have not read it, stop reading this book now, find a copy and read it.

Figure 5.10. (a) The malarial mosquito *Anopheles freeborni* (from the Public Health Image Library (PHIL) found at http://phil.cdc.gov/phil/default.asp thanks to Dr. James Gathany). (b) Egg rafts of the carrier of avian malaria *Culex laticinctus*, and (c) *Culex* attacking a host (both compliments of Dr. Leon Blaustein, Haifa University). (d) Ookinete of the human malaria parasite *Plasmodium falciparum* (top-right corner) within the basal region of the midgut wall of the mosquito vector *Anopheles stephensi*. The ookinete probably resides within the intercellular space between adjacent midgut cells, after having passed intracellularly through the midgut epithelial cell that exhibits abnormal dark staining (compliments of Dr. Luke Baton and Dr. Lisa Ranford-Cartwright, University of Glasgow).

- 2000–2010: The World Health Organization (WHO) embarks on a program called Roll Back Malaria, with the goal of reducing world-wide deaths by 50%.

Malaria is caused by amoeboid parasites *Plasmodium*; currently there are four main species that cause human malaria (*P. falciparum,*

P. malariae, P. ovale, P. vivax). The parasite itself has a complex life cycle and has been divided into more than ten separate steps (Oaks *et al*. 1991). For our purposes, the malarial cycle might be described as follows.

- An infected female mosquito seeks a blood meal so that she can make eggs. The sporozite form of the parasite migrates to the salivary glands of the mosquito.
- After entering a human host during a biting episode, the sporozites invade the liver cells and over the next 5–15 days, multiply into a new form (called merozites) which are released and invade red blood cells. The merozites reproduce within the red blood cell, ultimately rupturing it (with associated symptoms of fever and clinical indications of malaria).
- Some of the merozites differentiate into male and female sexual forms (gametocytes). These sexual forms are ingested by a different (potentially uninfected) mosquito during her blood meal. Once inside the mosquito, the gametes fuse to form a zygote, which migrates to the stomach of the mosquito and ultimately becomes an oocyst. Over the next week or so, the oocyst grows in the mosquito stomach, ultimately rupturing and releasing of the order of 10 000 sporozites which migrate to the salivary glands. And so the process goes.

There are more than 2500 species of mosquito in the world, but only the genus *Anopheles* transmits malaria; there are about 60 species in this genus. The mosquito life cycle consists of egg, larval, pupal and adult stages. Females require a blood meal for reproduction and deposit 200–1000 eggs in three or more batches, typically into relatively clean and still water. The development time from egg to adult is 7–20 days, depending upon species and environmental conditions. Adult survival is typically of the order of a month or so (especially under good conditions of high humidity and moderate temperature). The adults seek hosts via chemical cues that include plumes of carbon dioxide, body odors and warmth (Oaks *et al*. 1991).

There exists in the literature what one might call the "standard vector model" and we shall now derive it, using mosquitoes and humans as the motivation, but keeping in mind that these ideas are widely applicable. The key variables are the total population of humans and mosquitoes, H_T and M_T respectively, which are assumed to be approximately constant, and the population of infected humans and mosquitoes, H and M respectively. The malarial cycle is characterized by the following parameters.

$a =$ Biting rate of mosquitoes (bites/time).
$b =$ Fraction of bites by infectious mosquitoes on uninfected humans that lead to infections in humans.

$c =$ Fraction of bites by uninfected mosquitoes on infected humans that lead to parasites in the mosquito.

$r =$ Recovery rate of infected humans (rate at which the parasite is cleared).

$\mu =$ Mosquito death rate.

Examples of clearance rates of parasites are found in Anderson and May (1991; figures 14.2 and 14.3). To begin, we compute the basic reproductive rate of the disease. Imagine that one human becomes infected with the parasite. This individual is infectious for an interval that is roughly $1/r$. This infected human will thus be bitten a/r times and if we assume that the mosquitoes are uniformly distributed across hosts and that a mosquito only bites each human once, then the number of mosquitoes infected from biting this one infected human is $ac(M_T/H_T)(1/r)$. Each infected mosquito will make approximately $ab(1/\mu)$ infectious bites. Combining these, we conclude that the number of new cases is

$$R_0 = ac\frac{M_T}{H_T}\frac{1}{r}ab\frac{1}{\mu} = a^2bc\frac{M_T}{H_T}\left(\frac{1}{r}\right)\left(\frac{1}{\mu}\right) = \frac{a^2}{r\mu}bc\frac{M_T}{H_T} \qquad (5.29)$$

The last re-arrangement of terms in Eq. (5.29) makes the dimensionless combinations of parameters clear. In the mosquito literature, there is a tradition of using Z_0 for the basic reproductive rate. Perhaps the most important conclusion from this calculation is that the biting rate enters as a square, while all other parameters enter linearly. Thus, in general a given percentage reduction in the biting rate (e.g. by bed nets or by insect repellent) will have a much greater effect on the basic reproductive rate of the disease than a similar reduction in any of the other parameters. This was one of Ross's arguments for mosquito control as a means of malaria control.

We now construct the dynamics of infection. We begin with infected humans, $H(t)$, who come from interactions between infected mosquitoes, $M(t)$, and uninfected humans, $H_T - H(t)$. Assuming that transmission is characterized by mass action, thus depending upon the number of mosquitoes infected per human and the number of uninfected humans, and taking into account the clearance of parasites, we conclude that

$$\frac{dH}{dt} = ab\left(\frac{M}{H_T}\right)(H_T - H) - rH \qquad (5.30)$$

As in the computation of the basic reproductive rate, we have distributed infected mosquitoes across the human population. Mosquitoes become infected in a similar manner: transmission between infected humans and uninfected mosquitoes. The dynamics of infected mosquitoes become

$$\frac{dM}{dt} = ac\left(\frac{H}{H_T}\right)(M_T - M) - \mu M \qquad (5.31)$$

We will work with infected fractions of the human and mosquito populations. Dividing Eq. (5.30) by the total human population gives

$$\frac{d}{dt}\left(\frac{H}{H_T}\right) = ab\left(\frac{M_T}{H_T}\right)\left(\frac{M}{M_T}\right)\left(1 - \frac{H}{H_T}\right) - r\left(\frac{H}{H_T}\right) \qquad (5.32)$$

Note that in making this transition, I have rewritten M/H_T so that the fraction of infected mosquitoes appears on the right hand side of Eq. (5.32). If we divide Eq. (5.31) by the total mosquito population, we obtain

$$\frac{d}{dt}\left(\frac{M}{M_T}\right) = ac\left(\frac{H}{H_T}\right)\left(1 - \frac{M}{M_T}\right) - \mu\left(\frac{M}{M_T}\right) \qquad (5.33)$$

and we now work with variables $h(t)$ and $m(t)$ denoting the fraction of infected humans and fraction of infected mosquitoes respectively. From Eqs. (5.32) and (5.33), the dynamics of these infected fractions are

$$\begin{aligned} \frac{dh}{dt} &= ab\left(\frac{M_T}{H_T}\right)m(1 - h) - rh \\ \frac{dm}{dt} &= ach(1 - m) - \mu m \end{aligned} \qquad (5.34)$$

The steady state for the infected human population implies that $m = (r/ab)(H_T/M_T)(h/(1 - h))$ and this curve is shown in Figure 5.11a. Note that the slope of the tangent line to this curve at the origin (or, alternatively, the slope of the linear approximation to this curve) is $(r/ab)(H_T/M_T)$. The steady state for the infected mosquito population implies that $m = ach/(\mu + ach)$ and this curve is shown in Figure 5.11b. The slope of this curve at the origin is ac/μ. We understand the dynamics of the disease by putting the isoclines together, which I have done in three ways in Figures 5.11c–e. When the steady state determined by the intersection of the two isoclines is at a relatively high level of infection, MacDonald called the malaria "stable" (Anderson and May 1991, p. 397). When the steady state is at a lower level of infection, he called it "unstable" and it is possible for malaria to become extinct: if the mosquito isocline starts off below the human isocline, then the only steady state is the origin.

Malaria persists if the mosquito isocline rises faster than the human isocline at the origin. We can derive the condition for this to be true in terms of the slopes; in particular we must have $ac/\mu > (r/ab)(H_T/M_T)$ and combining these terms we conclude that malaria will persist if $(a^2bc/\mu r)(M_T/H_T) > 1$. Compare this with the computation that we did for the basic reproductive rate and you will see that they are the same: in

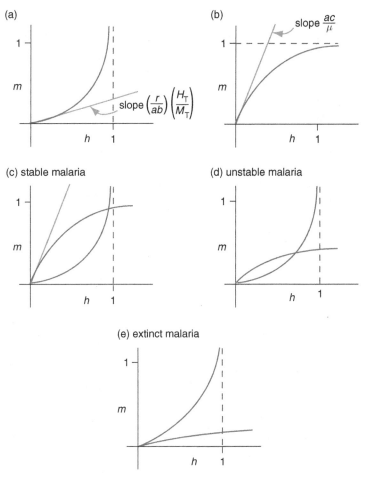

Figure 5.11. Analysis of the dynamics of malaria. (a) The isocline for infected humans, found by setting $dh/dt = 0$. (b) The isocline for infected mosquitoes, found by setting $dm/dt = 0$. Panels (c, d and e) show three ways that the isoclines can intersect.

this case the basic reproductive rate also gives us the condition for the persistence of the disease.

But the dynamics give us much more, particularly the ability to predict patterns of the transmission of disease. In Figure 5.12, I show a comparison of the predicted dynamics and those observed for three of the four *Plasmodium* parasites in infants following their birth. Given the simplicity of this model, the agreement is remarkably good. Anderson and May (1991) provide a variety of elaborations.

Helminth worms

We now turn to helminth worms (Figure 5.13), which have the following key properties. First, the parasites generally have complex life cycles, possibly using many hosts or having a free-living stage.

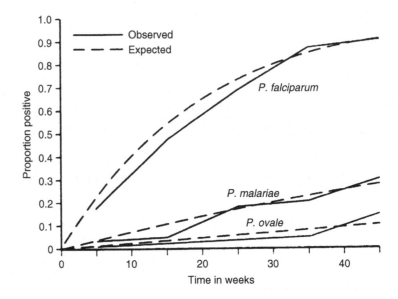

Figure 5.12. Comparison of the predicted and observed dynamics of malaria from Anderson and May (1991). Reprinted by permission.

Second, the actual number of parasites that a host harbors, rather than just the state of the host (infected or not), is important. Third, the number of parasites is most commonly overdispersed. That is, let $\tilde{W}$ denote the worm burden carried by a host and think of it as a random variable. We know from Chapter 3 that if worms were randomly distributed across hosts $\tilde{W}$ would have a Poisson distribution in which the mean and variance were equal. However, empirical data (an excellent review is given by Shaw and Dobson (1995)) show quite the contrary: that typically the variance of the worm burden exceeds the mean. So, the negative binomial distribution – which helped stabilize host–parasitoid interactions – will have an important role to play here.

In this section, I discuss three models of increasing complexity, each of which illustrates a different point and gives a sense of how we use the tools that have been developed. As before, I refer you to the original literature for more of the details.

The underlying host–worm model

We begin with a classic model due to Roy Anderson and Robert May (Anderson and May 1978, May and Anderson 1978). These papers contain eight different models (a basic model and then seven elaborations of it) and total nearly 50 pages, so in this section I can introduce only the basic model and one of its elaborations. But this should be sufficient to send you on your way.

Figure 5.13. (a) A female red grouse brooding chicks. Detailed field studies and experiments have shown that the parasites make their prey more vulnerable to predation (Hudson *et al.* 1992a, Packer *et al.* 2003). The parasites appear to interfere with the ability of females to control scent. (b) A young grouse chick infected with ticks *Ixodes ricinus*. These ticks can transmit the Louping ill virus that causes 80% mortality and reduces the growth rate of the grouse population (Hudson *et al.* 2002, Laurenson *et al.* 2003). (c) Experimental reductions in the intensity of nematode infection have shown that parasite removal results in an increase in clutch size, hatching success and survival of grouse chicks (Hudson 1986, Hudson *et al.* 1998, 2002). (d) Nematode worms (*Trichostrongylus tenuis*) burrowing into the caecal wall of red grouse. These nematodes have a major impact on the fitness of individual hosts to the extent that they cause morbidity, reduced fecundity and generate population cycles (Hudson *et al.* 1998, Dobson and Hudson 1992, Hudson *et al.* 1992b, 1998). Photo compliments of Pete Hudson.

Here, we consider parasites that do not reproduce within the host but which produce a transmission stage (eggs, spores, cysts) that passes out of the host before acquisition by the next host. To begin, we assume that the amount of time that the transmission stage is outside of hosts is very short (this assumption is relaxed in the next section). In that case, the dynamic variables are the population of hosts, $H(t)$, and the population of parasites, $P(t)$. These are treated as deterministic variables, so that the mean number of parasites per host is $P(t)/H(t)$, even though the underlying distribution of worms per host is stochastic. Watch how this is done.

We begin with the dynamics for the host, which we write as follows:

$$\frac{dH}{dt} = (a - b)H - \text{parasite induced mortality} \qquad (5.35a)$$

Here I keep the notation used by Anderson and May, so that a and b are the parasite-independent per capita rates of host reproduction and mortality. These are clearly not the same a and b that we used in the previous section, and as I wrestled with presentation of these different models, I decided that keeping original notation of primary sources was better, and would also keep you on your intellectual toes. You will see momentarily why they are separated.

We assume that the host per capita mortality rate due to parasites is proportional to the parasite burden of the host. Anderson and May (1978, figure 5.1) show data that are convincing enough for one to consider this a good starting assumption. We let $p(w)$ denote the probability that the worm burden at time t is w. If the per capita host mortality rate when the worm burden is w is αw, then the expected mortality rate for the host population is $\alpha H(t)\Sigma_{w=0}wp(w)$, where we understand that the upper limit of the sum is, in principle (but not in practice), infinite. We recognize the summation as the average number of worms per host, which is $P(t)/H(t)$, so that the mortality rate for the host population is thus $\alpha P(t)$ and the host dynamics are

$$\frac{dH}{dt} = (a-b)H - \alpha P \qquad (5.35b)$$

Parasite population growth involves production and transmission. At very high host densities, we assume that parasite production depends only upon the parasite per capita per host fecundity λ, so that parasite production at high host density is $\lambda H(t)\Sigma_{w=0}wp(w) = \lambda P(t)$. We correct this at lower host densities by assuming a type II functional response and thus assume that the net rate at which parasites are acquired by hosts is $\lambda P(t)[H(t)/(H_0 + H(t))]$.

There are three sources of mortality of the parasites. First, there is an intrinsic per capita mortality rate for the parasites, denoted by μ. Second, since hosts have an intrinsic per capita mortality rate b, the loss of parasites due to this intrinsic host mortality is $bH(t)\Sigma_{w=0}wp(w) = bP(t)$. Third, we have the parasite induced host mortality. Recall our assumption that the mortality rate for a host with w parasites is αw. When a host with worm burden w dies, the number of parasites dying is w, so that the rate of loss of parasites due to parasite induced mortality is $\alpha H(t)\Sigma_{w=0}w^2 p(w) = \alpha H(t)E(\tilde{W}^2)$. Combining parasite production and mortality we obtain the parasite dynamics

$$\frac{dP}{dt} = \lambda P\left(\frac{H}{H_0 + H}\right) - (b+\mu)P - \alpha HE(\tilde{W}^2) \qquad (5.36)$$

In order to make further progress, we have to specify the probability distribution for parasites per host. The simplest case would be one in which each of the parasites were randomly assigned to a host. Although I have already told you that overdispersion of parasites is common, let us work with this assumption for now. In that case $p(w)$ is a Poisson distribution, for which $E(\tilde{W}) = \text{Var}(\tilde{W}) = P(t)/H(t)$ and for which we then conclude that $E\left(\tilde{W}^2\right) = (P(t)/H(t)) + (P(t)/H(t))^2$. Substituting this expression into Eq. (5.36) we obtain

$$\frac{dP}{dt} = \lambda P\left(\frac{H}{H_0 + H}\right) - (b + \mu)P - \alpha H\left(\frac{P}{H} + \frac{P^2}{H^2}\right)$$

$$= P\left[\frac{\lambda H}{H_0 + H} - (b + \mu + \alpha) - \frac{\alpha P}{H}\right] \tag{5.37}$$

Equations (5.35b) and (5.37) need to be analyzed. By now, we know the drill: determine the steady states, examine the isoclines, and compute the eigenvalues of the system linearized around the steady states.

Exercise 5.9 (E/M)

Show that the steady states are

$$\bar{P} = \left(\frac{a - b}{\alpha}\right)\bar{H} \qquad \bar{H} = \frac{H_0(\mu + \alpha + a)}{\lambda - (\mu + \alpha + a)} \tag{5.38}$$

subject to the constraint that $\lambda > (\mu + \alpha + a)$.

We will not do the linearization around the steady state here, nor will I assign it as an exercise (although you might want to do it). Suffice to say: the eigenvalues are pure complex numbers, so that the steady state is neutrally stable and the dynamics are purely oscillatory. This point is illustrated in Figure 5.14a and should you remind you of our experience with the Nicholson–Bailey model when the fraction of hosts escaping parasitism was given by the zero term of the Poisson distribution.

We now turn to the case in which worms have a negative binomial distribution across hosts. Based on our experience with parasitoids, we expect that aggregation will stabilize the dynamics. We need to compute the second moment in Eq. (5.36), remembering that for the negative binomial distribution, the variance is $m + (m^2/k)$.

Exercise 5.10 (M)

Show that when worms have a negative binomial distribution across hosts, Eq. (5.37) is replaced by

$$\frac{dP}{dt} = P\left[\frac{\lambda H}{H_0 + H} - (b + \mu + \alpha) - \frac{\alpha(k + 1)P}{kH}\right] \tag{5.39}$$

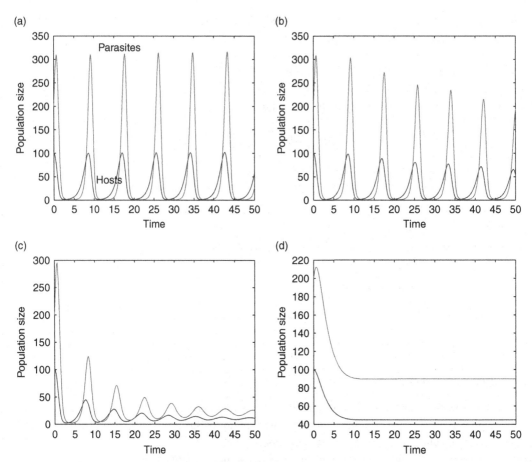

Figure 5.14. The basic worm model with a Poisson distribution (panel a) or negative binomial distribution of worms across hosts (panels b, c, and d). Common parameters are $a=2$, $b=1$, $\mu=1$, $H_0=5$, $\alpha=0.5$, and $\lambda=4$ and $k=100$ (panel b), 10 (panel c) or 1 (panel d). In each case the parasite population is the one with the larger maximum size.

In Figures 5.14b–d, I show the resulting dynamics for three values of k ($=100$, 10, and 1). Note that in each case the dynamics are stabilized. When one conducts an analysis of the linearized system (see Anderson and May 1978, Appendix A), a remarkable result emerges: the eigenvalues have negative real part as long as $\alpha/k > 0$; but both of these parameters are positive so we reach the conclusion that any level of aggregation – meaning any deviation from a completely random distribution – is sufficient to stabilize the dynamics.

The papers of Anderson and May (1978) and May and Anderson (1978) are well worth examining, to study the other various cases, for which you are now set. We, however, move on.

Accounting for the free-living stage

We will now develop, but not analyze, a model that accounts for the free-living stage of the parasite in more detail. It is based on, but not identical to, the work of Hudson *et al.* (1992a, b) and Hudson and Dobson (1997) which has all of the details, and additional models.

Until now, we have assumed that the transmission of adult parasites from hosts to hosts occurred with a very short intermediate stage, but now allow for a longer intermediate stage, understood to be the free-living eggs or larvae. The dynamical system that we consider is thus expanded to include hosts, $H(t)$, adult parasites, $P(t)$, and the free-living stage, which we denote by $W(t)$ to follow the notation in Dobson and Hudson (1992).

The dynamics of the hosts do not change, but the dynamics of the parasites do. We assume that the free-living stage, denoted by $W(t)$, is produced by adult parasites at constant per capita rate λ, have intrinsic natural mortality rate γ and are captured by hosts according to a type II functional response. In such a case, the dynamics of the free-living stage are

$$\frac{dW}{dt} = \lambda P - \gamma W - cW \frac{H}{H_0 + H} \tag{5.40}$$

The dynamics of the adult parasites still involve production and mortality. Mortality is unchanged from the previous case, but production is now determined by the uptake of the free-living stage and so the adult dynamics become

$$\frac{dP}{dt} = \frac{cWH}{H_0 + H} - (b + \mu + \alpha)P - \frac{\alpha(k+1)P^2}{kH} \tag{5.41}$$

Dobson and Hudson (1992) develop models that include larvae with arrested development (hypobiosis), show how to find the basic reproductive rate of the disease and analyze the dynamic properties of the model; I refer you to them for the details.

The broader ecological setting

Like other ecological interactions, host–parasite dynamics take place in a broader ecological setting. Grenfell (1988, 1992) constructed a scenario that connects parasite dynamics and grazing systems, with some interesting and remarkable conclusions. Here I describe the key features of a model that is similar, but not identical, to those used by Grenfell and, as before, you are encouraged to consult the original papers after this development.

We imbed the host–parasite dynamics in an herbivore–plant interaction. We allow $V(t)$ to denote the biomass of vegetation at time t
and assume that, in the absence of the herbivore, the vegetation grows
logistically. We assume that the herbivores consume vegetation
according to a type II functional response (Grenfell (1988, 1992)
assumed a type III functional response, which makes the analysis
much more complicated – and interesting – so I encourage you to
read his papers) and that herbivore per capita reproduction is proportional to this consumption. We assume that the free living stage of the
parasite can be ignored (Grenfell includes it), so that the simpler host–
parasite dynamics apply. When the plant–herbivore dynamics are
coupled to the host–parasite dynamics, we end up with the system
of equations

$$\frac{dV}{dt} = rV\left(1 - \frac{V}{K}\right) - \frac{eHV}{V_0 + V}$$

$$\frac{dH}{dt} = \frac{aHV}{V_0 + V} - bH - \alpha P \tag{5.42}$$

$$\frac{dP}{dt} = P\left[\frac{\lambda H}{H_0 + H} - (b + \mu + \alpha) - \frac{\alpha(k+1)P}{kH}\right]$$

In these equations, the new terms should be easily interpretable by you.
The question, of course, is what does one do with them now that one has
them?

Let us start by thinking about the steady states; we do so by holding the
herbivore population size fixed. Then the equation for vegetative biomass
is uncoupled from all of the others and the steady state of the vegetation is
determined by the solution of $rV(1 - (V/K)) = e\bar{H}V/(V_0 + V)$. The
graphical solution of this equation is shown in Figure 5.15. If we think
that e (the maximum per capita consumption rate of the herbivores) and
V_0 (the level of vegetative biomass at which 50% of this maximum is
reached) are fixed by the biology of the system, then steady states of the
vegetation are set by the steady state level of the herbivores, $\bar{H}$. The
origin, $V = 0$, is always a steady state but depending upon the values of
herbivores, there may be another steady state.

Figure 5.15. The graphical
solution of $rV(1 - (V/K)) = e\bar{H}V/(V_0 + V)$ for three values
of H.

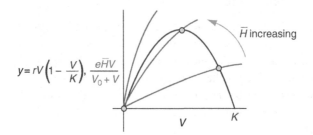

$$y = rV\left(1 - \frac{V}{K}\right), \frac{e\bar{H}V}{V_0 + V}$$

Exercise 5.11 (M/H)

Determine the stability of the steady states of V at the origin and at its positive value, when appropriate.

However, the herbivore population level cannot be manipulated at will, as I have just done. Rather it comes out of the simultaneous solution of the three nonlinear equations obtained by setting the left hand side of Eq. (5.42) equal to 0. We are not going to do that (once again, look at Grenfell's papers), but make the following observation: host populations will decline as parasite populations increase. Thus, the stability of the plant–herbivore system depends upon the parasites: the parasites function as top predator in this system! This means, for example, that one wants to think carefully about actions that will reduce parasite populations. For example, the indiscriminate application of antihelminth agents might have exactly the undesired effect of causing the entire system to crash.

Optimal immune responses

The immune system is a remarkable achievement of evolution and warrants its own careful analysis. Some starting points are Frank (2002), Schmid-Hempel and Ebert (2003) and Zuk and Stoehr (2002); also see Connections. In this section, we will explore two ways in which optimality theory may be applied to immunological response.

T-cell phenotypes in multiple infections

In response to an antigen, the body proliferates T-cells (Graham 2002). In a very broad brush, the proliferated cells are called T-helper cells and come in two forms (Mosmann and Sad 1996, Fishman and Perelson 1999): T-helper type 1 (Th_1) cells that generally work best against intracellular parasites or pathogens and T-helper type 2 (Th_2) cells that generally work best against extracelluar parasites or pathogens (Graham 2001). However, this simple description is not uniformly accepted (for a discussion of whether this is a reliable paradigm or dangerous dogma, see Romagnani (1996) and Allen and Maizels (1997)). It is agreed, however, that the two kinds of helper cells have some unique biochemical signals: interferon gamma (IFN-γ) is uniquely associated with Th_1 cells and interleukin4 (IL-4) is uniquely associated with Th_2 cells. Consequently, the ratio, with $[x]$ denoting the concentration of x,

$$t = \frac{[\text{IFN-}\gamma]}{[\text{IFN-}\gamma] + [\text{IL-4}]} \tag{5.43}$$

is a measure of the fraction of T-helper type 1 cells in response to an antigen.

Graham (2001) asks the following question: suppose that there are a wide variety of pathogens or parasites that might attack an organism, and that the ith disease evokes an optimal T-helper response t_i, understood to be the fraction of Th_1 cells. What will happen when the organism is simultaneously attacked by n different pathogens or parasites? This question is particularly interesting because of the generally held view of the emergent properties of the immune system. To begin, we will follow Graham's model, and then extend it. Suppose that the probability of surviving attack by the ith parasite or pathogen when the Th_1 response is t is

$$S_i(t) = \exp[-\alpha_i(t - t_i)^2] \qquad (5.44)$$

In this equation, α_i is a measure of the cost of deviations from the optimal response. The smaller this value, the less important are deviations from the optimal mix of T-helper cells. If we assume that the n different pathogens or parasites affect survival independently, then the probability of surviving all of them is

$$S(t) = \prod_{i=1}^{n} \exp[-\alpha_i(t - t_i)^2] \qquad (5.45)$$

and our objective is to maximize $S(t)$ by choosing t.

Exercise 5.12 (E)

Show that the optimal mixture of Th_1 cells is

$$t^* = \frac{\displaystyle\sum_{i=1}^{n} t_i\alpha_i}{\displaystyle\sum_{i=1}^{n} \alpha_i} \qquad (5.46)$$

This equation tells us two important things about the immune response. First, when a number of pathogens or parasites attack a host, the optimal response will be a mixture of the individual optimal responses. Second, this mixture will be weighted by the consequences of non-optimal response to each disease. Note that if one of the α_i is very large, then the optimal value of t will be very close to the response of Th_1 cells for that parasite or pathogen.

Here is an interesting extension of Graham's work, due to Steve Munch. There are some diseases that will kill a host, even if the host mounts the appropriately optimal response. Thus, we could generalize Eq. (5.44) to $S_i(t) = \beta_i\exp[-\alpha_i(t - t_i)^2]$ where the parameter β_i, with $0 < \beta_i \leq 1$, measures the probability of surviving the ith pathogen or parasite when the optimal response is applied.

Exercise 5.13 (E/M)

Show that the optimal response when confronted by n different pathogens or parasites is still given by Eq. (5.46). How is this result to be interpreted? Does this suggest that we should change the model in some way?

A trade-off between immune response and reproduction

All organisms are constrained by limited resources and this means that resources used to mount an immune response against a parasite or pathogen cannot be used for other things. In this section, we will explore a conceptual model of this trade-off using stochastic dynamic programming. The basic idea is due to Ruth Hamilton, who explored it in her honors thesis at the University of Edinburgh in the early 1990s. Some experimental evidence for this trade-off exists (Mosmann and Sad 1996, Sheldon and Verhulst 1996, Brunet et al. 1998, Lochmiller and Deerenberg 2000, Moret and Schmid-Hempel 2000, Rolff and Siva-Jothy 2002).

We characterize the organism under consideration by two variables: the resources $R(t)$ available for living life and reproduction, and the parasite burden $W(t)$ at the start of period t. In the absence of parasites, we assume that resources increase by an amount Y in each time period. Parasites can have at least three major effects on the organism. First, parasites may involve a metabolic cost since they use host resources, which we will assume to be at a fixed rate, a, per parasite. Second, parasites may increase the probability of mortality. Third, parasites may decrease reproductive output.

At any time period, the organism may use resources to kill parasites, at a cost of c per parasite, and at the same time may acquire new parasites while foraging; the net gain in resources will be denoted by Y. For simplicity (this is a conceptual model) we will assume that only one parasite can be acquired in each period. With these assumptions, the dynamics of host resources $R(t)$ when the host kills j parasites in period t are

$$R(t+1) = R(t) + Y - aW(t) - cj \qquad (5.47)$$

and the dynamics of the parasite burden $W(t)$ are

$$W(t+1) = \begin{array}{l} W(t) - j \text{ with probability } 1-p \\ W(t) - j + 1 \text{ with probability } p \end{array} \qquad (5.48)$$

We will assume that if resources drop below a critical value (r_c) or parasites exceed a maximum value (w_{max}), the organism dies. Resources are also constrained by a maximum value r_{max}.

To account for the effect of parasite burden on survival we will assume that the probability of surviving a single period when the parasite burden is $W(t)$ is $\exp(-m_0 - m_1 W(t))$. Again for simplicity of the model, let us assume that the organism is semelparous and reproduces at fixed time T. Reproduction at this time will be determined by both resources and parasite burden.

The approach we take is based on stochastic dynamic programming, in a somewhat more complicated version than the one of Chapter 4. We let $F(r, w, t)$ denote the maximum expected value of reproduction at time T, given that $R(t) = r$ and $W(t) = w$. At the final time, we assume that $F(r, w, T) = \Phi_1(w)\Phi_2(r)$ where $\Phi_1(w)$ is a decreasing function of w and $\Phi_2(r)$ is an increasing function of r.

For times previous to T, the appropriate equation of stochastic dynamic programming is

$$F(r, w, t) = \max_j[\exp(-m_0 - m_0 w)\{pF(r + Y - aw - cj, w - j + 1, t + 1) \\ + (1 - p)F(r + Y - aw - cj, w - j, t + 1)\}] \qquad (5.49)$$

which we now solve backwards in time, in the manner described in Chapter 4. The solution of this equation generates the expected lifetime reproductive success for an individual, given its resource level and parasite burden at any time. More importantly, the process of solving the equation generates the optimal immune response $j^*(r, w, t)$ for resource level $R(t) = r$ and parasite burden $W(t) = w$.

As before, the solution must be obtained numerically, so that one has to choose functional forms and parameter values. For the results reported here, I used $\Phi_1(w) = 1 - (w/w_{max})^{1.5}$, which is shown in Figure 5.16a, and $\Phi_2(r) = r^4/(r_0^4 + r^4)$, which is shown in Figure 5.16b and the following parameters: $r_c = 2$, $r_{max} = 25$, $r_0 = 0.4 r_{max}$, $w_{max} = 20$, $Y = 3$, $p = 0.1$, $m_0 = 0.05$, $m_1 = 3 m_0/w_{max}$, $T = 20$, $a = 0.2$ and $c = 0.8$. Perhaps the most important parameters for this discussion are the last two, which show that parasites have a moderate metabolic cost but are expensive to get rid of.

In Figures 5.16c and 5.16d, I show the optimal immune response for a parasite burden of 4, 8 or 12 at $t = 1$ (the beginning of the interval of interest) or $t = 18$ (close to reproduction). We draw a number of conclusions from these figures. First, the immune response is age and state dependent. Second, even though the organism can afford to eliminate some parasites, the predicted optimal response may be not to do so. Third, as age at reproduction is approached, we predict that the optimal immune response will weaken (as the organism conserves resources for reproduction). It is not clear that we could have made these predictions without the model. It is also not clear how robust these results are without further numerical exploration; hence the next exercise.

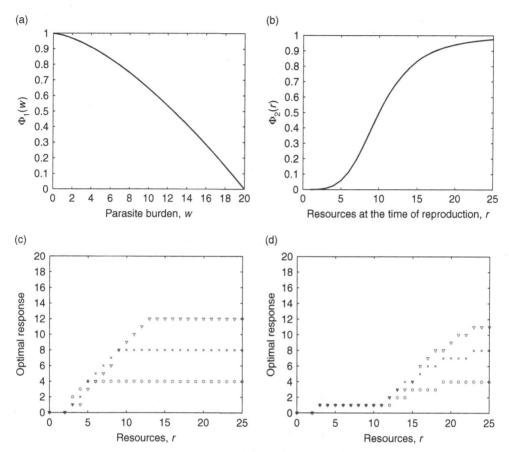

Figure 5.16. (a) The function $\Phi_1(w)$ used in the solution of the dynamic optimization model for immune response. (b) The function $\Phi_2(r)$ used in computation. (c) Optimal immune response as a function of parasite burden at $t=1$ for $W=12$ (triangles), 8 (crosses) or 4 (circles). (d) Optimal immune response at $t=18$. Other parameters are given in the text.

Exercise 5.14 (M/H)

Write a computer program to solve the stochastic dynamic programming equation. Then develop a forward Monte Carlo simulation for a population of individuals in which initial parasite burden is distributed in a negative binomial fashion with parameters $m=4$ and $k=1$. Use the simulation to predict the time dependence of the mean and variance of the predicted immune response.

Connections

General literature

The study of disease has a long and rich history; it is good to know some of that. Nuland (1993) is a fine discussion of disease and dying in general. On a more technical level, Anderson (1982) contains a good survey of material through to about 1980. In recent years, the evolutionary perspective to health and medicine has been framed as Darwinian medicine. Ewald (1994), Nesse and Williams (1994), Williams and Nesse (1991), and Stearns (1999) provide good introductions from different perspectives. We have focused on deterministic models, but there is a rich literature on stochastic epidemics, and there are differences in outcomes. For example, Rosà et al. (2003) show that while a deterministic model might predict oscillatory decay to a stable focus, a stochastic model could show sustained oscillations.

SIR, SIRS models

These models can become quite complex, for example when one takes age or stage structure into account (Keeling and Grenfell 1997, Hethcote 2000). But the basic kind of analyses that we conducted underlie more complicated situations. When we add demography, the SIR model also allows for vertical transmission of infection (that is, offspring may be born already infected). In such a case, we would modify the dynamics of infection in Eq. (5.18) to $dI/dt = rqI + bIS - vI - mI$ where q is the probability that the offspring of an infected parent is itself infected. The possibility of bioterrorism has made understanding these models and making them applicable even more important (Henderson 1999, Gani and Leach 2001, Meltzer et al. 2001, Enserink 2002, Meltzer 2003, Wynia and Gostin 2002). Some very interesting, and controversial (Enserink 2003), work is that of Edward Kaplan and colleagues (Kaplan et al. 2002, Kaplan and Wein 2003).

Evolution of virulence/evolution of resistance

This is again a topic with an enormous literature. Ewald (1994) and Frank (1996) provide great introductions to the subject. Other papers treating different aspects of the evolution of virulence include Antia et al. (1994), Lenski and May (1994), Antia et al. (1996), Antia and Lipschitz (1997), Mosquera and Adler (1998), Keeling and Grenfell (1997), Day (2001), Gandon et al. (2001), Day (2002a, b, c), Ferguson and Read (2002), Ganusov et al. (2002), Sylvain et al. (2002), Day (2003), Koella and Boëte (2003) and Day and Proulx (2004). Esteva and Vargas (2003) study the coexistence of different serotypes of dengue virus. A recent volume on the management of virulence is Dieckmann et al. (2002). On the other hand, the whole notion of the management of virulence is now being questioned (Ebert and Bull 2003).

Ecological applications of disease models

The recognition of the role of disease (either endemic or epidemic and thus catastrophic) as a regulator of natural populations has been slow in coming, perhaps because when disease regulates populations the dynamics are more complicated than simpler cases of autotrophic density dependence. However, good entries into the literature now exist through the papers of Grenfell and Hudson cited above, de Koeijer *et al.* (1998, 2004), and Dobson and Foutopoulos (2001). Species specific examples include papers by Barlow (1993, 2000), Dwyer and Elkinton (1993), White and Harris (1995), Smith *et al.* (1997), Bouloux *et al.* (1998), Begon *et al.* (1992, 1998, 1999), LoGiudice *et al.* (2003), Sauvage *et al.* (2003), Smith and Wilkinson (2002), Morgan *et al.* (2004) and Sumpter and Martin (2004).

Cultural and behavioral effects

Clearly social behavior and culture play a role in disease transmission, and the models that we have developed here can be extended to address them; two examples are the papers of Charles *et al.* (2002) and Tanaka *et al.* (2002).

Vector-based models

Anderson and May (1991) describe a number of elaborations of the basic model for malaria. Two interesting additional directions are these. First, recall that within the human host, the *Plasmodium* parasite has separate sexes. This suggests the possibility of local mate competition and sex ratios that deviate from 50:50. Read *et al.* (1992, 1995) recognized this possibility and applied an extension of Hamilton's sex ratio theory to *Plasmodium*. The development of resistance in *Plasmodium* to antimalarial drugs is another topic of considerable interest and opportunity for modeling (Lipp *et al.* 2002, Hastings 1997, 2001, Hastings and McKinnon 1998, Hastings and D'Allesandro 2000, Hastings *et al.* 2002) and opportunities exist for new kinds of approaches to drug therapy (Austin *et al.* 1998, Gardner 2000, 2001). Roitberg and Friend (1992) initiated a study of the behavioral ecology of the mosquito vector, using a combination of experiments and theory (mainly based on stochastic dynamic programming) to examine host seeking behavior in the vector. Ross's original papers (Ross 1916, Ross and Hudson 1917a, b) can be obtained at *JSTOR* and are well worth looking at. Catteruccia *et al.* (2003) model the effects of genetically modified mosquitoes.

Helminth parasites

More details about parasites with complex life cycles can be found in Esch *et al.* (1990).

Cholera

Cholera (and other water borne disease; see Brookhart *et al.* (2002)) is an extremely interesting disease for a number of reasons. The pathogen itself has a remarkable life history (Faruque *et al.* 2003, Reidl and Klose 2002, Yildiz and Schoolnik 1999), switching between two morphs (smooth and rugose) for reasons that we still do not understand. Second, the disease has a free-living stage, in stagnant water, that is key to transmission patterns (Codeço 2000). Third, the potential effects of climate change on cholera dynamics are profound (Lipp *et al.* 2002, Cottingham *et al.* 2003). Brookhart *et al.* (2002) also couple evolution of virulence and host demography in a different way than we did in the text.

Viral dynamics and AIDS

The same kinds of models that we used in this chapter apply to the study of viral dynamics, such as HIV. An excellent starting point are the articles by Perelson and Nelson (1999) and Callaway and Perelson (2002); the book by Nowak and May (2000) is also a good place to begin.

Optimal immune response

Understanding immune responses from an optimality perspective is going to be a growth area, since it bridges ecology, immunology, and epidemiology (Antia and Lipschitz 1997, Hellriegel 2001, Ahmed *et al.* 2002, Gardner and Thomas 2002, Rolff and Siva-Jothy 2002, 2003, Wegner *et al.* 2003). Shudo and Iwasa (2001) treat this question as a form of inducible defense (also see Gardner and Agrawal (2002)).

Prion disease kinetics

Prion diseases such as mad-cow disease (BSE) or Creutzfeldt–Jakob disease (CJD) are simultaneously infectious and heritable (Schwartz 2003). The agent in this case appears to be an improperly folded protein (Hur *et al.* 2002). The methods developed in this chapter apply to them too. Entries into the literature are the papers by Eigen (1996), Slepoy *et al.* (2001), Valleron *et al.* (2001), Ferguson *et al.* (2002), and Masel and Bergman (2003).

Space

We have ignored spatial aspects of disease transmission, or treated it very obliquely. Spatial considerations can be important, and sometimes paramount. Papers that show how spatial aspects of disease can be treated include those by White and Harris (1995), Hess (1996), Kirchner and Roy (1999), Caraco *et al.* (2001), and Fulford *et al.* (2002).

Stochastic epidemics

We have ignored stochastic effects in our study of the population biology of disease, but they can be important. Indeed, some of the classic papers in stochastic processes (Bailey 1953, Whittle 1955, Bartlett 1957) deal with stochastic epidemics. The book by Bharucha-Reid (1997 (1960)) and papers by Nasell (2002) and Rohani *et al.* (2002) are good starting points. O'Neill (2002) connects stochastic epidemics to modern Bayesian and Markov Chain Monte Carlo methods. These will be especially important when we must deal with new or re-emerging diseases for which the population parameters are not known and we simultaneously learn about the disease as we are dealing with it (Brookhart *et al.* 2002).

Chapter 6

An introduction to some of the problems of sustainable fisheries

There is general recognition that many of the world's marine and freshwater fisheries are overexploited, that the ecosystems containing them are degraded, and that many fish stocks are depleted and in need of rebuilding (for a review see the FAO report (Anonymous 2002)). There is also general agreement among scientists, the industry, the public and politicians that the search for sustainable fishing should receive high priority. To keep matters brief, and to avoid crossing the line between environmental science and environmentalism (Mangel 2001b), I do not go into the justification for studying fisheries here (but do provide some in Connections). In this chapter, we will investigate various single species models that provide intuition about the issues of sustainable fisheries. I believe that fishery management is on the verge of multispecies and ecosystem-based approaches (see Connections), but unless one really understands the single species approaches, these will be mysteries (or worse – one will do silly things).

The fishery system

Fisheries are systems that involve biological, economic and social/behavioral components (Figure 6.1). Each of these provides a distinctive perspective on the fishery, its goals, purpose and outputs. Biology and economics combine to produce outputs of the fishery, which are then compared with our expectations of the outputs. When the expectations and output do not match, we use the process of regulation, which may act on any of the biology, economics or sociology. Regulatory decisions constitute policy. Tony Charles (Charles 1992) answers the question "what is the fishery about?" with framework of three paradigms (Figure 6.2). Each of the paradigms shown in Figure 6.2 is a view of the fishery system, but according to different stakeholder groups.

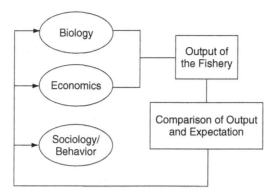

Figure 6.1. The fishery system consists of biological, economic and social/behavioral components; this description is due to my colleague Mike Healey (University of British Columbia). Biology and economics interact to produce outputs of the system, which can then be modified by regulation acting on any of the components. Quantitative methods can help us predict the response of the components to regulation.

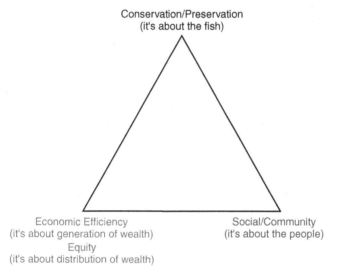

Figure 6.2. Tony Charles's view of "what the fishery is about" encompasses paradigms of conservation, economics and social/community. In the conservation perspective, the fishery is about preserving fish in the ocean and regulation should act to protect those fish. In the economic perspective, the fishery is about the generation of wealth (economic efficiency) and the distribution of that wealth (economic equity). In the social perspective, the fishery is about the people who fish and the community in which they live.

Indeed, a large part of the problem of fishery management is that these views often conflict.

It should be clear from these figures that the study of fisheries is inherently interdisciplinary, a word which regrettably suffers from terminological inexactitude (Jenkins 2001). My definition of interdisciplinary is this: one masters the core skills in all of the relevant disciplines (here, biology, economics, behavior, and quantitative methods). In this chapter, we will focus on biology and economics (and quantitative methods, of course) in large part because I said most of what I want to say about behavior in the chapter on human behavioral ecology in Clark and Mangel (2000); also see Connections.

The outputs of the fishery are affected by environmental uncertainty in the biological and operational processes (process uncertainty) and observational uncertainty since we never perfectly observe the system. In such a case, a natural approach is that of risk assessment (Anand 2002) in which we combine a probabilistic description of the states of nature with that of the consequences of possible actions and figure out a way to manage the appropriate risks. We will close this chapter with a discussion of risk assessment.

Stock and recruitment

Fish are a renewable resource, and underlying the system is the relationship between abundance of the spawning stock (reproductively active adults) and the number or biomass of new fish (recruits) produced. This is generally called the stock–recruitment relationship, and we encountered one version (the Ricker equation) of it in Chapter 2, in the discussion of discrete dynamical systems. Using S size of the spawning stock and R for the size of the recruited population, we have

$$R = aSe^{-bS} \tag{6.1}$$

where the parameters a and b respectively measure the maximum per capita recruitment and the strength of density dependence. Another commonly used stock–recruitment relationship is due to Beverton and Holt (1957)

$$R = \frac{aS}{b + S} \tag{6.2}$$

where the parameters a and b have the same general interpretations as before (but note that the units of b in Eq. (6.1) and in Eq. (6.2) are different) as maximum per capita reproduction and a measure of the strength of density dependence. When S is small, both Eqs. (6.1) and (6.2) behave according to $R \sim aS$, but when S is large, they behave very differently (Figure 6.3).

The Ricker and Beverton–Holt stock–recruitment relationships each have a mechanistic derivation. The Ricker is somewhat easier, so we start there. Each spawning adult makes a potential number of offspring, a, so that aS offspring are potentially produced by S spawning adults. Suppose that each offspring has probability per spawner p of surviving to spawning status itself. Then assuming independence, when there are S spawners the probability that a single offspring survives to spawning status is p^S. The number of recruits will thus be $R = aSp^S$. If we define $b = |\log(p)|$, then $p^S = \exp(-bS)$ and Eq. (6.1) follows directly, this is the traditional way of representing the Ricker stock–recruitment relationship (we could have left it as $R = aSp^S$).

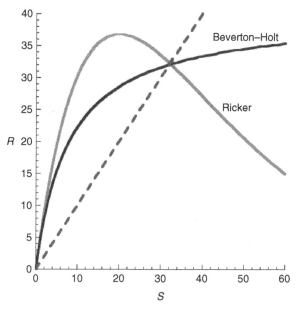

Figure 6.3. The Ricker and Beverton–Holt stock–recruitment relationships are similar when stock size is small but their behavior at large stock sizes differs considerably. I have also shown the 1:1 line, corresponding to $R = S$ (and thus a steady state for a semelparous species).

To derive the Beverton–Holt stock–recruitment relationship, let us follow the fate of a cohort of offspring from the time of spawning until they are considered recruits to the population at time T and let us denote the size of the cohort by $N(t)$, so that $N(0) = N_0$ is the initial number of offspring. If survival were density independent, we would write $dN/dt = -mN$ for which we know the solution at $t = T$ is $N(T) = N_0 e^{-mT}$. This is perhaps the simplest form of a stock–recruitment relationship once we specify the connection between S and N_0 (e.g. if we set $N_0 = fS$, where f is per-capita egg production, and $a = f e^{-mT}$, we then conclude $R = aS$).

We can incorporate density dependent survival by assuming that $m = m(N) = m_1 + m_2 N$ for which we then have the dynamics of N

$$\frac{dN}{dt} = -m_1 N - m_2 N^2 \tag{6.3}$$

and which needs to be solved with the initial condition $N(0) = N_0$.

Exercise 6.1 (M)

Use the method of partial fractions (that is, write $1/(m_1 N + m_2 N^2) = (A/N) + [B/(m_1 + m_2 N)]$ to solve Eq. (6.3) and show that

$$N(T) = \frac{e^{-m_1 T} N_0}{1 + (m_2/m_1)(1 - e^{-m_1 T}) N_0} \tag{6.4}$$

Now set $N_0 = fS$, make clear identifications of a and b from Eq. (6.2), and interpret them.

At this point, we can get a sense of how a fishery model might be formulated. Although in most of this chapter we will use discrete time formulations, let us use a continuous time formulation here with the assumptions of (1) a Beverton–Holt stock–recruitment relationship, and (2) a natural mortality rate M and a fishing mortality rate F on spawning stock biomass (we will shortly explore the difference between M and F, but for now simply think of F as mortality that is anthropogenically generated). The dynamics of the stock are

$$\frac{dN}{dt} = \frac{aN(t-T)}{b+N(t-T)} - MN - FN \tag{6.5}$$

This is a nonlinear differential-difference equation (owing to the lag between spawning and recruitment) and in general will be difficult to solve (which we shall not try to do). However, some simple explorations are worthwhile.

Exercise 6.2 (E)

The steady state population size satisfies $a\bar{N}/(b+\bar{N}) - M\bar{N} - F\bar{N} = 0$. Show that $\bar{N} = a/(M+F) - b$ and interpret this result. Also, show that the steady state yield (or catch, or harvest; all will be used interchangeably) from the fishery, defined as fishing mortality times population size will be $\bar{Y}(F) = F\bar{N} = F\big(a/(M+F)-b\big)$ and sketch this function.

There are other stock–recruitment relationships. For example, one due to John Shepherd (Shepherd 1982) introduces a third parameter, which leads to a single function that can transition between Ricker and Beverton–Holt shapes

$$R = \frac{aS}{1+(S/b)^c} \tag{6.6}$$

Here there is a third parameter c; note that I used the parameter b that characterizes density dependence in yet a different manner. I do this intentionally: you will find all sorts of functional relationships between stock and recruitment in the literature, with all kinds of different parametrizations. Upon encountering a new stock–recruitment relationship (or any other function for that matter), be certain that you fully understand the biological meaning of the parameters. A good starting point is always to begin with the units of the parameters and variables, to make certain that everything matches.

Each of Eqs. (6.1), (6.2), and (6.6) have the property that when S is small $R \sim aS$, so that when $S=0$, $R=0$. We say that this corresponds to a closed population, because if spawning stock size is 0, recruitment is 0. All populations are closed on the correct spatial scale (which might be global in the case of a highly pelagic species). However, on smaller spatial scales, populations might be open to immigration and emigration so that $R>0$ when $S=0$. In the late 1990s, it became fashionable in some quarters of marine ecology to assert that problems of fishery management were the result of the use of models that assume closed populations. Let us think about the difference between a model for a closed population model and a model for an open population:

$$\frac{dN}{dt} = rN\left(1 - \frac{N}{K}\right) \quad \text{or} \quad \frac{dN}{dt} = R_0 - MN \tag{6.7}$$

The equation on the left side is the standard logistic equation, for which $dN/dt = 0$ when $N=0$ or $N=K$. The equation on the right side is a simple model for an open population that experiences an externally determined recruitment R_0 and a natural mortality rate M.

Exercise 6.3 (E)

Sketch $N(t)$ vs t for an open population and think about how it compares to the logistic model.

For the open population model, dN/dt is maximum when N is small. Keep this in mind as we proceed through the rest of the chapter; it will not be hard to convince yourself that the assumption of a closed population is more conservative for management than that of an open population.

The Schaefer model and its extensions

In life, there are few things that "everybody knows," but if you are going to hang around anybody who works on fisheries, you must know the Schaefer model, which is due to Milner B. Schaefer, and its limitations (Maunder 2002, 2003). The original paper is hard to find, and since we will not go into great detail about the history of this model, I encourage you to read Tim Smith's wonderful book (Smith 1994) about the history of fishery science before 1955 (and if you can afford it, I encourage you to buy it). The Schaefer model involves a single variable $N(t)$ denoting the biomass of the stock, logistic growth of that biomass in the absence of harvest, and harvest proportional to abundance. We will use both continuous time (for analysis) and discrete time (for exercises) formulations:

$$\frac{dN}{dt} = rN\left(1 - \frac{N}{K}\right) - FN$$

$$N(t+1) = N(t) + rN(t)\left(1 - \frac{N(t)}{K}\right) - FN(t) \tag{6.8}$$

If you feel a bit uncomfortable with the lower equation in (6.8) because you know from Chapter 2 that it is not an accurate translation of the upper equation, that is fine. We shall be very careful when using the discrete logistic equation and thinking of it only as an approximation to the continuous one. On the other hand, for temperate species with an annual reproductive cycle, the discrete version may be more appropriate.

The biological parameters are r and K; we know from Chapter 2 that, in the absence of fishing, the population size that maximizes the growth rate is $K/2$ and that the growth rate at this population size is $rK/4$. When these are thought of in the context of fisheries we refer to the former as the population size giving maximum net productivity (MNP) and the latter as maximum sustainable yield (MSY), because if we could maintain the stock precisely at $K/2$ and then harvest the biological production, we can sustain the maximized yield. That is, if we then maintained the stock at MNP, we would achieve MSY. Of course, we cannot do that and these days MSY is viewed more as an upper limit to harvest than a goal (see Connections).

Exercise 6.4 (E/M)

Myers et al. (1997a) give the following data relating sea surface temperature (T) and r for a variety of cod Gadus morhua (Figure 6.4a; Myers et al. 1997b) stocks (each data point corresponds to a different spatial location). Construct a regression of r vs T. What explanation can you offer for the pattern? What implications are there for the management of "cod stocks"? You might want to check out Sinclair and Swain (1996) for the implication of these kind of data.

r (per year)	T (°C)	r (per year)	T (°C)
0.23	1.75	0.62	11.00
0.17	0.0	0.44	7.4
0.27	1.75	0.24	5.8
0.2	1.0	1.03	10.0
0.31	2.5	0.53	6.5
0.15	1.75	0.26	4.0
0.36	3.75	0.56	8.6
0.36	3.76	0.82	6.5
0.6	8.0	0.8	10.0
0.74	7.0	0.8	10.0
0.53	5.0		

There is a tradition of defining fishing mortality in Eqs. (6.8) as a function of fishing effort E and the effectiveness, q, of that effort in

(a)

(b)

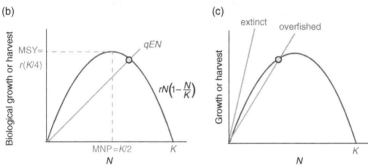

(c)

Figure 6.4. (a) Atlantic cod, *Gadus morhua*, perhaps a poster-child for poor fishery management (Hutchings and Myers 1994, Myers *et al.* 1997a, b). (b) Steady state analysis of the Schaefer model. I have plotted the biological production $rN(1 - (N/K))$ and the harvest on the same graph. The point of intersection is steady state population size. (c) As either effort or catchability increases, the line $y = qEN$ rotates counterclockwise and may ultimately lead to a steady state that is less than MNP, in which case the stock is considered to be overfished, in the sense that a larger stock size can lead to the same sustainable harvest. If qE is larger still, the only intersection point of the line and the parabola is the origin, in which case the stock can be fished to extinction.

removing fish (the catchability) so that $F = qE$. We already know that MSY is $rK/4$, but essentially all other population sizes will produce sustainable harvests (Figure 6.4b): as long as the harvest equals the biological production, the stock size will remain the same and the harvest will be sustainable. This is most easily seen by considering the steady state of Eqs. (6.8) for which $rN[1 - (N/K)] = qEN$. This equation has the solution $N = 0$, which we reject because it corresponds to extinction of the stock or solution $\overline{N} = K[1 - (qE/r)]$. We conclude that the steady state yield is

$$\overline{Y} = qE\overline{N} = qEK\left(1 - \frac{qE}{r}\right) \qquad (6.9)$$

which we recognize as another parabola (Figure 6.5) with maximum occurring at $E^* = r/2q$.

Exercise 6.5 (E)

Verify that, if $E = E^*$, then the steady state yield is the MSY value we determined from consideration of the biological growth function (as it must be).

Furthermore, note from Eqs. (6.8) that catch is $FN (= qEN)$, regardless of whether the stock is at steady state or not. Hence, in the Schaefer

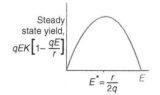

Figure 6.5. The steady state yield $\overline{Y} = qEK[1 - (qE/r)]$ is a parabolic function of fishing effort E.

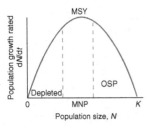

Figure 6.6. The acronym soup. Over the years, various reference points other than MSY (see Connections for more details) have developed. A stock is said to be in the range of optimal sustainable population (OSP) if stock size exceeds 60% of *K*, and to be depleted if stock size is less than 30%–36% of *K*.

model catch per unit effort (CPUE) is proportional to abundance and is thus commonly used as an indicator of abundance. This is based on the assumption that catchability is constant and that catch is proportional to abundance, neither of which need be true (see Connections) but they are useful starting points. In Figure 6.6, I summarize the variety of acronyms that we have introduced thus far, and add a new one (optimal sustainable population size, OSP).

Exercise 6.6 (M)

This multi-part exercise will help you cement many of the ideas we have just discussed. We focus on two stocks, the southern Gulf of St. Laurence, for which $r = 0.15$ and $K = 15\,234$ tons, and the faster growing North Sea stock for which $r = 0.56$ and $K = 185\,164$ tons (the data on r come from Myers *et al.* (1997a) cited above; the data on K come from Myers *et al.* (2001)). To begin, suppose that one were developing the fishery from an unfished state; we use the discrete logistic in Eqs. 6.8 and write

$$N(t+1) = N(t) + rN(t)\left(1 - \frac{N(t)}{K}\right) - C(t) \qquad (6.10)$$

where $C(t)$ is catch. Explore the dynamics of the Gulf of St. Laurence stock for a time horizon of 50 years, assuming that $N(0) = K$ and that (1) $C(t) = \text{MSY}$, or (2) $C(t) = 0.25N(t)$. Interpret your results. Now suppose that the stock has been overfished and that $N(0) = 0.2K$. What is the maximum sustainable harvest C_{max} associated with this overfished level? Fix the catch at $0, 0.1C_{max}, 0.2C_{max}$, up to $0.9C_{max}$ and compute the recovery time of the population from $N(0) = 0.2K$ to $N(t_{rec}) > 0.6K$. Make a plot of the recovery time as a function of the harvest level and try to interpret the social and institutional consequences of your plots. Repeat the calculations for the more productive North Sea stock. What conclusions do you draw? Now read the papers by Jeff Hutchings (Hutchings 2000, 2001) and think about them in the light of your work in this exercise.

Bioeconomics and the role of discounting

We now incorporate economics more explicitly by introducing the net revenue $R(E)$ (or economic rent or profit) which depends upon effort, the price p per unit harvest and the cost c of a unit of effort

$$R(E) = pY - cE = pqEN - cE \qquad (6.11)$$

In the steady state, for which $N = \bar{N} = K[1 - (qE/r)]$, we conclude that

$$\bar{R}(E) = pqEK\left(1 - \frac{qE}{r}\right) - cE \qquad (6.12)$$

We analyze this equation graphically (Figure 6.7), as we did with the steady state for population size, but in this case there is a bit more to talk

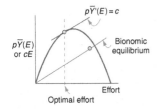

Figure 6.7. Steady state economic analysis of the net revenue from the fishery, which is composed of income $p\bar{Y}(E)$ and cost cE. When these are equal, the bionomic equilibrium is achieved; the value of effort that maximizes revenue is that for which the slope of the line tangent to the parabola is c.

about. First, we can consider the intersection of the parabola and the curve. At this intersection point $\bar{R}(E) = 0$ from which we conclude that the net revenue of the fishery is 0 (economists say that the rent is dissipated). H. Scott Gordon called this the "bionomic equilibrium" (Gordon 1954). It is a marine version of the famous tragedy of the commons, in which effort increases until there is no longer any money to be made.

Alternatively, we might imagine that somehow we can control effort, in which case we find the value of effort that maximizes the revenue. If we write the revenue as $\bar{R}(E) = p\bar{Y}(\mathrm{E}) - cE$ then the value of effort that maximizes revenue is the one that satisfies $p(\mathrm{d}/\mathrm{d}E)\bar{Y}(E) = c$, so that the level of effort that makes the line tangent to $p\bar{Y}(E)$ have slope c is the one that we want (Figure 6.7).

Exercise 6.7 (E/M)

Show that the bionomic level of effort (which makes total revenue equal to 0) is $E_b = (r/q)\left[1 - (c/pqK)\right]$ and that the corresponding population size is $N_b = \bar{N}(E) = c/pq$. What is frightening, from a biological perspective, about this deceptively beautiful equation? Does the former equation make you feel any more comfortable?

Next, we consider the dynamics of effort. Suppose that we assume that effort will increase as long as $R(E) > 0$, since people perceive that money can be made and that effort will decrease when people are losing money. Assuming that the rate of increase of effort and the rate of decrease of effort is the same, we might append an equation for the dynamics of effort to Eqs. (6.8) and write

$$
\frac{\mathrm{d}N}{\mathrm{d}t} = rN\left(1 - \frac{N}{K}\right) - qEN
$$
$$
\frac{\mathrm{d}E}{\mathrm{d}t} = \gamma(pqEN - cE)
$$

(6.13)

which can be analyzed by phase plane methods (and which will be *déjà vu* all over again if you did Exercise 2.12). One steady state of Eqs. (6.13) is $N = 0$, $E = 0$; otherwise the first equation gives the steady state condition $E = (r/q)[1 - (N/K)]$ and the second equation gives the condition $N = c/pq$. These are shown separately in Figure 6.8a and then combined. We conclude that if $K > c/pq$ (the condition for bionomic equilibrium and the economic persistence of the fishery), then the system will show oscillations of effort and stock abundance.

Now, you might expect that there are differences in the rate at which effort is added and at which effort is reduced. I agree with you and the following exercise will help sort out this idea.

Figure 6.8. Phase plane analysis of the dynamics of stock and effort. (a, b) The isoclines for population size and effort are shown separately. (c) If $K < c/pq$, the isoclines do not intersect and the fishery will be driven to economic extinction ($N = K$, $E = 0$). (d) If $K > c/pq$, then the isoclines intersect (at the bionomic equilibrium) and a phase plane analysis shows that the system will oscillate.

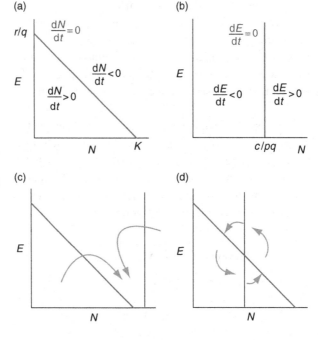

Exercise 6.8 (M)

In this exercise, you will explore the dynamics of the Schaefer model when the effort responds to profit. For simplicity, you will use parameter values chosen for ease of presentation rather than values for a real fishery. In particular, set $r = 0.1$ and $K = 1000$ (say tons, if you wish). Assume discrete logistic growth, written like this

$$N(t+1) = N(t) + rN(t)\left(1 - \frac{N(t)}{K}\right) - (1 - e^{-qE(t)})N(t) \tag{6.14}$$

where $E(t)$ is effort in year t and q is catchability. Set $q = 0.05$ and $E(0) = 0.2$ and assume that this is a developing fishery so that $N(0) = K$. (a) Use a Taylor expansion of $e^{-qE(t)}$ to show that this formulation becomes the Schaefer model in Eq. (6.8) when $qE(t) \ll 1$. Use this to explain the form of Eq. (6.14), rather than simply qEN for the harvest. (b) Next assume that the dynamics of effort are determined by profit and set

$$\Pi(t) = p(1 - e^{-qE(t)})N(t) - cE(t) \tag{6.15}$$

where $\Pi(t)$ is profit in year t; for calculations, set $p = 0.1$ and $c = 2$. Assume that in years when profit is positive effort increases by an amount ΔE_+ and that in years when profit is negative it decreases by an amount ΔE_-. For computations, set $\Delta E_+ = 0.2$ and $\Delta E_- = 0.1$, to capture the idea that fishing capacity is often irreversible (boats are more rather than less specialized). The effort dynamics are thus

$$E(t+1) = E(t) + \Delta E_+ \quad \text{if } \Pi(t) > 0$$

$$E(t+1) = E(t) \qquad\qquad \text{if } \Pi(t) = 0 \qquad\qquad (6.16)$$

$$E(t+1) = E(t) + \Delta E_- \quad \text{if } \Pi(t) < 0$$

Include the rule that if $E(t+1)$ is predicted by Eqs. (6.16) to be less than 0 then $E(t+1)=0$ and that if $E(t)=0$, then $E(t+1)=\Delta E_+$. Iterate Eqs. (6.15) and (6.16) for 100 years and interpret your results; using at least the following three plots: effort versus population size, catch versus time, and profit versus time. Interpret these plots. A more elaborate version of these kinds of ideas, using differential equations, is found in Mchich *et al.* (2002).

There is one final complication that we must discuss, whether we like its implications or not. This is the notion of discounting, which is the preference for an immediate reward over one of the same value but in the future (Souza, 1998). The basic concept is easy enough to understand: would you rather receive 100 dollars today or one year from today, given that you can do anything you want with that money between now and one year from today except spend it? It does not take much thinking to figure out that you'd take it today and put it in a bank account (if you are risk averse), a mutual fund (if you are less risk averse), or your favorite stock (if you really like to gamble). We can formalize this idea by introducing a rate δ at which future returns are devalued relative to the present in the sense that one dollar t years in the future is worth $e^{-\delta t}$ dollars today. That is, all else being equal, when the discount rate is greater than 0 you would always prefer rewards now rather than in the future. Thus, discounting compounds the effects of the tragedy of the commons.

Let us now think about the problem of harvesting a renewable resource when the returns are discounted. We will conduct a fairly general analysis, following the example of Colin Clark (Clark 1985, 1990). Instead of logistic dynamics, we assume a general biological growth function $g(N)$, and instead of $C(t)=qEN(t)$ we assume a general harvest function $h(t)$, so that the dynamics for the stock are $dN/dt = g(N) - h(t)$. A harvest $h(t)$ obtained in the time interval t to $t+dt$ years in the future has a present-day value $h(t)e^{-\delta t}dt$, so that the present value, PV, of all future harvest is

$$PV = \int_0^\infty h(t)e^{-\delta t}dt \qquad\qquad (6.17)$$

and our goal is to find the pattern of harvest that maximizes the present value, given the stock dynamics. In light of those dynamics, we write $h(t) = g(N) - (dN/dt)$ so that the present value becomes

(a)

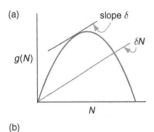

(b)

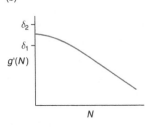

Figure 6.9. (a) The condition $g'(N) - \delta N$ maximizes the present value of harvest as long as $g'(0)$ is sufficiently big. If it is not (as for δ_2 in panel (b), drawn for a $g(N)$ that may not be logistic) then the optimal behavior, in terms of maximizing present value, is to drive the stock to extinction.

$$PV = \int_0^\infty \left(g(N) - \frac{dN}{dt} \right) e^{-\delta t} dt$$

We integrate by parts according to

$$\int_0^\infty \frac{dN}{dt} e^{-\delta t} dt = N(t) e^{-\delta t} \Big|_0^\infty + \int_0^\infty \delta N(t) e^{-\delta t} dt$$

from which we conclude that the present value is

$$PV = \int_0^\infty (g(N) - \delta N) e^{-\delta t} dt + N(0) \tag{6.18}$$

We maximize the present value by maximizing $g(N) - \delta N$ over N; the condition for maximization is $(d/dN)\{g(N) - \delta N\} = 0$ so that $(d/dN)g(N) = g'(N) = \delta$. In fact, if you look back to the previous section, just above Exercise 6.7 and to Figure 6.7 you see that this is basically the same kind of condition that we had previously reached: the present value is maximized when the stock size is such that the tangent line of the biological growth curve has slope δ (Figure 6.9a). Since we know that $g'(N)$ is a decreasing function of N, we recognize that this argument makes sense only if $g'(0) > \delta$. But what if that is not true, as for example in the case of whales or rockfish, where $g'(0) \sim r$ may be 0.04–0.08 and the discount rate may be much higher (say even 12% or 15%)? Then the optimal behavior, in terms of present value, is to take everything as quickly as possible (drive the stock to extinction). This result was first noted by Colin Clark in 1973 (Clark 1973) using methods of optimal control theory. In his book on mathematical bio-economics (Clark 1990, but the first edition published in 1976) he uses calculus of variations and the Euler–Lagrange equations, and in his 1985 book on fishery modeling (Clark 1985), Colin uses the method of integration by parts that we have done here.

In a more general setting, we would be interested in discounting a stream of profits, not harvest, so our starting point would be

$$PV = \int_0^\infty (p - c(N))h(t) e^{-\delta t} dt \tag{6.19}$$

where p is the price received per unit harvest and $c(N)$ is the cost of a unit of harvest when stock size is N. The same kind of calculation leads to a more elaborate condition (Clark 1990).

There is yet another way of thinking about this question, which I discovered while teaching this material in 1997, and which led to a paper with some of students from that class (Mangel et al. 1998) and which makes a good exercise.

Exercise 6.9 (E/M)

If a fishery develops on a stock that is previously unfished, we may assume that the initial biomass of the stock is $N(0) = K$. A sustainable steady state harvest that maintains the population size at N_s will remove all of the biological production, so that if h is the harvest, we know

$$h = rN_s\left(1 - \frac{N_s}{K}\right) \qquad (6.20)$$

(a) Show that, in general, solving Eq. (6.20) for N_s leads to two steady states, one of which is dynamically unstable; to do this, it may be helpful to analyze the dynamical system $N(t + 1) = N(t) + rN(t)\left[1 - (N(t)/K)\right] - h$ graphically. (b) Now envision that the development of the fishery consists of two components. First, there is a "bonus harvest" in which the stock is harvested from K to N_s, which for simplicity we assume takes place in the first year. Second, there is the sustainable harvest in each subsequent year, given by Eq. (6.20). The harvest in year t after the bonus harvest is discounted by the factor $1/(1 + \delta)^t$. (This is the common representation of discounting in discrete time models. To connect it with what we have done before, note that $(1 + \delta)^{-t} = e^{-t\log(1+\delta)} \approx e^{-\delta t}$ when δ is small.) Combining these, the present value $PV(N_s)$ of choosing the value N_s for the steady state population size is

$$PV(N_s) = K - N_s + \sum_{t=1}^{\infty} \frac{1}{(1 + \delta)^t} h \qquad (6.21)$$

Now we can factor h out of the summation and then you should verify that

$$\sum_{t=1}^{\infty} \frac{1}{(1 + \delta)^t} = \frac{1}{\delta}$$

so that Eq. (6.21) becomes

$$PV(N_s) = K - N_s + \frac{1}{\delta}rN_s\left(1 - \frac{N_s}{K}\right) \qquad (6.22)$$

(c) Show that the value of N_s that maximizes $PV(N_s)$ is $(K/2)[1 - (\delta/r)]$ and interpret the result. Compare this with the condition following Eq. (6.18). (d) In order to illustrate Eq. (6.22), use the following data (Clark 1990; pp. 47–49, 65).

Species	r	K
Antarctic fin whale	0.08	400 000 whales
Pacific halibut	0.71	80.5×10^6 kg
Yellowfin tuna	2.61	134×10^6 kg

Determine the maximum value of $PV(N_s)$ as δ varies by making a matrix in which columns are labeled by the value of δ, rows are labeled by N_s/K and the entry of matrix is $PV(N_s)$. Let δ vary between 0.01 and 0.21 in steps of 0.04 and let N_s vary between 0 and K in steps of $0.1K$. You may also want to measure population size in handy units, such as 1000 whales or 10^6 kg, or as a fraction of the carrying capacity. Interpret your results.

Age structure and yield per recruit

The models that we have discussed thus far are called production models because they focus on removing the "excess production" associated with biological growth. But that production has thus far been treated in an exceedingly simple manner. We will now change that. Models that incorporate individual growth play a crucial role in modern fishery management, so we shall spend a bit of time showing that connection. Let us return to Eq. (2.13) and explicitly write a, for age, instead of t so that $L(a)$ represents length at age a and $W(a)$ represents weight at age a, still assumed to be given allometrically. Imagine that we follow a single cohort of fish, with initial numbers $N(0) = R$. In the absence of fishing mortality, the number of individuals at any other age is given by $N(a) = Re^{-Ma}$.

When following a population with overlapping generations, we introduce $N(a, t)$ as the number of individuals of age a at time t, and $F(a)$ as the fishing mortality of individuals of age a. The dynamics of all age classes except the youngest are

$$N(a + 1, \; t + 1) = e^{-(M+F(a))}N(a, t) \tag{6.23}$$

since next year's 10 year olds, for example, must come from this year's 9 year olds. We assume that $p_m(a)$ is the probability that an individual of age a is mature and reproductively active, and an allometric relationship between length at age $L(a)$ and egg production $(=cL(a)^b$, with c and b constants). The total number of eggs produced in a particular year is

$$E(t) = \sum_a p_m(a)cL(a)^b N(a, t) \tag{6.24}$$

and we append the dynamics of the youngest age class $N(0, t + 1) = N_0(E(t))$, where $N_0(E(t))$ is the relationship between the number of eggs produced by spawning adults and the number of individuals in the youngest age class. For example, in analogy to the Beverton–Holt recruitment function for we have $N_0(0, t + 1) = \alpha E(t)/[1 + \beta E(t)]$ and in analogy to the Ricker recruitment $N_0(0, \; t + 1) = \alpha E(t)e^{-\beta E(t)}$; in both cases the parameters α and β require new interpretations from the ones that we have given previously. For example, the parameter α is now a measure of egg to juvenile survival when population size is low and the parameter β is still a measure of the effects of density dependence.

In light of Eq. (6.23), the number of fish of age a that died in year t is $N(a, t)(1 - e^{-(M+F(a))})$, and if we assume that the natural and anthropogenic components are in proportion to the contribution of total mortality $m + F(a)$ owing to each, we conclude that a fraction $M/[M + F(a)]$ of the fish are lost owing to natural mortality and a fraction $F(a)/[M + F(a)]$ of the fish are taken by the fishery. Thus, the yield of fish of age a in year t is

$$Y(a, t) = \frac{F(a)}{M + F(a)} N(a,t)(1 - e^{-(M+F(a))})W(a) \qquad (6.25)$$

where $W(a)$ is the weight of fish of age a; the total yield in year t is $Y(t) = \sum_{a=0}^{a_{max}} Y(a, t)$, where a_{max} is the maximum age to which fish live (for most of this chapter, I will not write the upper limit).

Very often, we assume "knife-edge" fishing mortality, so that $F(a) = 0$ if a is less than the age a_r at which fish are recruited to the fishery and $F(a) = F$, a constant, for ages greater than or equal to the age of recruitment to the fishery. Note, too, that there are now two kinds of recruitment – to the population (at age 0) and to the fishery (at age a_r).

Yield per recruit

Let us now follow the fate of a single cohort through time. Why would we want to do this? Part of the answer is that we are much less certain about stock and recruitment relationships than we are about survival from one age class to the next. So, wouldn't it be nice if we could learn a lot about sustaining fisheries by simply looking at cohort dynamics and not stressing about the stock–recruitment relationship? That, at least, is the hope.

When we follow a single cohort, age a and time t are identical, if we start the time clock at age 0, for which we fix $N(0) = N_0$, assumed to be a known constant. The dynamics of the cohort are exceedingly simple, since $N(a + 1) = N(a)e^{-M-F(a)}$ and if individuals are recruited to the fishery at age a_r and fishing mortality is knife-edge at level F the yield from this cohort is

$$Y(a_r, F) = \sum_{a=a_r} \frac{F}{M + F} N(a)(1 - e^{-(M+F)})W(a) \qquad (6.26)$$

Intuition tells us (and you will confirm in an exercise below) that yield as a function of F will look like Figure 6.10. When F is small, we expect that yield will be an increasing function of fishing effort (from a Taylor expansion of the exponential). As F increases, fewer individuals reach high age (and large weight), so that yield declines. The slope of the yield versus effort curve will be largest at the origin and very often you will

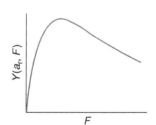

Figure 6.10. The yield from a cohort as a function of fishing effort.

encounter rules for setting fishing mortality that are called $F_{0.x}$, which means to choose F so that the slope of the tangent line of the yield versus effort curve is $0.x$ times the value of the slope at the origin.

Since $p_m(a)$ is the probability that an individual of age a is mature, the number of spawners when the fishing mortality is F and the age of recruitment to the fishery is a_r is $S(a_r, F) = \sum_a p_m(a)N(a)$ and the spawning stock biomass produced by this cohort is $SSB(a_r, F) = \sum_a p_m(a)W(a)N(a)$ (note that F and a_r are actually "buried" in $N(a)$). The number of spawners and the spawning stock biomass that we have just constructed will depend upon the initial size of the cohort. Consequently, it is common to divide these values by the initial size of the cohort and refer to the spawners per recruit or spawning stock biomass per recruit.

In the early 1990s, W. G. Clark (Clark 1991, 2002) noted that some of the biggest uncertainty in fishery management arises in the spawner recruit relationship. Clark proceeded to simulate a number of different stock recruitment relationships and studied how the long term yield was related to the fishing mortality F. In the course of this work, he used the spawning potential ratio, which is the value of F that makes $SSB(F)$ a specified fraction of $SSB(0)$. For many fast growing stocks, a $SSB(F)$ of 0.35 or 0.40 (that is, 35% or 40%) is predicted to produce maximum long term yields while for slower growing stocks the value is closer to 55% or 60% (MacCall 2002).

Exercise 6.10 (E/M)

Imagine a stock with von Bertalanffy growth with parameters $k = 0.25 \, \text{yr}^{-1}$, $L_\infty = 50 \, \text{cm}$, $t_0 = 0$, $M = 0.1 \, \text{yr}^{-1}$, and a length weight allometry $W = 0.01 \, L^3$, where W is measured in grams. Assume that no fish lives past age 10. With knife-edge dynamics for recruitment to the fishery, the dynamics of the cohort are

$$N(0) = R$$
$$N(a+1) = N(a)e^{-M} \qquad \text{for } a = 0, 1, 2, \ldots a_r - 1$$
$$N(a+1) = N(a)e^{-M-F} \qquad \text{for } a = a_r \text{ to } 9 \qquad (6.27)$$
$$Y(a) = \frac{F}{M+F}(1 - e^{-M-F})N(a)W(a) \qquad \text{for } a > a_r$$

Assume that $N_0 = 500\,000$ individuals. Compute the total yield (in metric tons $= 1000 \, \text{kg}$) per recruit assuming that fish are recruited to the fishery at age 2, 3, or 4. Make three separate plots of yield vs fishing effort for the three different ages of recruitment to the fishery. Pick one of these ages and construct a table of age vs number of individuals in the presence or absence of fishing. Next compute the number of spawners per recruit and spawning stock biomass per recruit, assuming that all individuals mature at age 3. Now convert your code to a time dependent problem for the number of fish of age a at time t, $N(a, t)$, by assuming that recruitment $N(0, t)$ is a Beverton–Holt function of spawning stock biomass $S(t-1)$ according to $N(0,t) = 3S(t-1)/[1 + 0.002S(t-1)]$ and repeat the previous calculations.

Salmon are special

Salmon life histories are somewhat different than most fish life histories, and a separate scientific jargon has grown up around salmon life histories (fisheries science has its own jargon that is distinctive from ecology although the same problems are studied, and salmon biology has its own jargon that is somewhat distinctive from the rest of fisheries science). Eggs are laid by adults returning from some time in the ocean in nests, called redds, in freshwater. In general (for all Pacific salmon, but not necessarily for steelhead trout or Atlantic salmon) adults die shortly after spawning and how long an adult stays alive on the spawning ground is itself an interesting question (McPhee and Quinn 1998). Eggs are laid in the fall and offspring emerge the following spring, in stages called aelvin, fry, and parr. Parr spend some numbers of years in freshwater and then, in general, migrate to the ocean before maturation. A Pacific salmon that returns to freshwater for reproduction after one sea winter or less is called a jack; an Atlantic salmon that returns early is called a grilse. Salmon life histories are thus described by the notation $X \cdot Y$ meaning X years in freshwater and Y years in seawater.

When individuals die after spawning, we use dynamics that connect the number of spawners in one generation, $S(t)$, with the number of spawners in the next generation, $S(t+1)$. In the simplest case all individuals from a cohort will return at the same time and using the Ricker stock–recruitment relationship we write

$$S(t+1) = aS(t)e^{-bS(t)} \qquad (6.28)$$

In this case (Figure 6.11) the steady state population size at which $S(t+1) = S(t)$ satisfies $1 = ae^{-b\bar{S}}$ (see Exercise 6.11 below) and the stock that can be harvested for a sustainable fishery is the difference $S(t+1) - S(t)$, keeping the stock size at $S(t)$. Thus, the maximum sustainable yield occurs at the stock size at which the difference $S(t+1) - S(t)$ is maximized (also shown in Figure 6.11).

Salmon fisheries can be managed in a number of different ways. In a fixed harvest fishery, a constant harvest H, is taken thus allowing $S(t) - H$ fish to "escape" up the river for reproduction. The dynamics are then $S(t+1) = a(S(t) - H)e^{-b(S(t)-H)}$. In a fixed escapement fishery, a fixed number of fish E is allowed to "escape" the fishery and return to spawn. The harvest is then $S(t) - E$ as long as this is positive

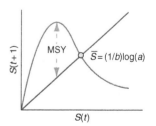

Figure 6.11. The Ricker stock–recruitment function is used when characterizing the dynamics of salmonid stocks.

and zero otherwise. With a policy based on a constant harvest fraction, a fraction q of the returning spawners are taken, making the spawning stock $(1-q)$ and the dynamics become $S(t+1) = a(1-q)$ $S(t)e^{-b(1-q)S(t)}$. More details about salmon harvesting can be found in Connections.

Exercise 6.11 (M)

This is a long and multi-part exercise. (a) Show that the steady state of Eq. (6.28) satisfies $\bar{S} = (1/b)\log(a)$. For computations that follow, choose $a = 6.9$ and $b = 0.05$. (b) Draw the phase plane showing $S(t)$ (x-axis) vs $S(t)$ (y-axis) and use cob-webbing to obtain a graphical characterization of the data. (If you do not recall cob-webbing from your undergraduate days, see Gotelli (2001)). (c) Next, numerically iterate the dynamics, starting at an initial value of your choice, for 20 years, to demonstrate the dynamic behavior of the system. (d) Show that Eq. (6.28) can be converted to a linear regression of recruits per spawner of the form $\log[S(t+1)/S(t)] = \log(a) - bS(t)$ so that a plot of $S(t)$ (x-axis) vs $\log[S(t+1)/S(t)]$ (y-axis) allows one to estimate log(a) from the intercept and b from the slope. (e) My colleague John Williams proposed that Eq. (6.28) could be modified for habitat quality by rewriting it as $S(t+1) = ah(t)S(t)$ $\exp(-bS(t)/h(t))$, where $h(t)$ denotes the relative habitat, with $h(t) = 1$ corresponding to maximum habitat in year t. What biological reasoning goes into this equation? What are the alternative arguments? (f) You will now conduct a very simple power analysis (Peterman 1989, 1990a, b) for habitat improvement. Assume that habitat has been reduced to 20% of its original value and that habitat restoration occurs at a rate of 3% per year (so that $h(t+1) = 1.03h(t)$, until $h(t) = 1$ is reached). Find the steady state population size if habitat is reduced to 20% of its original value. Starting at this lower population size, increase the habitat by 3% each year (without ever letting it exceed 1) and assume that the population is observed with uncertainty, so that the 95% confidence interval for population size is $0.5S(t)$ to $1.5S(t)$. Use this plot to determine how long it will be before you can confidently state that the habitat improvement is having the positive effect of increasing the population size of the stock. Interpret your result. See Korman and Higgins (1997) and Ham and Pearsons (2000) for applications similar to these ideas.

Incorporating process uncertainty and observation error

Thus far, we have discussed deterministic models. In this section, I discuss some aspects of stochastic models, and offer one exercise to give you a flavor of them. More details – and a more elaborate version of the exercise – can be found in Hilborn and Mangel (1997).

Stochastic effects may enter through the population dynamics (process uncertainty) or through our observation of the system (observation error). For example, if we assumed that biological production, but not catch, were subject to process uncertainty and that this uncertainty had a log-normal distribution, then the Schaefer model, Eq. (6.10), would be modified to

$$N(t+1) = N(t) + rN(t)\left(1 - \frac{N(t)}{K}\right)e^{Z_p} - qEN(t) \qquad (6.29)$$

where Z_p is a normally distributed random variable with mean 0 and standard deviation σ_p. Our index of abundance is still catch per unit effort, but this is now observed with error, so that we have an index of abundance

$$I(t) = qN(t)e^{Z_{obs}} \qquad (6.30)$$

where Z_{obs} is a normally distributed random variable with mean 0 and standard deviation σ_{obs}.

Exercise 6.12 (E)

Referring to Chapter 3 and the properties of the log-normal distribution, explain why Eq. (6.30) produces a biased index of abundance, in the sense that $E\{I(t)\} > qN(t)$. Explain why a better choice in Eq. (6.30) is that Z_{obs} is a normally distributed random variable with mean $-(1/2)(\sigma_{obs})^2$ and standard deviation σ_{obs}. Would this cause you to change the form of Eq. (6.29)?

One of the great quantitative challenges in fishery management is to figure out practicable means of analysis of models such as Eqs. (6.29) and (6.30) (or their extensions; see Connections). The following exercise, which is a simplification of the analysis in Hilborn and Mangel (1997, chapter 10) will give you a flavor of how the thinking goes. Modern Bayesian methods allow us to treat process uncertainty and observation error simultaneously, but that is the subject for a different book (see, for example, Gelman *et al.* (1995), West and Harrison (1997)).

Exercise 6.13 (M)

The Namibian fishery for two species of hake (*Merluccius capensis* and *M. paradoxus*) was managed by the International Commission for Southeast Atlantic Fisheries (ICSEAF) from the mid 1960s until about 1990. Your analysis will be concerned with the data from the period up to and including ICSEAF management. Hake were fished by large ocean-going trawlers primarily from Spain, South Africa, and the (former) Soviet Union. Adults are found in large schools, primarily in mid-water. While both species are captured in the fishery, the fishermen are unable to distinguish between them and they are treated as a single stock for management purposes. The fishery developed essentially without any regulation or conservation. Catch per unit effort (CPUE), measured in tons of fish caught per hour, declined dramatically until concern was expressed by all the nations fishing this stock. The concern about the dropping CPUE led to the formation of ICSEAF and subsequent reductions in catch. After catches were reduced, the CPUE began to increase. In the data used in this analysis, CPUE is the catch per hour of a standardized class of Spanish trawlers. Such standardized analysis is used to avoid bias due to increasing gear efficiency or differences in fishing pattern by different classes or nationalities of vessels. The data are as follows.

Year	CPUE	Catch (thousands of tons)
1965	1.78	94
1966	1.31	212
1967	0.91	195
1968	0.96	383
1969	0.88	320
1970	0.9	402
1971	0.87	366
1972	0.72	606
1973	0.57	378
1974	0.45	319
1975	0.42	309
1976	0.42	389
1977	0.49	277
1978	0.43	254
1979	0.4	170
1980	0.45	97
1981	0.5	91
1982	0.53	177
1983	0.58	216
1984	0.64	229
1985	0.66	211
1986	0.65	231
1987	0.63	223

(a) To get a sense of the issues, make plots of CPUE vs year (remembering that CPUE is an index of abundance), catch vs year, and cumulative catch vs year.
(b) You are going to use a Schaefer model without process uncertainty but with observation error to analyze the data. That is, we assume that the biological dynamics are given by Eq. (6.10). Ray Hilborn and I treat the case in which both r and K are unknown, but here we will assume that r is known from other sources and is $r = 0.39$. However, carrying capacity K is unknown. Assume that the index of abundance is CPUE and is proportional to biomass; the predicted index

of abundance is $I_{pre}(t) = qN(t)$, where q is the catchability coefficient. As with r, Hilborn and I consider the case in which q also has to be determined. To make life easier for you, assume that $q = 0.000\,45$. However, the index $I_{pre}(t)$ is not observed. Rather, the observed CPUE is $CPUE(t) = I_{pre}(t)e^{Z(t)}$ where $Z(t)$ is normally distributed with mean 0 and standard deviation σ. (c) Show that $Z(t) = \log\{CPUE(t) - \log(I_{pre}(t))\}$ so that the log-likelihood of a single deviation $Z(t)$ is $L(t) = -\log(\sigma) - (1/2)\log(2\pi) - (Z(t)^2/2\sigma^2)$. The total log-likelihood for a particular value of K is the sum of the single year log-likelihoods $L_T(K|\text{data}) = \sum_{t=1965}^{1987} L(t)$, where I have emphasized the dependence of the likelihood for K on the data. (d) Compute the total log-likelihood associated with different values of carrying capacity K, as K ranges from 2650 to 2850 in steps of 10. To do this, use Eq. (6.10) to determine $N(t)$ for each year, assuming that the population started at K in 1965. Find the value of K that makes the total log-likelihood the largest. Denote this value by K^* and the associated total log-likelihood by L_T^*; it is the best point estimate. Make a plot of L_T (x-axis) vs K (y-axis) and show K^* and L_T^*. (e) From Chapter 3, we know that the 95% confidence interval for the carrying capacity are the values of K for which the total log-likelihood $L_T = L_T^* - 1.96$. Use your plot from part (d) to find these confidence intervals. (Note: if you look in Hilborn and Mangel (1997), you will see that the confidence intervals are much broader. This is caused by admitting uncertainty in r and q, and having to determine σ as part of the solution. But don't let that worry you. We also used the negative log-likelihood, which is minimized, rather than the likelihood, which is maximized.) (f) At this point, you should have estimates for the 95% confidence interval for carrying capacity. Now suppose that the management objective is to keep the population within the optimal sustainable region, in which $N(t) > 0.6\,K$ from 1988 to 2000 (assume that you were doing this work in 1988). Determine the catch limit that you would apply to achieve this goal. Hint: How do you determine the population size in 1987?

The theory of marine reserves

No-take marine reserves (or marine protected areas), in which all forms of catch are prohibited, are gaining increasing attention as conservation and management tools. Rather than provide a comprehensive review, I point you to recent issues of the *Bulletin of Marine Science* (**66**(3), 2000) and *Ecological Applications* (**13**(1) (Supplement), 2003). A summary of these is that there is general agreement that no-take marine reserves are likely to be effective tools for conservation, but it is still not clear if they will enhance fishery catches, either in the short-term or the long-term (Mangel 1998, 2000a, b, c, Botsford *et al.* 2001, Lockwood *et al.* 2002).

In this section, we analyze a relatively simple model for reserves, because it allows us to use a variety of our tools and to see things in a new way. Other modeling approaches are discussed in Connections.

Figure 6.12. A model for marine reserves involves a habitat that is divided into a reserve zone and a harvest zone. In the harvest zone, a fraction u of the stock is taken by the fishery. Following fishing, the stocks in the two zones merge for reproduction.

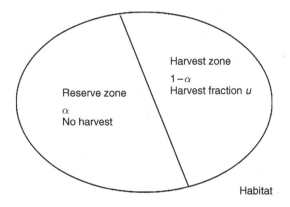

Envision a stock that grows logistically, again in discrete time, in a known habitat. Rather than fishing in the entire habitat, we set aside a fraction α of it as a reserve in which there is no fishing. We then allow a fraction u of the stock in the non-reserve area to be taken by the fishery (Figure 6.12). If $N(t)$ is the size of the stock at the start of fishing season t, then after fishing the stock size in the reserve is $\alpha N(t)$ and in the fishing region is $(1 - \alpha)(1 - u)N(t)$. Hence the total stock after fishing but before reproduction is $\alpha N(t) + (1 - \alpha)(1 - u)N(t) = [1 - u(1 - \alpha)]N(t)$. With logistic dynamics, we have

$$
\begin{aligned}
N(t+1) = {}& [1 - u(1 - \alpha)]N(t) \\
& + r[1 - u(1 - \alpha)]N(t)\left(1 - \frac{[1 - u(1 - \alpha)]N(t)}{K}\right)
\end{aligned}
\tag{6.31}
$$

To begin, as always, we ask about the steady state.

Exercise 6.14 (E)

Show that the steady state of Eq. (6.31) is given by

$$
\bar{N} = \frac{K}{1 - u(1 - \alpha)}\left[1 - \frac{u(1 - \alpha)}{r(1 - u(1 - \alpha))}\right]
\tag{6.32}
$$

Note that we have already learned something valuable about the system: $I = u(1 - \alpha)$ is an invariant for the marine reserve in the sense that the steady state takes the same value, regardless of individual values of u and α as long as the product remains the same. Thus for example, if we have a large reserve (making α big) then we can allow a high fraction of take in the harvest zone and vice versa (i.e. a higher fraction of a smaller available population in the fishing region). This observation, also noted by Hastings and Botsford (1999), suggests that there is an equivalence between protecting area and reducing catch.

One objective for the design of a reserve could be that the steady state given by Eq. (6.32) is a fixed fraction f of the carrying capacity. The value of that fraction is not something that can be set by quantitative analysis; it is a policy decision. For example, for a relic population $f=0.1$ (or even 0.05 – definition of relic is an open topic); we might want to ensure that the population is at worst depleted and set $f=0.35$ or we might want to ensure that the population is in its optimal sustainable range and set $f=0.6$. If we set $\bar{N}=fK$ and solve the resulting equation for f, we obtain

$$f = \frac{1}{1-u(1-\alpha)}\left[1 - \frac{u(1-\alpha)}{r(1-u(1-\alpha))}\right]$$

We should actually like to solve this equation for the reserve fraction, hence obtaining $\alpha(f)$, which is the fraction of habitat needed to be reserve to maintain the population steady state at fK, once f is specified. This can be done (Mangel 1998); you might set it as an optional exercise. One interesting question arises if we set $f=0$; why we will do this becomes clear momentarily.

Exercise 6.15 (E)

Set $f=0$ in the previous equation and show that

$$\alpha(0) = \frac{u(r+1)-r}{u(r+1)} \tag{6.33}$$

We conclude from Eq. (6.33) that if the reserve fraction is greater than $\alpha(0)$ then the steady state stock size will be greater than 0. Interpret Eq. (6.33) for the case of very large r. (Of course, to assert that we sustain a fish stock as long as one individual remains is kind of a silly idea; we should like many more individuals than 1.) Interpret the case of modest or small r.

There are a number of ways that one can present the information contained in Eq. (6.32) (Figure 6.13).

There are also many ways in which stochastic effects could enter into what we have done. One possibility is that the catch fraction in the harvest region is not fixed but is a random variable $U(t)$. An example of the distribution of this random variable is shown in Figure 6.14a; the mean and mode of the catch fraction are about 0.25, but the actual fraction varies from about 0.1 to 0.45. This should remind us that in operational situations such as fisheries, fishing mortality can be targeted but it cannot be controlled (Mangel 2000b).

When there are stochastic effects, the whole notion of sustainability must change and we have to think in terms of probabilities (Mangel 2000a). We understand now that the population size after fishing but before

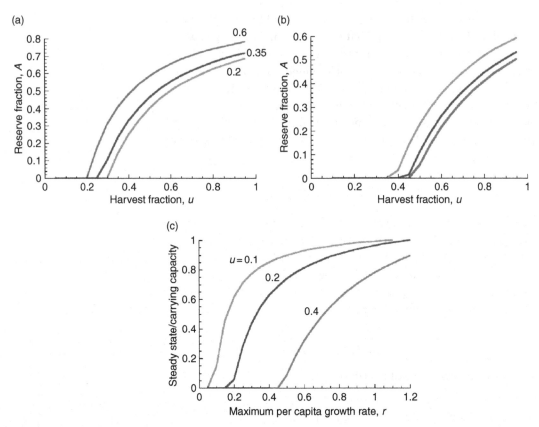

Figure 6.13. The reserve fraction needed to achieve steady state population sizes that are 20%, 35%, or 60% of carrying capacity as a function of the harvest fraction outside of the maximum per capita growth rate $r=0.5$ (panel a) or $r=1$ (panel b). (c) An alternative way to view the information is to fix reserve size (say at 20%) and see how steady state population size varies with maximum per capita growth rate and harvest fraction.

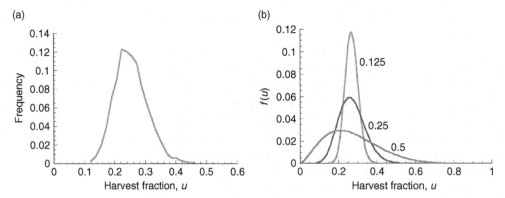

Figure 6.14. (a) The distribution of fishing mortality on herring *Clupea harengus* (from Patterson (1999)). (b) The beta density with mean 0.25 and three different values of the coefficient of variation.

reproduction, $\alpha N(t) + (1 - \alpha)(1 - U(t))N(t) = [1 - U(t)(1 - \alpha)]\,N(t)$, is a random variable because harvest fraction $U(t)$ is a random variable. Let us fix the reserve fraction at α, a time horizon T, and a critical population size N_c and define

$$p(n, t|\alpha, N_c) = \Pr\{N(s) \text{ exceeds } N_c \text{ for all } s, t \leq s \leq T|$$
$$\text{reserve fraction } \alpha \text{ and that } N(t) = n\} \tag{6.34}$$

For example, in the case of a developing fishery, we might assume that the population starts at carrying capacity and as a target that we do not want it to fall below 60% of carrying capacity, so that $N_c = 0.6\,K$. For the case of an exploited fishery, we might change the goal so that the population does not fall below 35% of carrying capacity, so that $N_c = 0.35\,K$.

We evaluate the probability defined in Eq. (6.33) by methods similar to, and actually easier than, stochastic dynamic programming. We have the end condition

$$p(n, T|\alpha, N_c) = \begin{array}{ll} 1 & \text{if } n \geq N_c \\ 0 & \text{otherwise} \end{array} \tag{6.35}$$

At any earlier time t, when $N(t) = n$, assume that the random variable $U(t)$ takes the value u (which will occur with probability determined by the density function that describes $U(t)$). Then the population size at the start of time period $t + 1$ will be $g[(1 - u(1 - \alpha))n]$ where $g(\cdot)$ is given by the right hand side of Eq. (6.31) and we are interested in the probability of staying above the critical level, starting at this new time and new population size. (In the next chapter, we will call this reasoning "thinking along sample paths.") Since the value of $U(t) = u$ is determined by a probability density we have

$$p(n, t|\alpha, N_c) = E_u\{p(g[(1 - u(1 - \alpha))n], t + 1|\alpha, N_c)\} \tag{6.36}$$

In order to make Eq. (6.36) operational, we need to pick a distribution for $U(t)$. Since $U(t)$ is a catch fraction, a natural choice for the probability distribution is the beta density, some examples of which are shown in Figure 6.14b.

In year t, when $N(t) = n$ and $U(t) = n$, the harvest is $(1 - \alpha)un$, so that if we define $C(n, t|\alpha, N_c)$ to be the accumulated catch between t and T, given that $N(t) = n$ and the reserve fraction α, reasoning similar to that leading to Eq. (6.36) gives us

$$C(n, t|\alpha, N_c) = E_u\{(1 - \alpha)un + C(g[(1 - u(1 - \alpha))n], t + 1|\alpha, N_c)\} \tag{6.37}$$

The first term on the right hand side is a linear function of u; we take its expectation and write (using $\bar{u}$ to denote the mean $U(t)$)

$$C(n, t|\alpha, N_c) = (1 - \alpha)\bar{u}n + E_u\{C(g[(1 - u(1 - \alpha))n], t + 1|\alpha, N_c)\} \tag{6.38}$$

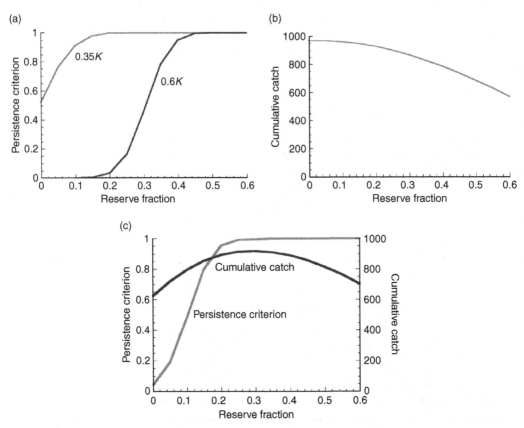

Figure 6.15. A marine reserve with fluctuations in harvest fraction over a planning horizon of 100 years. (a) The persistence criterion p (K, $1|\alpha$, N_c) for critical values of population size 0.6 K or 0.35 K, mean catch fraction 0.2 and coefficient of variation 50%. (b) The cumulative catch under these circumstances. (c) The persistence criterion p (0.35 K, $1|\alpha$, 0.2 K) and the cumulative catch when the mean catch fraction is 0.27 and the coefficient of variation is 50%. Based on Mangel (2000a).

In Figure 6.15, I show results of calculations using Eqs. (6.36) and (6.37) (also see Mangel (2000a) where other results are given). Note that a marine reserve is predicted to sometimes, but not always, increase catch. However, a reserve is always predicted to improve the probability of long-term persistence of the stock.

Risk analysis as a framework in fishery systems

Risk analysis (Anand 2002) is the appropriate decision tool when the science is ambiguous (almost always true in environmental problem solving) and this is especially true for fisheries (Rosenberg and Restrepo 1994, Tyutyunov et al. 2002). To illustrate the idea as simply as possible, imagine a stock for which we know that $r = 0.12$ yr^{-1} but that carrying capacity is uncertain, known to be either 1000 mt, with probability p, or 2000 mt, with probability $1 - p$. In the former case, the MSY harvest is 30 mt/yr and in the latter case, the MSY harvest is 60 mt/yr. Now the average value of carrying capacity is $\bar{K} = 1000p + 2000(1 - p) = 1000(2 - p)$ and the average of the MSY harvest is $30p + 60(1 - p) = 30(2 - p)$. However, it is pretty clear, because of the simplicity of this problem, that if we had to choose a value of the harvest rate, it would be nonsensical to choose the average of the MSY harvests. If the true carrying capacity is 1000 mt, then choosing the average of the MSY harvests will overfish the stock as long as $p > 0$. On the other hand, if the true carrying capacity is 2000 mt, applying the average of the MSY harvest will cause the loss of sustainable yield as long as $p < 1$. Whether it is better to overfish the stock or lose sustainable yield is a question that cannot be answered by quantitative methods alone (and may not even be within the purview of quantitative methods (Ludwig *et al.* 2001)), but it is clear that averaging uncertain values as a means of attaining a "consensus" value for action is arbitrary (Mangel *et al.* 1993).

The procedures of risk analysis recognize that one must be explicit about the potential states of nature, the actions one takes, and the consequences of those actions. In the simplest case we have two states of nature S_1, S_2; two possible actions, A_1, A_2; and values $V_{ij} = V(S_i, A_j)$ that accrue when the state of nature is S_i and the action is A_j. These are best summarized in a table; if p is the probability that the state of nature is S_1, then the table looks like this.

	State of nature		
	S_1	S_2	
Action	Probability p	$1-p$	Average value
A_1	$V(S_1, A_1)$	$V(S_2, A_1)$	$pV(S_1, A_1) + (1 - p)V(S_2, A_1)$
A_2	$V(S_1, A_2)$	$V(S_2, A_2)$	$pV(S_1, A_2) + (1 - p)V(S_2, A_2)$

In this table, we have explicitly laid out the states of nature, the probability of their occurrence and the actions. Each entry in the last column is the value of a particular action given the state of nature. We are then able to compute the average values of different actions; for example, the average value of action A_1 will be $pV(S_1, A_1) + (1 - p)V(S_2, A_1)$. The hard work, of course, is filling in the entries in the table. For example, to estimate the probability of the alternative states of nature we can use methods of likelihood theory or Bayesian analysis. To evaluate the effects of alternative actions with uncertainty, we can use Monte Carlo forward simulations.

Connections

The fishery system

It helps to think broadly about the fishery system (Cole-King 1993, Okey 2003). The recent books of Jennings *et al.* (2001) and Hart and Reynolds (2002) are great starting points. Some other entry points include Gulland (1977, 1988), Wooster (1988), Norse (1993), King (1995), Olver *et al.* (1995), Pikitch *et al.* (1997), McAllister *et al.* (1999), Cochrane (2000), Corkett (2002), and Rosenberg (2003). Zabel *et al.* (2003) introduced the notion of ecologically sustainable yield (ESY), defined as the maximum yield of fish that an ecosystem can sustain without shifting states in the sense of a system with multiple steady states. Other community-level metrics such as species richness, evenness, or community resiliency could also be used as ESY targets. However, because communities change in response to natural processes in ways that we do not fully understand, and we can never predict the behavior of communities with absolute certainty, we can and should improve our understanding of the bounds of expected community behavior and define ESY within the limits of their predictability. Smith (1994) provides a history of fishery science before 1955. It is also helpful to learn about some particular fishery systems, such as the Northeast Pacific (Trumble 1998), or the Atlantic groundfish fishery (Boreman *et al.* 1997). For more quantitative approaches to the subject, the classic work is Beverton and Holt (1957); updates are Quinn and Deriso (1999) and Walters and Martell (2004). Some years ago, the Marine Fisheries Section of the American Fisheries Society republished Ray Beverton's 1951 lectures on the use of theoretical models in the study of the dynamics of exploited fish (Beverton 1994).

Models and data

In thinking about the issues of fishery management, Don Ludwig (Ludwig 1995) recognized that we will have to use models for management, and that the data associated with the fishery system will have both process uncertainty and observation error. This caused him to raise two paradoxes:

(1) management for sustained yield cannot be optimal;
(2) effective management models cannot be realistic.

Each paradox is caused by the interaction of data and models and we still lack complete resolution of those paradoxes (see Mangel *et al.* (2001) for some thoughts about resolution).

Fisherman behavior

Models of the behavior of fishermen are important to know about. I have not included them here because most of what I would write is contained in Chapter 7 of Clark and Mangel (2000). I especially like the work of Abrahams and Healey (1993), Gillis et al. (1995), Gillis (1999, 2003), and Babcock and Pitcher (2000); other nice papers include Healey (1985), Healey and Morris (1992), Holland and Sutinen (1999), Vestergaard (1996) and Vestergaard et al. (2003). One of the most important reasons for understanding behavior, as Gillis and his colleagues and Vestergaard argue, is to get a sense of the nature of discarding, which causes additional and often unreported mortality on stock (Perkins and Edwards 1995, Crowder and Murawski 1998, Harris and Dean 1998, Stratoudakis et al. 1998). Gillis and Peterman (1998) discuss how the behavior of fishing vessels affects the interpretation of CPUE. Anderson (1991a, b) discusses individual transferable quotas. When most broadly interpreted, behavior should also include that of scientists (Starr et al. 1998).

Stock, recruitment, and catchability

As mentioned in the text, there are many other stock–recruitment relationships (a recent review – with both diagnosis and prognosis – is by Needle (2002)). In the alpha-logistic or z-logistic, we find $N(t+1) = N(t) + rN(t)(1 - (N(t)/K)^z)$, where typically $z > 1$ for mammals and $z < 1$ for small fish. As an exercise, you might want to make sketches of the biological production in each of these cases. One can build more complicated density dependence into stock–recruitment relationships (Bjorksted 2000) given the appropriate life history information. Schnute and Kronland (1996) describe a management oriented approach to stock–recruitment relationships. Some of my favorite papers deal with stock–recruitment models for Pacific sardine *Sardinops sagax* (see Jacobson and MacCall (1995), and Jacobson et al. (2001)), and swordfish (Prager 2002). Amazingly, there is actually still argument from some quarters that there is no relationship between spawning stock size and recruitment (e.g. that the main factors driving recruitment are abiotic, such as climate, or non-autotrophic, such as food web interactions) or that the relationship is extremely weak (Marshall et al. 1998); also see Hennemuth et al. (1980), Leggett and Deblois (1994), Rickman et al. (2000), and Chen et al. (2002). Brodziak et al. (2001) give a nice summary of the debate and a very convincing reply to the charges. In recent years, stock–recruitment relationships have been parametrized by the biomass in the absence of fishing (B_0) and the "steepness," defined to be the fraction of the maximum number of recruits when the spawning stock biomass is $0.2B_0$. One generalization of the Beverton–Holt stock recruitment curve is the "hockey-stick:" a piecewise linear relationship that rises linearly until it flattens as a horizontal line (Barrowman and Myers 2000). Another involves "depensation," in which the line $R = S$ intersects the stock–recruitment relationship $R = f(S)$ at more than one point, so that there is an unstable steady state between the origin and a high stable steady state (or even more steady states). Depensation has been proposed as a possible cause for the lack of recovery of the northern cod *Gadus morhua* (Shelton and Healey 1999). The obituary of Ricker (Beamish 2002) is very interesting. Age-structured models are commonly used in fishery management as means of estimating stock abundances and setting management levels; an example is found in Matsuda and Nishimori (2003). A good starting point for more general approaches is the extended survivors analysis described by John Shepherd (Shepherd 1999). Jacobson *et al.* (2002) describe ways to estimate the fishing mortality that generates MSY in any stock assessment model. There is a growing literature applying life history concepts more directly to fishery related problems. Good entry points are Jennings *et al.* (1998), Denney *et al.* (2002), Frisk *et al.* (2001), King and McFarlane (2003). We have assumed that catch per unit effort is proportional to abundance, but there are both theoretical and empirical reasons that it might not be (Harley *et al.* 2001). For example, if the catch is constrained by operational considerations to be a fixed amount (say 10 mt) and we recognize that in a small amount of time the fraction of the stock taken is $1 - e^{-qE\Delta t}$ then it becomes clear that it will appear that q depends upon the total stock size (taking 10 mt from a stock with biomass of 100, 1000, or 1 000 000 mt represents very different fractions). Ray Beverton discusses this issue in depth with excellent examples in his lecture series http://spo.nwr.noaa.gov/ BevertonLectures1994. In his new book (Clark 2006), Colin Clark argues that we would be more conservative to assume that catch were independent of N, so that $C = qE$. An alternative, which to my knowledge has not been investigated, is to think of catch as a functional response so that $C(N) = c_{max}qEN/(qEN + C_0)$. Williams (2002) discusses the effects of unaccounted discard and incorrectly specified natural mortality on estimates of spawners per recruit and on the harvest policies based on spawners per recruit. We have ignored spatial aspects of stock, recruitment and harvesting, but the reaction diffusion models that we discussed in Chapter 2 apply; an excellent starting point is MacCall (1990). Schnute and Richards (2001) have an interesting discussion on the role of models (and the abuse of models) in stock assessment; whether we want to become fishmeticians doing fishmetic is a different question; also see Mangel *et al.* (2001).

Targets, thresholds, and reference points

Perhaps a generation ago, MSY was viewed as a "target for management." We are much wiser than that now (Maunder 2002). Whether or not MSY should be viewed as a target, reference point or limit for management is a topic that can be addressed by quantitative means; some entry points are Thompson (1993), Nakken *et al.* (1996), Schnute and Richards (1998), Overholtz (1999), Bradford *et al.* (2000), Caddy (2002), Hilborn *et al.* (2002), Ulrich and Marchal (2002), Koeller (2003) and Prager *et al.* (2003). A recent issue of the *Bulletin of Marine Science* (**70**(2), 2002) is focussed on targets and thresholds.

Bioeconomics

We have just barely touched on bioeconomics, through our introduction of the discount rate, bionomic equilibrium and Exercises 6.8 and 6.9. The subject is very important. Corkett (2002) argues that bioeconomics is essential for making fishery stock assessment a falsifiable science; the problem of excess capacity for catching fish is perhaps the most significant factor leading to overfishing (Figure 6.16). The classic text in bioeconomics (and indeed, the one that got the field going) is Clark (1990; this is the second edition; the first published in the mid 1970s). Clark (1985) is also superb. As with understanding the specifics of fishery systems, it is also good to understand specific bioeconomic models of fishery systems such as New England groundfish (Overholtz *et al.* 1995), the southern bluefin tuna fishery (McDonald *et al.* 2002), US silver hake fisheries (Helser *et al.* 1996), English channel artisinal fisheries (Ulrich *et al.* 2002a), North Sea flatfish fishery (Ulrich *et al.* 2002b), traditionally managed Fijian fisheries (Jennings and Polunin 1996), the Gulf of Mexico red snapper fishery (Gillig *et al.* 2001), or the role of individual transferable quotas (McGarvey 2003).

The role of Bayesian methods

For even the simplest model of the fishery, we have seen that parameters are confounded (e.g. MSY $= rK/4$) and Bayesian methods provide the natural way for dealing with problems in which parameters are confounded and there is prior information (in fisheries, for example on similar stocks elsewhere). There is an excellent and growing literature

Figure 6.16. The overcapacity ratio, defined as the actual capacity of the fishery divided by the estimated long-run sustainable capacity, of various historical fisheries (data from Clark 1990) and of the USA as a whole in 2002 (D. Fluharty, personal communication).

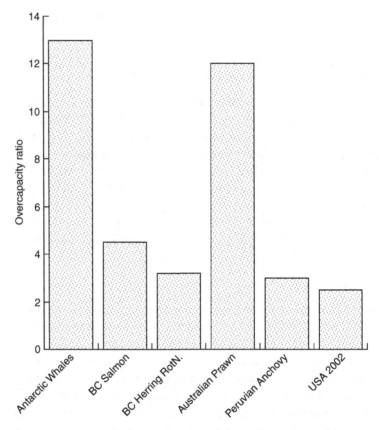

on Bayesian methods in stock assessment and fishery management. Good starting points include Thompson (1992), McAllister *et al.* (1994), Ellison (1996), Kinas (1996), McAllister (1996), McAllister and Ianelli (1997), McAllister and Kirkwood (1998a, b, 1999), Meyer and Millar (1999a, b), Patterson (1999), Wade (2000), Adkison and Zhenming (2001), Hammond and O'Brien (2001), Harley and Myers (2001), McAllister *et al.* (2001), Dorn (2002), Millar (2002), Rivot *et al.* (2001), and Chen *et al.* (2003). Liermann and Hilborn (1997) apply Bayesian methods to the analysis of depensation in the stock–recruitment relationship.

More about salmon

There is so much more that I would like to tell you about salmon that it could be another book. These are remarkable fish because of their diadromy (migration between freshwater and marine environments; see McDowall (1988)) and semelparity (in general, just a single reproductive event, the major exception being Atlantic salmon *Salmo salar* L.). If one extends from the salmon to the salmonids (thus including, for example, trout and charr), iteroparity is much more common. Salmonids are fished both commercially and recreationally. Because of their interest as sport fish, salmon are described in a number of well written books for the lay person; some of my favorites are Ade (1989), Stolz and Schnell (1991), Behnke (1992, 2002), Watson (1993, 1999), and Greer (1995), Although not about salmon, John McPhee's book (McPhee 2002) about shad, which are also diadromous, is a great joy. If you can find Malloch (1994), which is a reprint of a 1909 publication, it is well worth reading. For more academic treatments on Atlantic salmon consult Mills (1989), and on Pacific salmon consult Groot and Margolis (1991), Pearcy (1992) and Groot *et al.* (1995). Some of the most interesting questions involving salmon are those relating to the diversity of life histories and early maturation (usually males) in freshwater before migration to the ocean. Fisher *et al.* (1991) discuss the integration of fishery and water resource management.

Models with process uncertainty and observation noise

Taking into account process uncertainty and observation noise is a difficult task, somewhat beyond the goal of this chapter. Schnute (1993) lays out some of the questions via clear and simple examples, but the methodology of solution rapidly becomes difficult. Entry points for methods and for some applications include Reed and Simons (1996), Patterson (1998), Patterson *et al.* (2001), de Valpine and Hastings (2002), Hinrichsen (2002), Kehler *et al.* (2002), Tang and Wang (2002), Cooper *et al.* (2003), Lindley (2003), and Mesnil (2003). As Sinclair *et al.* (2002a) note, size-selective mortality, density and temperature form a tangled bank when one tries to understand length at age, which as we have discussed is fundamental for age-structured management models. Millar and Meyer (2000) and Schnute and Kronlund (2002) use an explicit Bayesian approach for this problem.

Marine reserves

There is a growing literature on both theoretical (Apollorio 1994, Guenette *et al.* 1998, Horwood *et al.* 1998, Guenette and Pitcher 1999, Pezzey *et al.* 2000, Soh *et al.* 2000, Lindholm *et al.* 2001, Council 2001, Acosta 2002, Apostolaki *et al.* 2002, Brooks 2002, and

Lockwood *et al.* 2002) and empirical (McClanahan and Kauna-Arara 1996, Edgar and Barrett 1999, Jennings 2000, Mosquera *et al.* 2000, Paddack and Estes 2000, Sanchez Lizaso *et al.* 2000, Cote *et al.* 2001, Halpern and Warner 2002, Fanshawe *et al.* 2003, and Shears and Babcock 2003) aspects of marine reserves. As we have discussed, one of the important issues with marine reserves is economic (associated with the suite of questions concerning foregone catch, the effects of reserves on yield and displaced fishing effort). There is less literature on this question (Farrow and Sumaila 2002, Rudd *et al.* 2003) but the topic is important (Sanchirico *et al.* 2003).

Fishing as an agent of selection

We have ignored the evolutionary response of stocks to fishing, but there is a growing literature that fishing acts as a clear agent of selection and that responses can be rapid (Cardinale and Modin 1999, Cardinale and Arrhenius 2000, Law 2000, Hutchings 2000, 2001, Sadovy 2001, and Conover and Munch 2002).

Ecosystem-based approaches to fishery management

We have focussed on single species models because understanding them is essential if one wants to move forward to models based on community or ecosystem concepts, as we surely must (Sherman and Alexander 1989, Sherman *et al.* 1990, 1991, 1993, Richards and Maguire 1998, Murawski 2000, Pitcher 2000, Link *et al.* 2002, Pitcher *et al.* 2002, Sinclair *et al.* 2002b, and Christensen *et al.* 2003). In the future, fishery management will likely take an ecosystem-based approach (Pikitch *et al.* 2004). We are accumulating some theoretical and empirical (Gislason 1994, Daskalov 2002) knowledge about ecosystem effects of fishing and ecosystem approaches to management. Furthermore, management based on an ecosystem approach explicitly recognizes the role of climate in the production of fish (for examples, see Healey (1990), Mullan (1993), Bakun (1996), Klyashtorin (1998), Kuikka *et al.* (1999), Fiksen and Slotte (2002), Swansburg *et al.* (2002), and Williams (2003)).

In the late 1990s, I served on the Ecosystem Advisory Panel which sent to Congress *Ecosystem-based Fishery Management. A Report to Congress by the Ecosystem Advisory Panel* (available, among other places at my website: http://www.soe.ucsc.edu/~msmangel/ and http://www.soe.ucsc.edu/~msmangel/eprints_topical.htm). An executive summary of our conclusions is as follows.

Basic ecosystem characteristics and operating principles

(1) Prediction

- Prediction is limited. Uncertainty and indeterminacy are fundamental characteristics of the dynamics of complex adaptive systems. It is not possible to predict the behaviors of these systems with absolute certainty, regardless of the amount of scientific effort invested. We can, however, find the boundaries of expected behavior and improve our understanding of the underlying dynamics. Thus, ecosystems are not totally predictable, but they are not totally unpredictable either. There are limits to their predictability.

(2) Resilience

- Thresholds are real. Ecosystems are finite and exhaustible. But, they usually have a high buffering capacity and are fairly resilient to stress. Often, as we begin to apply a stress to an ecosystem, its structure and behavior may at first not change noticeably. Only after a critical threshold is passed does the system begin to deteriorate rapidly. Since there is little change initially in behavior with increasing stress, these thresholds are very difficult to predict before they are reached. The nonlinear dynamics which cause this kind of behavior are a basic characteristic of ecosystems.

- Changes can be irreversible. When an ecosystem is radically altered, it may never return to its original condition, even after the stress is removed. This phenomenon (called hysteresis) is common in many complex adaptive systems.

- Diversity is important to ecosystem functioning. The diversity of components at the individual, species, and landscapes scales strongly affects ecosystem behavior. Although the overall productivity of ecosystems may not change significantly when particular species are added or removed, their resilience may be affected.

(3) Space and time linkages

- Multiple scales interact. Ecosystems cannot be understood from the perspective of a single time, space, or complexity scale. At minimum, both the next larger scale and the next lower scale of interest must be considered when effects of perturbations are analyzed.

- Components are strongly linked. The components within ecosystems are linked by flows of materials, energy, and information in complex patterns. The impacts of disrupting these patterns are highly variable and poorly understood.

- Ecosystem boundaries are open. Ecosystems are thermodynamically open, far from equilibrium systems, and cannot be adequately understood without knowledge of their boundary conditions, energy flows, and internal cycling of nutrients and other materials. Environmental variability can alter spatial boundaries and energy inputs to ecosystems.

- Ecosystems change with time. Ecosystems change with time in response to natural and anthropogenic influences. Different components of ecosystems change at different rates and can influence the overall structure of the ecosystem itself. The human component of ecosystems (especially technology and institutions) changes rapidly, far outstripping the capacity for change of other components of the ecosystem.

Social goals

- Apply a precautionary approach. Because predictability is limited and we now live in a world where humans are an important component of almost all ecosystems, it is reasonable to assume that human activities will impact ecosystems at several scales. We should reverse the current burden of proof and presume that adverse impacts will occur, unless and until it can be shown otherwise.
- Purchase "insurance". To guard against uncertain adverse impacts we should purchase "insurance" of various kinds, ranging from the physical insurance of marine protected areas to the financial insurance of environmental bonds.
- Make local goals compatible with global objectives. Changing human behavior is most easily accomplished by changing the local incentives which individuals face to be consistent with broader social goals. The lack of consistency between local incentives and global goals is the root cause of many "social traps," including those in fisheries management. Changing incentives is complex and must be accomplished in culturally appropriate ways. For example, in western market economies, changing market incentives may be appropriate, but this will not generally be the case for other cultures.
- Promote participation, fairness and equity in policy and management. Policies that are developed and implemented with the full participation and consideration of all stakeholders, including the interests of future generations, are more likely to be – and to be perceived as – fair and equitable.
- Treat actions as experiments. Management actions and policies are analogous to experiments and should be based upon hypotheses about the ecosystem response. This requires close monitoring of results to determine to what extent the hypotheses hold.

Risk analysis

Francis and Shotton (1997) offer a general review of the notions of risk in fishery management; Pielke and Conant (2003) another view based on "best-practices"; Hutchings *et al.* (1997) discuss the role of science and government control of information. One must deeply understand uncertainty and its implications, at a variety of levels, to proceed with effective risk analyses. The work of Dovers and colleagues (Dovers and Handmer 1995; Dovers *et al.* 1996) is a good starting point. Other explicit treatments of risk analysis in fisheries include McAllister and Peterman (1992a, b), Punt and Hilborn (1997), MacCall (1998), Robb and Peterman (1998), Schnute *et al.* (2000), van Oostenbrugge *et al.* (2001), Jonzen *et al.* (2002) and Myers *et al.* (2002).

Chapter 7

The basics of stochastic population dynamics

In this and the next chapter, we turn to questions that require the use of all of our tools: differential equations, probability, computation, and a good deal of hard thinking about biological implications of the analysis. Do not be dissuaded: the material is accessible. However, accessing this material requires new kinds of thinking, because funny things happen when we enter the realm of dynamical systems with random components. These are generally called stochastic processes. Time can be measured either discretely or continuously and the state of the system can be measured either continuously or discretely. We will encounter all combinations, but will mainly focus on continuous time models. Much of the groundwork for what we will do was laid by physicists in the twentieth century and adopted in part or wholly by biologists as we moved into the twentyfirst century (see, for example, May (1974), Ludwig (1975), Voronka and Keller (1975), Costantino and Desharnais (1991), Lande *et al.* (2003)). Thus, as you read the text you may begin to think that I have physics envy; I don't, but I do believe that we should acknowledge the source of great ideas. Both in the text and in Connections, I will point towards biological applications, and the next chapter is all about them.

Thinking along sample paths

To begin, we need to learn to think about dynamic biological systems in a different way. The reason is this: when the dynamics are stochastic, even the simplest dynamics can have more than one possible outcome. (This has profound "real world" applications. For example, it means that in a management context, we might do everything right and still not succeed in the goal.)

To illustrate this point, let us reconsider exponential population growth in discrete time:

$$X(t+1) = (1+\lambda)X(t) \tag{7.1}$$

which we know has the solution $X(t) = (1+\lambda)^t X(0)$. Now suppose that we wanted to make these dynamics stochastic. One possibility would be to assume that at each time the new population size is determined by the deterministic component given in Eq. (7.1) and a random, stochastic term $Z(t)$ representing elements of the population that come from "somewhere else." Instead of Eq. (7.1), we would write

$$X(t+1) = (1+\lambda)X(t) + Z(t) \tag{7.2}$$

In order to iterate this equation forward in time, we need assumptions about the properties of $Z(t)$. One assumption is that $Z(t)$, the process uncertainty, is normally distributed with mean 0 and variance σ^2. In that case, there are an infinite number of possibilities for the sequence $\{Z(0), Z(1), Z(2), \ldots\}$ and in order to understand the dynamics we should investigate the properties of a variety of the trajectories, or sample paths, that this equation generates. In Figure 7.1, I show ten such trajectories and the deterministic trajectory.

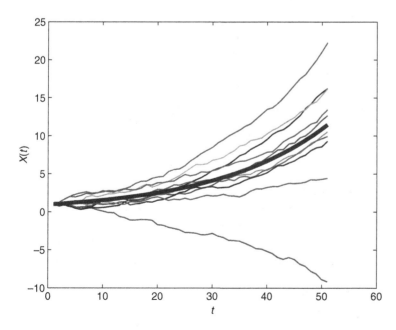

Figure 7.1. Ten trajectories (thin lines) and the deterministic trajectory (thick line) generated by Eq. (7.2) for $X(1) = 1$, $\lambda = 0.05$ and $\sigma = 0.2$.

Note that in this particular case, the deterministic trajectory is predicted to be the same as the average of the stochastic trajectories. If we take the expectation of Eq. (7.2), we have

$$E\{X(t+1)\} = E\{(1+\lambda)X(t)\} + E\{Z(t)\} = (1+\lambda)E\{X(t)\} \qquad (7.3)$$

which is the same as Eq. (7.1), so that the deterministic dynamics characterize what the population does "on average." This identification of the average of the stochastic trajectories with the deterministic trajectory only holds, however, because the underlying dynamics are linear. Were they nonlinear, so that instead of $(1+\lambda)X(t)$, we had a term $g(X(t))$ on the right hand side of Eq. (7.2), then the averaging as in Eq. (7.3) would not work, since in general $E\{g(X)\} \neq g(E\{X\})$.

The deterministic trajectory shown in Figure 7.1 accords with our experience with exponential growth. Since the growth parameter is small, the trajectory grows exponentially in time, but at a slow rate. How about the stochastic trajectories? Well, some of them are close to the deterministic one, but others deviate considerably from the deterministic one, in both directions. Note that the largest value of $X(t)$ in the simulated trajectories is about 23 and that the smallest value is about -10. If this were a model of a population, for example, we might say that the population is extinct if it falls below zero, in which case one of the ten trajectories leads to extinction. Note that the trajectories are just a little bit bumpy, because of the relatively small value of the variance (try this out for yourself by simulating your own version of Eq. (7.2) with different choices of λ and σ^2).

The transition from Eq. (7.1) to Eq. (7.2), in which we made the dynamics stochastic rather than deterministic, is a key piece of the art of modeling. We might have done it in a different manner. For example, suppose that we assume that the growth rate is composed of a deterministic term and a random term, so that we write $X(t+1) = (1+\lambda(t))X(t)$, where $\lambda(t) = \bar{\lambda} + Z(t)$, and understand $\bar{\lambda}$ to be the mean growth rate and $Z(t)$ to be the perturbation in time of that growth rate. Now, instead of Eq. (7.2), our stochastic dynamics will be

$$X(t+1) = (1+\bar{\lambda})X(t) + Z(t)X(t) \qquad (7.4)$$

Note the difference between Eq. (7.4) and Eq. (7.2). In Eq. (7.4), the stochastic perturbation is proportional to population size. This slight modification, however, changes in a qualitative nature the sample paths (Figure 7.2). We can now have very large changes in the trajectory, because the stochastic component, $Z(t)$, is amplified by the current value of the state, $X(t)$.

Which is the "right" way to convert from deterministic to stochastic dynamics – Eq. (7.2) or Eq. (7.4)? The answer is "it depends." It depends

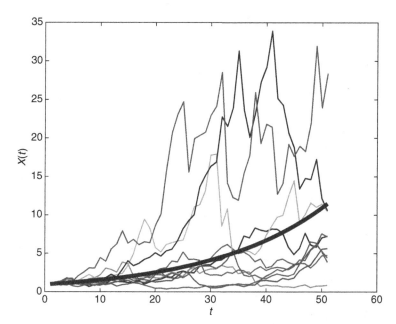

Figure 7.2. Ten trajectories and the deterministic trajectory generated by Eq. (7.4) for the same parameters as Figure 7.1.

upon your understanding of the biology and on how the random factors enter into the biological dynamics. That is, this is a question of the art of modeling, at which you are becoming more expert, and which (the development of models) is a life-long pursuit. We will put this question aside mostly, until the next chapter when it returns with a vengeance, when new tools obtained in this chapter are used.

Brownian motion

In 1828 (Brown 1828), Robert Brown, a Scottish botanist, observed that a grain of pollen in water dispersed into a number of much smaller particles, each of which moved continuously and randomly (as if with a "vital force"). This motion is now called Brownian motion; it was investigated by a variety of scientists between 1828 and 1905, when Einstein – in his miraculous year – published an explanation of Brownian motion (Einstein 1956), using the atomic theory of matter as a guide. It is perhaps hard for us to believe today but, at the turn of the last century, the atomic theory of matter was still just that – considered to be an unproven theory. Fuerth (1956) gives a history of the study of Brownian motion between its report and Einstein's publication. Beginning in the 1930s, pure mathematicians got hold of the subject, and took it away from its biological and physical origins; they tend to call Brownian motion a Wiener process, after the brilliant Norbert Wiener who began to mathematize the subject.

Figure 7.3. A set of four times $\{t_1, t_2, t_3, t_4\}$ with non-overlapping intervals. A key assumption of the process of Brownian motion is that $W(t_2) - W(t_1)$ and $W(t_4) - W(t_3)$ are independent random variables, no matter how close t_3 is to t_2.

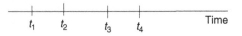

In compromise, we will use $W(t)$ to denote "standard Brownian motion," which is defined by the following four conditions:

(1) $W(0) = 0$;

(2) $W(t)$ is continuous;

(3) $W(t)$ is normally distributed with mean 0 and variance t;

(4) if $\{t_1, t_2, t_3, t_4\}$ represent four different, ordered times with $t_1 < t_2 < t_3 < t_4$ (Figure 7.3), then $W(t_2) - W(t_1)$ and $W(t_4) - W(t_3)$ are independent random variables, no matter how close t_3 is to t_2. The last property is said to be the property of independent increments (see Connections for more details) and is a key assumption.

In Figure 7.4, I show five sample trajectories, which in the business are described as "realizations of the stochastic process." They all start at 0 because of property (1). The trajectories are continuous, forced by property (2). Notice, however, that although the trajectories are continuous, they are very wiggly (we will come back to that momentarily).

For much of what follows, we will work with the "increment of Brownian motion" (we are going to convert regular differential equations of the sort that we encountered in previous chapters into stochastic differential equations using this increment), which is defined as

$$dW = W(t + dt) - W(t) \qquad (7.5)$$

Exercise 7.1 (M)

By applying properties (1)–(4) to the increment of Brownian motion, show that:

(1) $E\{dW\} = 0$;

(2) $E\{dW^2\} = dt$;

(3) dW is normally distributed;

(4) if $dW_1 = W(t_1 + dt) - W(t_1)$ and $dW_2 = W(t_2 + dt) - W(t_2)$ where $t_2 > t_1 + dt$ then dW_1 and dW_2 are independent random variables (for this last part, you might want to peek at Eqs. (7.29) and (7.30)).

Now, although Brownian motion and its increment seem very natural to us (perhaps because have spent so much time working with normal random variables), a variety of surprising and non-intuitive results emerge. To begin, let's ask about the derivative dW/dt. Since $W(t)$ is a random variable, its derivative will be one too. Using the definition of the derivative

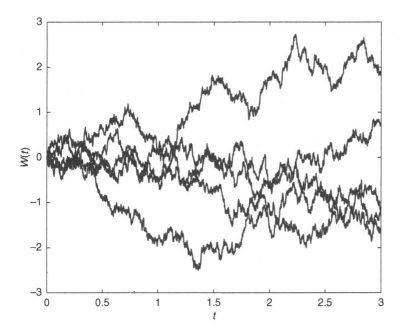

Figure 7.4. Five realizations of standard Brownian motion.

$$\frac{dW}{dt} = \lim_{dt \to 0} \frac{W(t + dt) - W(t)}{dt} \tag{7.6}$$

so that

$$E\left\{\frac{dW}{dt}\right\} = \lim_{dt \to 0} E\left\{\frac{W(t + dt) - W(t)}{dt}\right\} = 0 \tag{7.7}$$

and we conclude that the average value of dW/dt is 0. But look what happens with the variance:

$$E\left\{\left(\frac{dW}{dt}\right)^2\right\} = \lim_{dt \to 0} E\left\{\frac{(W(t + dt) - W(t))^2}{dt^2}\right\} = \lim_{dt \to 0} \frac{dt}{dt^2} \tag{7.8}$$

but we had better stop right here, because we know what is going to happen with the limit – it does not exist. In other words, although the sample paths of Brownian motion are continuous, they are not differentiable, at least in the sense that the variance of the derivative exists. Later in this chapter, in the section on white noise (see p. 261), we will make sense of the derivative of Brownian motion. For now, I want to introduce one more strange property associated with Brownian motion and then spend some time using it.

Suppose that we have a function $f(t,W)$ which is known and well understood and can be differentiated to our hearts' content and for which we want to find $f(t + dt, w + dW)$ when dt (and thus $E\{dW^2\}$)

is small and t and $W(t) = w$ are specified. We Taylor expand in the usual manner, using a subscript to denote a derivative

$$
\begin{aligned}
f(t + dt, \ w + dW) = &f(t, w) + f_t dt + f_w dW \\
&+ \frac{1}{2} \left\{ f_{tt} \, dt^2 + 2f_{tw} dt dW + f_{ww} \, dW^2 \right\} + o(dt^2) \\
&+ o(dt dW) + o(dW^2)
\end{aligned}
\tag{7.9}
$$

and now we ask "what are the terms that are order dt on the right hand side of this expression?" Once again, this can only make sense in terms of an expectation, since $f(t + dt, w + dW)$ will be a random variable. So let us take the expectation and use the properties of the increment of Brownian motion

$$
E\{f(t + dt, \ w + dW)\} = f(t, w) + f_t dt + \frac{1}{2} f_{ww} dt + o(dt)
\tag{7.10}
$$

so that the particular property of Brownian motion that $E\{dW^2\} = dt$ translates into a Taylor expansion in which first derivatives with respect to dt and first and second derivatives with respect to dW are the same order of dt. This is an example of Ito calculus, due to the mathematician K. Ito; see Connections for more details. We will now explore the implications of this observation.

The gambler's ruin in a fair game

Many – perhaps all – books on stochastic processes or probability include a section on gambling because, let's face it, what is the point of studying probability and stochastic processes if you can't become a better gambler (see also Dubins and Savage (1976))? The gambling problem also allows us to introduce some ideas that will flow through the rest of this chapter and the next chapter.

Imagine that you are playing a fair game in a casino (we will discuss real casinos, which always have the edge, in the next section) and that your current holdings are $X(t)$ dollars. You are out of the game when $X(t)$ falls to 0 and you break the bank when your holdings $X(t)$ reach the casino holdings C. If you think that this is a purely mathematical problem and are impatient for biology, make the following analogy: $X(t)$ is the size at time t of the population descended from a propagule of size x that reached an island at time $t = 0$; $X(t) = 0$ corresponds to extinction of the population and $X(t) = C$ corresponds to successful colonization of the island by the descendants of the propagule. With this interpretation, we have one of the models for island biogeography of MacArthur and Wilson (1967), which will be discussed in the next chapter.

Since the game is fair, we may assume that the change in your holdings are determined by a standard Brownian motion; that is, your holdings at time t and time $t + dt$ are related by

$$X(t + dt) = X(t) + dW \tag{7.11}$$

There are many questions that we could ask about your game, but I want to focus here on a single question: given your initial stake $X(0) = x$, what is the chance that you break the casino before you go broke? One way to answer this question would be through simulation of trajectories satisfying Eq. (7.11). We would then follow the trajectories until $X(t)$ crosses 0 or crosses C and the probability of breaking the casino would be the fraction of trajectories that cross C before they cross 0. The trajectories that we simulate would look like those in Figure 7.4 with a starting value of x rather than 0. This method, while effective, would be hard pressed to give us general intuition and might require considerable computer time in order for us to obtain accurate answers. So, we will seek another method by thinking along sample paths.

In Figure 7.5, I show the $t - x$ plane and the initial value of your holdings $X(0) = x$. At at time dt later, your holdings will change to $x + dW$, where dW is normally distributed with mean 0 and variance dt. Suppose that, as in the figure, they have changed to $x + w$, where we can calculate the probability of dW falling around w from the normal distribution. What happens when you start at this new value of holdings? Either you break the bank or you go broke; that is, things start over exactly as before except with a new level of holdings. But what happens between 0 and dt and after dt are independent of each other because of the properties of Brownian motion. Thus, whatever happens after dt is determined solely by your holdings at dt. And those holdings are normally distributed.

To be more formal about this, let us set

$$u(x) = \Pr\{X(t) \text{ hits } C \text{ before it hits } 0 | X(0) = x\} \tag{7.12}$$

(which could also be recognized as a colonization probability, using the metaphor of island biogeography) and recognize that the argument of the previous paragraph can be summarized as

$$u(x) = E_{dW}\{u(x + dW)\} \tag{7.13}$$

where E_{dW} means to average over dW. Now let us Taylor expand the right hand side of Eq. (7.13) around x:

$$u(x) = E_{dW}\left\{u(x) + dW u_x + \frac{1}{2}(dW)^2 u_{xx} + o((dW)^2)\right\} \tag{7.14a}$$

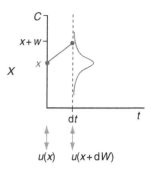

Figure 7.5. To compute the probability $u(x)$ that $X(t)$ crosses C before 0, given $X(0) = x$ we recognize that, in the first dt of the game, holdings will change from x to $x + w$, where w has a normal distribution with mean 0 and variance dt. We can thus relate $u(x)$ at this time to the average of $u(x + dW)$ at a slightly later time (later by dt).

and take the average over dW, remembering that it is normally distributed with mean 0 and variance dt:

$$u(x) = u(x) + \frac{1}{2}u_{xx}dt + o(dt) \qquad (7.14b)$$

The last two equations share the same number because I want to emphasize their equivalence. To finish the derivation, we subtract $u(x)$ from both sides, divide by dt and let $dt \rightarrow 0$ to obtain the especially simple differential equation

$$u_{xx} = 0 \qquad (7.15)$$

which we now solve by inspection. The second derivative is 0, so the first derivative of $u(x)$ is a constant $u_x = k_1$ and thus $u(x)$ is a linear function of x

$$u(x) = k_2 + k_1 x \qquad (7.16)$$

We will find these constants of integration by thinking about the boundary conditions that $u(x)$ must satisfy.

From Eq. (7.12), we conclude that $u(0)$ must be 0 and $u(C)$ must be 1 since if you start with $x = 0$ you have hit 0 before C and if you start with C you have hit C before 0. Since $u(0) = 0$, from Eq. (7.16) we conclude that $k_2 = 0$ and to make $u(C) = 1$ we must have $k_1 = 1/C$ so that $u(x)$ is

$$u(x) = \frac{x}{C} \qquad (7.17)$$

What is the typical relationship between your initial holdings and those of a casino? In general $C \gg x$, so that $u(x) \sim 0$ – you are almost always guaranteed to go broke before hitting the casino limit.

But, of course, most of us gamble not to break the bank, but to have some fun (and perhaps win a little bit). So we might ask how long it will be before the game ends (i.e., your holdings are either 0 or C). To answer this question, set

$$T(x) = \text{average amount of time in the game, given } X(0) = x \qquad (7.18)$$

We derive an equation for $T(x)$ using logic similar to that which took us to Eq. (7.15). Starting at $X(0) = x$, after dt the holdings will be $x + dW$ and you will have been in the game for dt time units. Thus we conclude

$$T(x) = dt + E_{dW}\{T(x + dW)\} \qquad (7.19)$$

and we would now proceed as before, Taylor expanding, averaging, dividing by dt and letting dt approach 0. This question is better left as an exercise.

Exercise 7.2 (M)

Show that $T(x)$ satisfies the equation $-1 = (1/2)\, T_{xx}$ and that the general solution of this equation is $T(x) = -x^2 + k_1 x + k_2$. Then explain why the boundary conditions for the equation are $T(0) = T(C) = 0$ and use them to evaluate the two constants. Plot and interpret the final result for $T(x)$.

The gambler's ruin in a biased game

Most casinos have a slight edge on the gamblers playing there. This means that on average your holdings will decrease (the casino's edge) at rate m, as well as change due to the random fluctuations of the game. To capture this idea, we replace Eq. (7.11) by

$$dX = X(t + dt) - X(t) = -m\,dt + dW \qquad (7.20)$$

Exercise 7.3 (E/M)

Show that dX is normally distributed with mean $-m\,dt$ and variance $dt + o(dt)$ by evaluating $E\{dX\}$ and $E\{dX^2\}$ using Eq. (7.20) and the results of Exercise 7.1.

As before, we compute $u(x)$, the probability that $X(t)$ hits C before 0, but now we recognize that the average must be over dX rather than dW, since the holdings change from x to $x + dX$ due to deterministic ($m\,dt$) and stochastic (dW) factors. The analog of Eq. (7.13) is then

$$u(x) = E_{dX}\{u(x + dX)\} = E_{dX}\{u(x - m\,dt + dW)\} \qquad (7.21)$$

We now Taylor expand and combine higher powers of dt and dW into a term that is $o(dt)$

$$u(x) = E_{dX}\left\{ u(x) + (-m\,dt + dW)u_x + \frac{1}{2}(-m\,dt + dW)^2 u_{xx} + o(dt) \right\} \qquad (7.22)$$

We expand the squared term, recognizing that $O(dW^2)$ will be order dt, take the average over dX, divide by dt and let $dt \to 0$ (you should write out all of these steps if any one of them is not clear to you) to obtain

$$\frac{1}{2}u_{xx} - m u_x = 0 \qquad (7.23)$$

which we need to solve with the same boundary conditions as before $u(0) = 0$, $u(C) = 1$. There are at least two ways of solving Eq. (7.23). I will demonstrate one; the other uses the same method that we used in Chapter 2 to deal with the von Bertalanffy equation for growth.

Let us set $w = u_x$, so that Eq. (7.23) can be rewritten as $w_x = 2mw$, for which we immediately recognize the solution $w(x) = k_1 e^{2mx}$, where k_1 is a constant. Since $w(x)$ is the derivative of $u(x)$ we integrate again to obtain

$$u(x) = k_2 e^{2mx} + k_3 \tag{7.24}$$

where k_2 and k_3 are constants and, to be certain that we are on the same page, try the next exercise.

Exercise 7.4 (E)

What is the relationship between k_1 and k_2?

When we apply the boundary condition that $u(0) = 0$, we conclude that $k_3 = -k_2$, and when we apply the boundary condition $u(C) = 1$, we conclude that $k_2 = 1/(e^{2mC} - 1)$. We thus have the solution for the probability of reaching the limit of the casino in a biased game:

$$u(x) = \frac{e^{2mx} - 1}{e^{2mC} - 1} \tag{7.25}$$

and now things are very bleak: the chance that you win is, for almost all situations, vanishingly small (Figure 7.6).

Once again, we can ask about how long you can stay in the game and, possibly, about connections between the biased and fair gambles. I leave both of these as exercises.

Figure 7.6. When the game is biased, the chance of reaching the limit of the casino before going broke is vanishingly small. Here I show $u(x)$ given by Eq. (7.25) for $m = 0.1$ and $C = 100$. Note that if you start with even 90% of the casino limit, the situation is not very good. Most of us would start with $x \ll C$ and should thus just enjoy the game (or develop a system to reduce the value of m, or even change its sign.)

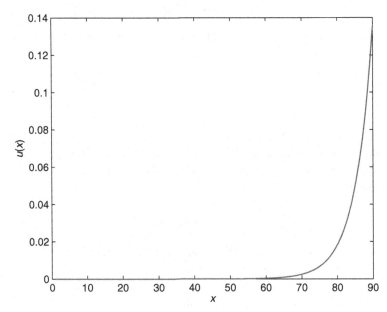

Exercise 7.5 (M/H)

Derive the equation for $T(x)$, the mean time that you are in the game when dX is given by Eq. (7.20). Solve this equation for the boundary conditions $T(0) = T(C) = 0$.

Exercise 7.6 (E/M)

When m is very small, we expect that the solution of Eq. (7.25) should be close to Eq. (7.17) because then the biased game is almost like a fair one. Show that this is indeed the case by Taylor expansion of the exponentials in Eq. (7.25) for $m \to 0$ and show that you obtain our previous result. If you have more energy after this, do the same for the solutions of $T(x)$ from Exercises 7.5 and 7.2.

Before moving on, let us do one additional piece of analysis. In general, we expect the casino limit C to be very large, so that $2mC \gg 1$. Dividing numerator and denominator of Eq. (7.25) by e^{2mC} gives

$$u(x) = \frac{e^{-2m(C-x)} - e^{-2mC}}{1 - e^{-2mC}} \approx e^{-2m(C-x)} \qquad (7.26)$$

with the last approximation coming by assuming that $e^{-2mC} \ll 1$. Now let us take the logarithm to the base 10 of this approximation to $u(x)$, so that $\log_{10}(u(x)) = -2m(C - x)\log_{10}e$. I have plotted this function in Figure 7.7, for $x = 10$ and $C = 50$, 500, or 1000. Now, $C = 1000, x = 10$, and $m = 0.01$ probably under-represents the relationship of the bank of a casino to most of us, but note that, even in this case, the chance of reaching the casino limit before going broke when $m = 0.01$ is about 1 in

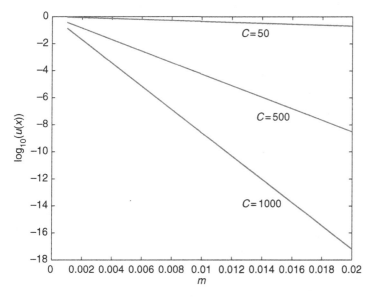

Figure 7.7. The base 10 logarithm of the approximation of $u(x)$, based on Eq. (7.26) for $x = 10$ and $C = 50$, 500, or 1000, as a function of m.

a billion. So go to Vegas, but go for a good time. (In spring 1981, my first year at UC Davis, I went to a regional meeting of the American Mathematical Society, held in Reno, Nevada, to speak in a session on applied stochastic processes. Many famous colleagues were they, and although our session was Friday, they had been there since Tuesday doing, you guessed it, true work in applied probability. All, of course, claimed positive gains in their holdings.)

The transition density and covariance of Brownian motion

We now return to standard Brownian motion, to learn a little bit more about it. To do this, consider the interval $[0, t]$ and some intermediate time s (Figure 7.8). Suppose we know that $W(s) = y$, for $s < t$. What can be said about $W(t)$? The increment $W(t) - W(s) = W(t) - y$ will be normally distributed with mean 0 and variance $t - s$. Thus we conclude that

$$\Pr\{a \leq W(t) \leq b\} = \frac{1}{\sqrt{2\pi(t-s)}} \int_a^b \exp\left(-\frac{(x-y)^2}{2(t-s)}\right) dx \qquad (7.27)$$

Note too that we can make this prediction knowing only $W(s)$, and not having to know anything about the history between 0 and s. A stochastic process for which the future depends only upon the current value and not upon the past that led to the current value is called a Markov process, so that we now know that Brownian motion is a Markov process.

The integrand in Eq. (7.27) is an example of a transition density function, which tells us how the process moves from one time and value to another. It depends upon four values: $s, y, t,$ and x, and we shall write it as

$$q(x,t,y,s)dx = \Pr\{x \leq W(t) \leq x + dx | W(s) = y\}$$
$$= \frac{1}{\sqrt{2\pi(t-s)}} \exp\left(-\frac{(x-y)^2}{2(t-s)}\right) dx \qquad (7.28)$$

This equation should remind you of the diffusion equation encountered in Chapter 2, and the discussion that we had there about the strange properties of the right hand side as t decreases to s. In the next section all of this will be clarified. But before that, a small exercise.

Figure 7.8. The time s divides the interval 0 to t into two pieces, one from 0 to just before s (s_-) and one from just after s (s_+) to t. The increments in Brownian motion are then independent random variables.

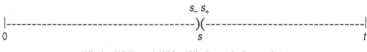

$$s_- \ s_+$$
$$|\text{-----------------------}\rangle(\text{-----------------------}|$$
$$0 \qquad\qquad\qquad\qquad s \qquad\qquad\qquad\qquad t$$

$W(s_-) - W(0)$ and $W(t) - W(s_+)$ are independent random variables

Exercise 7.7 (E/M)

Show that $q(x, t, y, s)$ satisfies the differential equation $q_t = (1/2)q_{xx}$. What equation does $q(x, t, y, s)$ satisfy in the variables s and y (think about the relationship between q_t and q_s and q_{xx} and q_{yy} before you start computing)?

Keeping with the ordering of time in Figure 7.8, let us compute the covariance of $W(t)$ and $W(s)$:

$$\begin{aligned}
E\{W(t)W(s)\} &= E\{(W(t) - W(s))W(s)\} + E(\{W(s)^2\}) \\
&= E\{(W(t) - W(s))(W(s) - 0)\} + s \qquad (7.29) \\
&= s
\end{aligned}$$

where the last line of Eq. (7.29) follows because $W(s) - W(0)$ and $W(t) - W(s)$ are independent random variables, with mean 0. Suppose that we had interchanged the order of t and s. Our conclusion would then be that $E\{W(t)W(s)\} = t$. In other words

$$E\{W(t)W(s)\} = \min(t, s) \qquad (7.30)$$

and we are now ready to think about the derivative of Brownian motion.

Gaussian "white" noise

The derivative of Brownian motion, which we shall denote by $\xi(t) = dW/dt$, is often called Gaussian white noise. It should already be clear where Gaussian comes from; the origin of white will be understood at the end of this section, and the use of noise comes from engineers, who see fluctuations as noise, not as the element of variation that may lead to selection; Jaynes (2003) has a particularly nice discussion of this point. We have already shown the $E\{\xi(t)\} = 0$ and that problems arise when we try to compute $E\{\xi(t)^2\}$ in the usual way because of the variance of Brownian motion (recall the discussion around Eq. (7.8)). So, we are going to sneak up on this derivative by computing the covariance

$$E\{\xi(t)\xi(s)\} = \frac{\partial}{\partial t}\frac{\partial}{\partial s} E\{W(t)W(s)\} \qquad (7.31)$$

Note that I have exchanged the order of differentiation and integration in Eq. (7.31); we will do this once more in this chapter. In general, one needs to be careful about doing such exchanges; both are okay here (if you want to know more about this question, consult a good book on advanced analysis). We know that $E\{W(t)W(s)\} = \min(t, s)$. Let us think about this covariance as a function of t, when s is held fixed, as if it were just a parameter (Figure 7.9)

$$\chi(t, s) = \min(t, s) = \begin{array}{l} t \text{ if } t < s \\ s \text{ if } t \geq s \end{array} \qquad (7.32)$$

Figure 7.9. (a) The covariance function $\chi(t,s) = E\{W(t)W(s)\} = \min(t,s)$, thought of as a function of t with s as a parameter. (b) The derivative of the covariance function is either 1 or 0 with a discontinuity at $t = s$. (c) We approximate the derivative by a smooth function $\chi_n(t,s)$, which in the limit has the discontinuity. (d) The approximate derivative is the tail of the cumulative Gaussian from $t = s$.

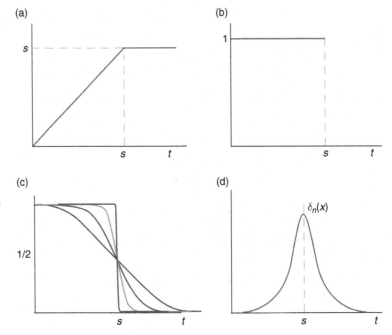

Now the derivative of this function will be discontinuous; since the derivative is 1 if $t < s$, and is 0 if $t > s$, there is a jump at $t = s$ (Figure 7.9). We are going to deal with this problem by using the approach of generalized functions described in Chapter 2 (and in the course of this, learn more about Gaussians).

We will replace the derivative $(\partial/\partial t)\chi(t,s)$ by an approximation that is smooth but in the limit has the discontinuity. Define a family of functions

$$\chi_n(t,s) = \frac{\sqrt{n}}{\sqrt{2\pi}} \int\limits_{(t-s)}^{\infty} \exp\left(-\frac{nx^2}{2}\right) dx$$

which we recognize as the tail of the cumulative distribution function for the Gaussian with mean 0 and variance $1/n$. That is, the density is

$$\delta_n(x) = \frac{\sqrt{n}}{\sqrt{2\pi}} \exp\left(-\frac{nx^2}{2}\right)$$

We then set

$$\frac{\partial}{\partial t}\chi(t,s) = \lim_{n\to\infty} \frac{\sqrt{n}}{\sqrt{2\pi}} \int\limits_{(t-s)}^{\infty} \exp\left(-\frac{nx^2}{2}\right) dx = \lim_{n\to\infty} \frac{\partial}{\partial t}\chi_n(t,s) \quad (7.33)$$

When $t = s$, the lower limit of the integral is 0, so that the integral is 1/2. To understand what happens when t does not equal s, the following exercise is useful.

Exercise 7.8 (E)

Make the transformation $y = x\sqrt{n}$ so that the integral in Eq. (7.33) is the same as

$$\frac{1}{\sqrt{2\pi}} \int_{\sqrt{n}(t-s)}^{\infty} \exp\left(-\frac{y^2}{2}\right) dy \qquad (7.34)$$

The form of the integral in expression (7.34) lets us understand what will happen when $t \neq s$. If $t < s$, the lower limit is negative, so that as $n \to \infty$ the integral will approach 1. If $t > s$, the lower limit is positive so that as n increases the integral will approach 0. We have thus constructed an approximation to the derivative of the correlation function.

Equation (7.31) tells us what we need to do next. We have constructed an approximation to $(\partial/\partial t)\chi(t, s)$, and so to find the covariance of Gaussian white noise, we now need to differentiate Eq. (7.33) with respect to s. Remembering how to take the derivative of an integral with respect to one of its arguments, we have

$$\frac{\partial}{\partial t}\frac{\partial}{\partial s}\chi(t,s) = \lim_{n \to \infty}\frac{\sqrt{n}}{\sqrt{2\pi}}\exp\left(-\frac{n(t-s)^2}{2}\right) = \lim_{n \to \infty}\delta_n(t-s) \quad (7.35)$$

Now, $\delta_n(t-s)$ is a Gaussian distribution centered not at 0 but at $t = s$ with variance $1/n$. Its integral, over all values of t, is 1 but in the limit that $n \to \infty$ it is 0 everywhere except at $t = s$, where it is infinite. In other words, the limit of $\delta_n(t-s)$ is the Dirac delta function that we first encountered in Chapter 2 (some $\delta_n(x)$ are shown in Figure 7.10).

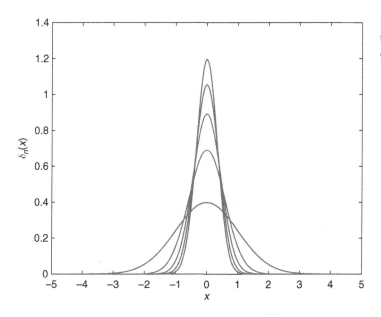

Figure 7.10. The generalized functions $\delta_n(x)$ for $n = 1, 3, 5, 7,$ and 9.

Figure 7.11. The spectrum of
the covariance function given
by Eq. (7.36) is completely
flat so that all frequencies are
equally represented. Hence
the spectrum is "white." In the
natural world, however, the
higher frequencies are less
represented, leading to a
fall-off of the spectrum.

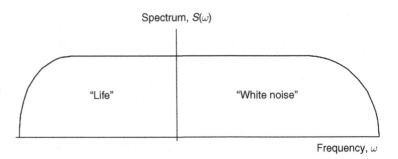

This has been a tough slog, but worth it, because we have shown that

$$E\{\xi(t)\xi(s)\} = \delta(t - s) \tag{7.36}$$

We are now in a position to understand the use of the word "white" in the description of this process. Historically, engineers have worked interchangeably between time and frequency domains (Kailath 1980) because in the frequency domain tools other the ones that we consider are useful, especially for linear systems (which most biological systems are not). The connection between the time and frequency (Stratonovich 1963) is the spectrum $S(\omega)$ defined for the function $f(t)$ by

$$S(\omega) = \int e^{-i\omega t} f(t) dt \tag{7.37}$$

where the integral extends over the entire time domain of $f(t)$. In our case then, we set $s = 0$ for simplicity, since Eq. (7.36) depends only on $t - s$; the spectrum of the covariance function given by Eq. (7.36) is then

$$S(\omega) = \int e^{-i\omega t} \delta(t) dt = 1 \tag{7.38}$$

where the last equality follows because the delta function picks out $t = 0$, for which the exponential is 1. The spectrum of Eq. (7.36) is thus flat (Figure 7.11): all frequencies are equally represented in it. Well, that is the description of white light and this is the reason that we call the derivative of Brownian motion white noise. In the natural world, the covariance does not drop off instantaneously and we obtain spectra with color (see Connections).

The Ornstein–Uhlenbeck process and stochastic integrals

In our analyses thus far, the dynamics of the stochastic process have been independent of the state, depending only upon Brownian motion. We will now begin to move beyond that limitation, but do it appropriately slowly. To begin, recall that if $X(t)$ satisfies the dynamics

$dX/dt = f(X)$ and K is a stable steady state of this system, so that $f(K) = 0$, and we consider the behavior of deviations from the steady state $Y(t) = X(t) - K$ then, to first order, $Y(t)$ satisfies the linear dynamics $dY/dt = -|f'(K)|Y$, where $f'(K)$ is the derivative of $f(X)$ evaluated at K. We can then define a relaxation parameter $\beta = |f'(K)|$ so that the dynamics of Y are given by

$$\frac{dY}{dt} = -\beta Y \tag{7.39}$$

We call β the relaxation parameter because it measures the rate at which fluctuations from the steady state return (relax) towards 0. Sometimes this parameter is called the dissipation parameter.

Exercise 7.9 (E)

What is the relaxation parameter if $f(X)$ is the logistic $rX(1 - (X/K))$? If you have the time, find Levins (1966) and see what he has to say about your result.

We fully understand the dynamics of Eq. (7.39): it represents return of deviations to the steady state: which ever way the deviation starts (above or below K), it becomes smaller. However, now let us ask what happens if in addition to this deterministic attraction back to the steady state, there is stochastic fluctuation. That is, we imagine that in the next little bit of time, the deviation from the steady state declines because of the attraction back towards the steady state but at the same time is perturbed by factors independent of this decline. Bjørnstadt and Grenfell (2001) call this process "noisy clockwork;" Stenseth *et al.* (1999) apply the ideas we now develop to cod, and Dennis and Otten (2000) apply them to kit fox.

We formulate the dynamics in terms of the increment of Brownian motion, rather than white noise, by recognizing that in the limit $dt \to 0$, Eq. (7.39) is the same as $dY = -\beta Y\, dt + o(dt)$ and so our stochastic version will become

$$dY = -\beta Y\, dt + \sigma dW \tag{7.40}$$

where σ is allowed to scale the intensity of the fluctuations. The stochastic process generated by Eq. (7.40) is called the Ornstein–Uhlenbeck process (see Connections) and contains both deterministic relaxation and stochastic fluctuations (Figure 7.12). Our goal is to now characterize the mixture of relaxation and fluctuation.

To do so, we write Eq. (7.40) as a differential by using the integrating factor $e^{\beta t}$ so that

$$d(e^{\beta t}Y) = \sigma e^{\beta t}dW \tag{7.41}$$

Figure 7.12. Five trajectories of the Ornstein–Uhlenbeck process, simulated for $\beta = 0.1$, $dt = 0.01$, $q = 0.1$, and $Y(0)$, uniformly distributed between -0.01 and 0.01. We see both the relaxation (or dissipation) towards the steady state $Y = 0$ and fluctuations around the trajectory and the steady state.

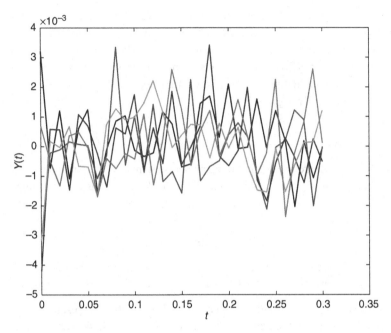

and we now integrate from 0 to t:

$$e^{\beta t} Y(t) - Y(0) = \int_0^t \sigma e^{\beta s} dW(s) \tag{7.42}$$

We have created a new kind of stochastic entity, an integral involving the increment of Brownian motion. Before we can understand the Ornstein–Uhlenbeck process, we need to understand that stochastic integral, so let us set

$$G(t) = \int_0^t \sigma e^{\beta s} dW(s) \tag{7.43}$$

Some properties of $G(t)$ come to us for free: it is normally distributed and the mean $E\{G(t)\} = 0$. But what about the variance? In order to compute the variance of $G(t)$, let us divide the interval $[0, t]$ into pieces by picking a large number N and setting

$$t_j = \frac{t}{N} j \qquad dW_j = W(t_{j+1}) - W(t_j) \qquad j = 0, \ldots N \tag{7.44}$$

so that we can approximate $G(t)$ by a summation

$$G(t) = \lim_{N \to \infty} \sum_{j=0}^{N} \sigma e^{\beta t_j} dW_j \tag{7.45}$$

We now square $G(t)$ and take its expectation, remembering that $E\{dW_i dW_j\} = 0$ if $i \neq j$ and equals dt if $i = j$, so that all the cross terms vanish when we take the expectation, and we see that

$$\text{Var}\{G(t)\} = \lim_{N \to \infty} \sum_{j=0}^{N} \sigma^2 e^{2\beta t_j}\, dt_j = \int_0^t \sigma^2 e^{2\beta s} ds = \sigma^2 \left\{ \frac{e^{2\beta t} - 1}{2\beta} \right\} \quad (7.46)$$

We can now rewrite Eq. (7.42) as

$$Y(t) = e^{-\beta t} Y(0) + e^{-\beta t} G(t) \quad (7.47)$$

and from this can determine the properties of $Y(t)$.

Exercise 7.10 (E/M)

Using the results we have just derived, confirm that (i) $Y(t)$ is normally distributed, (ii) $E\{Y(t)\} = e^{-\beta t} Y(0)$, and (iii) $\text{Var}\{Y(t)\} = [\sigma^2(1 - e^{-2\beta t})]/2\beta$.

Note that when t is very large $\text{Var}\{Y(t)\} \sim \sigma^2/2\beta$, which is a very interesting result because it tells us how fluctuations, measured by σ, and dissipation, measured by β, are connected to create the variance of $Y(t)$. In physical systems, this is called the "fluctuation–dissipation" theorem and another piece of physical insight (the Maxwell–Boltzmann distribution) allows one to determine σ for physical systems (see Connections). Given the result of Exercise 7.10, we can also immediately write down the transition density for the Ornstein–Uhlenbeck process $q(x, t, y, s)dx$ defined to be the probability that $x \leq Y(t) \leq x + dx$ given that $Y(s) = y$. It looks terribly frightening, but is simply a mathematical statement of the results of Exercise 7.10 in which $Y(0)$ is replaced by $Y(s) = y$:

$$q(x, t, y, s) = \frac{1}{\sqrt{2\pi \left(\frac{\sigma^2(1 - e^{-2\beta(t-s)})}{2\beta} \right)}} \exp\left[-\frac{(x - e^{-\beta(t-s)} y)^2}{\frac{2\sigma^2(1 - e^{-2\beta(t-s)})}{2\beta}} \right] \quad (7.48)$$

I intentionally did not cancel the 2s in the constant or the denominator of the exponential, so that we can continue to carry along the variance intact. If we wait a very long time, the dependence on the initial condition disappears, but we still have a probability distribution for the process. Let us denote by $\bar{q}(x)$ the limit of the transition density given by Eq. (7.48) when s is fixed and $t \to \infty$. This is

$$\bar{q}(x) = \frac{1}{\sqrt{2\pi \left(\frac{\sigma^2}{2\beta} \right)}} \exp\left[-\frac{x^2}{2 \left(\frac{\sigma^2}{2\beta} \right)} \right] \quad (7.49)$$

Figure 7.13. The set up for the
study of "escape from a
domain of attraction." The
phase line for the Ornstein–
Uhlenbeck process has a
single, stable steady state at
the origin. We surround the
origin by an interval [A, B],
where $A < 0$ and $B > 0$, and
assume that $Y(0)$ is in this
interval. Because the long time
limit of $q(x,t,y,s), \bar{q}(x)$, is
positive outside of [A, B]
(Eq. (7.49)), escape from the
interval is guaranteed.

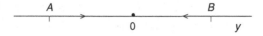

Figure 7.13. The set up for the study of "escape from a domain of attraction." The phase line for the Ornstein–Uhlenbeck process has a single, stable steady state at the origin. We surround the origin by an interval [A, B], where $A < 0$ and $B > 0$, and assume that $Y(0)$ is in this interval. Because the long time limit of $q(x,t,y,s), \bar{q}(x)$, is positive outside of [A, B] (Eq. (7.49)), escape from the interval is guaranteed.

Perhaps the most interesting insight from Eq. 7.49 pertains to "escapes from domains of attraction" (which we will revisit in the next chapter). The phase line for the deterministic system the underlies the Ornstein–Uhlenbeck process has a single steady state at the origin (Figure 7.13). Suppose that we start the process at some point $A \leq y \leq B$. Equations (7.48) and (7.49) tell us that there is always positive probability that $Y(t)$ will be outside of the interval [A, B]. In other words, the Ornstein–Uhlenbeck process will, with probability equal to 1, escape from [A, B]. As we will see in Chapter 8, how it does this becomes very important to our understanding of evolution and conservation.

When Ornstein and Uhlenbeck did this work, they envisioned that $Y(t)$ was the velocity of a Brownian particle, experiencing friction (hence the relaxation proportional to velocity) and random fluctuations due to the smaller molecules surrounding it. We need to integrate velocity in order to find position, so if $X(t)$ denotes the position of this particle

$$X(t) = X(0) + \int_0^t Y(s)\mathrm{d}s \tag{7.50}$$

and now we have another stochastic integral to deal with. But that is the subject for a more advanced book (see Connections).

General diffusion processes and the backward equation

We now move from the specific – Brownian motion, the Ornstein–Uhlenbeck process – to the general diffusion process. To be honest, two colleagues who read this book in draft suggested that I eliminate this section and the next. Their argument was something like this: "I don't need to know how my computer or car work in order to use them, so why should I have to know how the diffusion equations are derived?" Although I somewhat concur with the argument for both computers and cars, I could not buy it for diffusion processes. However, if you want to skip the details and get to the driving, the key equations are Eqs. (7.53), (7.54), (7.58), and (7.79).

The route that we follow is due to the famous probabilist William Feller, who immigrated to the USA from Germany around the time of the Second World War and ended up in Princeton. Feller wrote two beautiful books about probability theory and its applications (Feller

1957, 1971) which are simply known as Feller Volume 1 and Feller Volume 2; when I was a graduate student there was apocrypha that a faculty member at the University of Michigan decided to spend a summer doing all of the problems in Feller Volume 1 and that it took him seven years. Steve Hubbell, whose recent volume (Hubbell 2001) uses many probabilistic ideas, purchased Feller's home when he (Hubbell) moved to Princeton in the 1980s and told me that he found a copy of Feller Volume 1 (first edition, I believe) in the basement. A buyer's bonus!

We imagine a stochastic process $X(t)$ defined by its transition density function

$$\Pr\{y+z \leq X(s+dt) \leq y+z+dy | X(s) = z\} = q(y+z, s+dt, z, s)dy \tag{7.51}$$

so that $q(y+z, s+dt, z, s)dy$ tells us the probability that the stochastic process moves from the point z at time s to around the point $y+z$ at time $s+dt$. Now, clearly the process has be somewhere at time $t+dt$ so that

$$\int q(y+z, s+dt, z, s)dy = 1 \tag{7.52}$$

where the integral extends over all possible values of y.

A diffusion process is defined by the first, second, and higher moments of the transitions according to the following

$$\int q(y+z, s+dt, z, s)y\,dy = b(z, s)dt + o(dt)$$

$$\int q(y+z, s+dt, z, s)y^2 dy = a(z, s)dt + o(dt) \tag{7.53}$$

$$\int q(y+z, s+dt, z, s)y^n dy = o(dt) \text{ for } n \geq 3$$

In Eqs. (7.53), y is the size of the transition, and we integrate over all possible values of this transition. The second line in Eqs. (7.53) tells us about the variance, and the third line tells us that all higher moments are $o(dt)$. This description clearly does not fit all biological systems, since in many cases there are discrete transitions (the classic example is reproduction). But in many cases, with appropriate scaling (see Connections) the diffusion approximation, as Eq. (7.53) is called, is appropriate. In the last section of this chapter, we will investigate a process in which the increments are caused by a Poisson process rather than Brownian motion. The art of modeling a biological system consists in understanding the system well enough that we can choose appropriate forms for $a(X, t)$ and $b(X, t)$. In the next chapter, we will discuss this artistry in more detail, but before we create new art, we need to understand how the tools work.

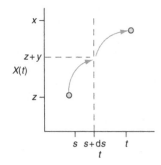

Figure 7.14. The process $X(t)$ starts at the value z at time s. To reach the vicinity of the value x at time t, it must first transition from z to a some value y at time $s+ds$ and then from y to the vicinity of x in the remaining time (ds is not to scale).

A stochastic process that satisfies this set of conditions on the transitions is also said to satisfy the stochastic differential equation

$$dX = b(X,t)dt + \sqrt{a(X,t)}dW \tag{7.54}$$

with infinitesimal mean $b(X,\ t)dt + o(dt)$ and infinitesimal variance $a(X,\ t)dt + o(dt)$. Symbolically, we write that given $X(t)=x$, $E\{dX\} = b(x,t) + o(dt)$, $\text{Var}\{dX\} = a(x,t)dt + o(dt)$ and, of course, dX is normally distributed. We will use both Eqs. (7.53) and Eq. (7.54) in subsequent analysis, but to begin will concentrate on Eqs. (7.53).

Let us begin by asking: how does the process get from the value z at time s to around the value x at time t? It has to pass through some point $z+y$ at intermediate time $s+ds$ and then go from that point to the vicinity of x at time t (Figure 7.14). In terms of the transition function we have

$$q(x,t,z,s)dx = \int q(x,t,y+z,s+ds)q(y+z,s+ds,z,s)dydx \tag{7.55}$$

This equation is called the Chapman–Kolmogorov equation and sometimes simply "The Master Equation." Keeping Eqs. (7.53) in mind, we Taylor expand in powers of y and ds:

$$q(x,t,z,s) = \int \left[q(x,t,z,s) + q_s(x,t,z,s)ds + q_z(x,t,z,s)y + \frac{1}{2}q_{zz}(x,t,z,s)y^2 \right.$$
$$\left. + O(y^3) \right] q(y+z,s+ds,z,s)dy \tag{7.56}$$

and now we proceed to integrate, noting that integral goes over y but that by Taylor expanding, we have made all of the transition functions to depend only upon x, so that they are constants in terms of the integrals. We do those integrals and apply Eqs. (7.53)

$$q(x,t,z,s) = q(x,t,z,s) + ds\Big\{q_s(x,t,z,s) + b(z,s)q_z(x,t,z,s)$$
$$+ \frac{1}{2}a(z,s)q_{zz}(x,t,z,s)\Big\} + o(ds) \tag{7.57}$$

We now subtract $q(x,\ t,\ z,\ s)$ from both sides, divide by ds, and let ds approach 0 to obtain the partial differential equation that the transition density satisfies in terms of z and s:

$$q_s(x,t,z,s) + b(z,s)q_z(x,t,z,s) + \frac{1}{2}a(z,s)q_{zz}(x,t,z,s) = 0 \tag{7.58}$$

Equation (7.58) is called the Kolmogorov Backward Equation. The use of "backward" refers to the variables z and s, which are the starting value and time of the process; in a similar manner the variables x and t are called "forward" variables and there is a Kolmogorov Forward

Equation (also called the Fokker–Planck equation by physicists and chemists), which we will derive in a while. In the backward equation, x and t are carried as parameters as z and s vary.

Equation (7.58) involves one time derivative and two spatial derivatives. Hence we need to specify one initial condition and two boundary conditions, as we did in Chapter 2. For the initial condition, let us think about what happens as $s \rightarrow t$? As these two times get closer and closer together, the only way the transition density makes sense is to guarantee that the process is at the same point. In other words $q(x, t, z, t) = \delta(x - z)$. As in Chapter 2, boundary conditions are specific to the problem, so we defer those until the next chapter.

Very often of course, we are not just interested in the transition density, but we are interested in more complicated properties of the stochastic process. For example, suppose we wanted to know the probability that $X(t)$ exceeds some threshold value x_c, given that $X(s) = z$. Let us call this probability $u(z, s, t|x_c)$ and recognize that it can be found from the transition function according to

$$u(z, s, t|x_c) = \int_{x_c}^{\infty} q(x, t, z, s) dx \qquad (7.59)$$

and now notice that with t treated as a parameter then $u(z, s, t|x_c)$ viewed as a function of z and s will satisfy Eq. (7.58), as long as we can take those derivatives inside the integral. (Which we can do. As I mentioned earlier, one should not be completely cavalier about the processes of integration and differentiation, but everything that I do in this book in that regard is proper and justified.) What about the initial and boundary conditions that $u(z, s, t|x_c)$ satisfies? We will save a discussion of them for the next chapter, in the application of these ideas to extinction processes.

We can also find the equation for $u(z, s, t|x_c)$ directly from the stochastic differential equation (7.54), by using the same kind of logic that we did for the gambler's ruin. That is, the process starts at $X(s) = z$ and we are interested in the probably that $X(t) > x_c$. In the first bit of time ds, the process moves to a new value $z + dX$, where dX is given by Eq. (7.54) and we are then interested in the probability that $X(t) > x_c$ from this new value. The new value is random so we must average over all possible values that dX might take. In other words

$$u(z, s, t|x_c) = E_{dX}\{u(z + dX, s + ds, t|x_c)\} \qquad (7.60)$$

and the procedure from here should be obvious: Taylor expand in powers of dX and dt and then take the average over dX.

Exercise 7.11 (E/M)

Do the Taylor expansion and averaging and show that

$$u_s(z, s, t|x_c) + b(z, s)u_z(z, s, t|x_c) + \frac{1}{2}a(z, s)u_{zz}(z, s, t|x_c) = 0 \qquad (7.61)$$

It is possible to make one further generalization of Eq. (7.59), in which we integrated the "indicator function" $I(x) = 1$ if $x > x_c$ and $I(x) = 0$ otherwise over all values of x. Suppose, instead, we integrated a more general function $f(x)$ and defined $u(z, s, t)$ by

$$u(z, s, t) = \int f(x)q(x, t, z, s)dx \qquad (7.62)$$

for which we see that $u(z, s, t)$ satisfies Eq. (7.61). If we recall that $q(x, s, z, s) = \delta(z - x)$, then it becomes clear that $u(z, t, t) = f(z)$; more formally we write that $u(z, s, t) \to f(z)$ as $s \to t$ and we will defer the boundary conditions until the next chapter.

We will return to backward variables later in this chapter (with discussion of Feyman–Kac and stochastic harvesting equations) but now we move on to the forward equation.

The forward equation

We now derive the forward Kolmogorov equation, which describes the behavior of $q(x, t, z, s)$ as a function of x and t, treating z and s as parameters. This derivation is long and there are a few subtleties that we will need to explore. The easy way out, for me at least, would simply be to tell you the equation and cite some other places where the derivation could be found. However, I want you to understand how this tool arises.

Our starting point is the Chapman–Kolmogorov equation, which I write in a slightly different form than Eq. (7.56) (Figure 7.15)

$$q(x, t + dt, z, s) = \int q(y, t, z, s)q(x, t + dt, y, t)dy \qquad (7.63)$$

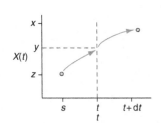

Figure 7.15. The transition process for the forward equation. From the point $X(s) = z$, the process moves to value y at time t and then from that value to the vicinity of x at time $t + dt$. Note the difference between this formulation and that in Figure 7.14: in the former figure the small interval of time occurs at the beginning (with the backward variable). In this figure, the small interval of time occurs near the end (with the forward variable); here dt is not to scale.

That is: to be around the value x at time $t + dt$, the process starts at z at time s and moves from there to the value y at time t; from y at time t the process then has to move to the vicinity of x in the next dt.

Now we know that we are going to want the derivative of the transition density with respect to t, so let us subtract $q(x, t, z, s)$ from both sides of Eq. (7.63)

$$q(x, t + dt, z, s) - q(x, t, z, s) = \int q(y, t, z, s)q(x, t + dt, y, t)dy - q(x, t, z, s) \qquad (7.64)$$

Now here comes something subtle and non-intuitive (in the "why did you do that?" with answer "because I learned to" sense). Suppose that

$h(x)$ is a function for which we can find derivatives and which goes to 0 as $|x| \to \infty$. We multiply both sides of Eq. (7.64) by $h(x)$ and integrate over x:

$$\int (q(x, t+dt, z, s) - q(x, t, z, s))h(x)dx = \int \left[\int q(y, t, z, s)q(x, t+dt, y, t)dy \right.$$
$$\left. -q(x, t, z, s) \right] h(x)dx \qquad (7.65)$$

Now we divide both sides of Eq. (7.65) by dt and let dt approach 0. The left hand side (LHS) becomes

$$\text{LHS} = \lim_{dt \to 0} \frac{1}{dt} \int (q(x, t+dt, z, s) - q(x, t, z, s))h(x)dx \qquad (7.66)$$

and taking the limit inside the integral, we recognize the derivative of q with respect to t:

$$\text{LHS} = \int q_t(x, t, z, s)h(x)dx \qquad (7.67)$$

and now "all" we have to do is deal with the right hand side (RHS). We begin by Taylor expansion of $h(x)$ around the intermediate point y, using subscript notation so that $h_y(y)$ is the first derivative of $h(x)$ evaluated at the point $x = y$. The Taylor expansion of $h(x)$ is

$$h(x) = h(y) + h_y(y)(x-y) + \frac{1}{2}h_{yy}(y)(x-y)^2 + O((x-y)^3) \qquad (7.68)$$

The right hand side of Eq. (7.65) is now (with multiplication intended from the top line to the bottom one):

$$\text{RHS} = \lim_{dt \to 0} \frac{1}{dt} \int \left[\int q(y, t, z, s)q(x, t+dt, y, t)dy - q(x, t, z, s) \right]$$
$$\left\{ h(y) + h_y(y)(x-y) + \frac{1}{2}h_{yy}(y)(x-y)^2 + O((x-y)^3) \right\} dx \qquad (7.69)$$

Now, we do the following: we multiply through all of the terms and re-order them. In addition, we will replace the Taylor expansion by $h(x)$ itself when we multiply by $q(x, t, z, s)$. The result of this process is a long expression, but an intelligible one:

$$\text{RHS} = \lim_{dt \to 0} \frac{1}{dt} \int \left\{ h(y)q(y, t, z, s) \int q(x, t+dt, y, t) \right\} dxdy$$
$$+ \lim_{dt \to 0} \frac{1}{dt} \int \left\{ h_y(y)q(y, t, z, s) \int (x-y)q(x, t+dt, y, t) \right\} dxdy$$
$$+ \lim_{dt \to 0} \frac{1}{dt} \int \left\{ \frac{1}{2}h_{yy}(y)q(y, t, z, s) \int (x-y)^2 q(x, t+dt, y, t) \right\} dxdy$$
$$+ \lim_{dt \to 0} \frac{1}{dt} \int \left\{ q(y, t, z, s) \int O((x-y)^3)q(x, t+dt, y, t) \right\} dxdy$$
$$- \lim_{dt \to 0} \frac{1}{dt} \int q(x, t, z, s)h(x)dx \qquad (7.70)$$

Now let us consider the terms on the right hand side of Eq. (7.70). First, we know that $\int q(x, t + dt, y, t)\}dx = 1$ because from y at time t, the process must move to somewhere at time $t + dt$, so that when we integrate over all values of x, the result is 1. Thus, the first term on the right hand side of Eq. (7.70) and the fifth term are the same, except that in the former the variable of integration is y and in the latter it is x. So these terms cancel. The other three terms are defined in terms of the transition moments for the diffusion process given by Eqs. (7.53) in a slightly different form

$$\int (x - y)q(x, t + dt, y, t)\}dx = b(y, t)dt + o(dt)$$

$$\int (x - y)^2 q(x, t + dt, y, t)\}dxdy = a(y, t) + o(dt) \qquad (7.71)$$

$$\int O((x - y)^3)q(x, t + dt, y, t)\}dx = o(dt)$$

Although they appear to be different, Eqs. (7.53) and (7.71) are really the same, since they both deal with the average of the transition. In Eqs. (7.53), we move from value z to value $y + z$, so that the transition size is y. In Eqs. (7.71), we move from y to x, so that the transition size is $x - y$.

Now, we can clearly write the LHS, Eq. (7.67), as

$$\int q_t(y, t, z, s)h(y)dy$$

through the simple change of variables of replacing x by y as the integration variable. We have already agreed that the first and fifth terms of Eq. (7.70) cancel, so that if we divide through by dt and apply Eq. (7.71), in Eq. (7.70), we are left with

$$\text{RHS} = \lim_{dt \to 0} \left\{ \int h_y(y)q(y, t, z, s)b(y, t)dy \right.$$
$$\left. + \int \frac{1}{2}h_{yy}(y)q(y, t, z, s)a(y, t)dy + \frac{o(dt)}{dt} \right\} \qquad (7.72)$$

and the term $o(dt)/dt$ will disappear as dt goes to 0. We now equate the LHS and the RHS:

$$\int q_t(y, t, z, s)h(y)dy = \int \left[h_y(y)q(y, t, z, s)b(y, t) + \frac{1}{2}h_{yy}(y)q(y, t, z, s)a(y, t) \right] dy \qquad (7.73)$$

and this is a good, but not especially useful, equation, because we have $h(y)$ on the left hand side, but its derivatives on the right hand side.

We will deal with this difficulty by integrating by parts, recalling the basic formula $\int_{-\infty}^{\infty} u \, dv = uv|_{-\infty}^{\infty} - \int_{-\infty}^{\infty} v \, du$. I have included the limits of integration here because they will be important. However, for simplicity, I am going to suppress the indices on $h(y)$, $q(y, t, z, s)$ and $b(y, t)$

in what follows. Thus, integrating the term involving $b(y, t)$ by parts and using the notation $(qb)_y = (\partial / \partial y)(qb)$ we have

$$\int_{-\infty}^{\infty} h_y qb = hqb|_{-\infty}^{\infty} - \int_{-\infty}^{\infty} h(qb)_y dy \qquad (7.74)$$

and, for the term involving $a(y, t)$, we integrate by parts not once, but twice:

$$\int_{-\infty}^{\infty} h_{yy} qa = h_y qa|_{-\infty}^{\infty} - \int_{-\infty}^{\infty} h_y (qa)_y dy$$

$$= h_y qa|_{-\infty}^{\infty} - \left[h(qa)_y|_{-\infty}^{\infty} - \int_{-\infty}^{\infty} h(qa)_{yy} dy \right] \qquad (7.75)$$

If we now assume that $h(y)$ and its derivative approach 0 as $|y| \to \infty$, Eqs. (7.74) and (7.75) simplify (and putting the indices back in):

$$\int_{-\infty}^{\infty} h_y(y) q(y, t, z, s) b(y, t) dy = - \int_{-\infty}^{\infty} h(y)(q(y, t, z, s)b(y, t))_y dy$$

$$\int_{-\infty}^{\infty} h_{yy}(y) q(y, t, z, s) a(y, t) dy = \int_{-\infty}^{\infty} h(y)(q(y, t, z, s)a(y, t))_{yy} dy \qquad (7.76)$$

We return to Eq. (7.71) and conclude

$$\int q_t(y, t, z, s) h(y) dy = \int \left[-h(y)(q(y, t, z, s)b(y, t))_y \right.$$

$$\left. + \frac{1}{2} h(y)(q(y, t, z, s)a(y, t))_{yy} \right] dy \qquad (7.77)$$

so that the entire equation now only involves $h(y)$. In fact, we could bring everything to the same side of the equation and factor $h(y)$ through to obtain

$$\int h(y) \left[q_t(y, t, z, s) + (q(y, t, z, s)b(y, t))_y - \frac{1}{2}(q(y, t, z, s)a(y, t))_{yy} \right] dy = 0 \qquad (7.78)$$

but this equation is supposed to hold for almost any choice of $h(y)$ – remember that we made minimal assumptions about it. We thus conclude that $q(y, t, z, s)$ satisfies (formally we say "in a weak sense") the equation

$$q_t(y, t, z, s) = \frac{1}{2}(q(y, t, z, s)a(y, t))_{yy} - (q(y, t, z, s)b(y, t))_y \qquad (7.79)$$

If we replace y by x, the form of the equation remains unchanged.

As before, we have the initial condition that $q(y, t, z, s) \rightarrow \delta(y - z)$ as $s \rightarrow t$, and the boundary condition that $q(y, t, z, s) \rightarrow 0$ as $|y| \rightarrow \infty$. The forward equation is often used in population genetics (see Connections). In the mathematical literature, Eqs. (7.59) and (7.81) are said to be adjoints of each other (Haberman (1998) is a good source for more mathematical detail). We shall employ both backward and forward equations in the next chapter. Before we get to that next chapter, however, I would like to show a couple more backward equations, which have various interesting uses. And before that, I offer an exercise which I hope may clarify some of the differences between backward and forward equations.

Exercise 7.12 (M)

Remember that in Chapter 2 we reached the diffusion equation as the limit of a random walk. Let us reconsider such a random walk. Thus, $X(t)$ represents the position of the process at time t on a lattice with separation ε between sites. We assume that steps take place in time interval Δt and let $r(x)$ and $l(x)$ represent the probabilities of moving to the right or left in the next interval, given that $X(t) = x$. To get to a backward equation, let

$$u(x,t) = \Pr\{\text{the process has left } [A, B] \text{ by time } t | X(0) = x\} \qquad (7.80)$$

(a) Show that $u(x, t)$ satisfies the equation

$$u(x,t) = (1 - r(x) - l(x))u(x, t - \Delta t) + r(x)u(x + \varepsilon, t - \Delta t) \\ + l(x)u(x - \varepsilon, t - \Delta t) \qquad (7.81)$$

and be certain that you can explain why the right hand side involves $t - \Delta t$.
(b) To get to the forward equation, let $v(x, t) = \Pr\{X(t) = x\}$. Show that $v(x, t)$ satisfies the equation

$$v(x, t + \Delta t) = (1 - r(x) - l(x))v(x, t) + r(x - \varepsilon)v(x - \varepsilon, t) \\ + l(x + \varepsilon)v(x + \varepsilon, t) \qquad (7.82)$$

(c) Now Taylor expand Eqs. (7.81) and (7.82) to first order in Δt and to second order in ε and compare those results with the backward and forward equations that we have derived. A word of warning: this exercise is heuristic and ignoring all of the higher order terms in the Taylor expansion is not a generally safe thing to do, rather one needs to carefully take limits as we did in Chapter 2.

The Feynman–Kac (stochastic survival) and stochastic harvesting equations

Here are two more examples of backward equations. The first, although fully general, applies to a foraging animal with stochastic mortality. The second arises in natural resource management and bioeconomics (Clark 1985, 1990).

Suppose we have some function $\psi(x)$ (the interpretation of which will follow almost immediately, in Exercise 7.13, but for now simply think of it as positive) and we define a function $u(x, t, T)$ by

$$u(x, t, T) = E\left\{ \exp\left[-\int_t^T \psi(X(s))ds \right] \middle| X(t) = x \right\} \qquad (7.83)$$

where we understand the expectation to be over the sample paths of $X(s)$, starting from $X(t) = x$, satisfying the stochastic differential equation $dX = b(X, t)dt + \sqrt{a(X, t)}dW$. When $t = T$ the integral will be 0, we conclude that $u(x, T, T) = 1$ and that $u(x, t, T) < 1$ otherwise.

If we break the integral into a piece from t to $t + dt$ and then a piece from $t + dt$ to T, we have

$$\exp\left[-\int_t^T \psi(X(s))ds \right] = \exp\left[-\int_t^{t+dt} \psi(X(s))ds \right] \exp\left[-\int_{t+dt}^T \psi(X(s))ds \right] \qquad (7.84)$$

and if we now Taylor expand the first term on the right hand side of Eq. (7.84) we have

$$\exp\left[-\int_t^T \psi(X(s))ds \right] = [1 - \psi(X(t))dt + o(dt)]\exp\left[-\int_{t+dt}^T \psi(X(s))ds \right] \qquad (7.85)$$

The expectation over paths beginning at $X(t)$ can be broken into two pieces: first the expectation of paths that go from $X(t) = x$ to values of $X(t + dt) = x + dX$ and then the expectation of paths starting at $x + dX$. In other words, we have shown that

$$u(x, t, T) = E_{dX}\{[1 - \psi(x)dt + o(dt)]u(x + dX, t + dt, T)\} \qquad (7.86)$$

which we now Taylor expand and average in the usual manner to obtain a differential equation.

Exercise 7.13 (M)

Do the Taylor expansion and averaging of Eq. (7.86) to show that $u(x, t, T)$ satisfies the differential equation (with variables suppressed)

$$u_t + b(x, t)u_x + \frac{1}{2}a(x, t)u_{xx} - \psi(x)u = 0 \qquad (7.87)$$

Now add the interpretation that $X(t)$ indicates the position of an individual following the stochastic differential equation given above and that

$$\Pr\{\text{being killed in the next } dt | X(t) = x\} = \psi(x)dt + o(dt) \qquad (7.88)$$

so that $u(x, t, T)$ represents the probability of surviving from t to T, given that $X(t) = x$. (If you don't want to think of this as position, think of $X(t)$ as energy reserves, with death by starvation in the next dt determined by $\psi(x)$, or any other analogy that works for you.) Use the method of thinking along sample paths to get directly to Eq. (7.87) (hint: to survive from t to T, the individual must first survive from t to $t + dt$ and then from $t + dt$ to T).

Equations (7.83) and (7.87) are called the Feynman–Kac formula. In 1948, when Richard Feynman presented his path integral formulation of quantum mechanics (which involves the Schroedinger equation, also a diffusion-like equation), Mark Kac recognized that path integrals could thus be used to solve the usual diffusion equation (see Connections).

There is also associated forward equation for the probability density of the process:

$$f(y,s,x,t)dy = \Pr\{y \le X(s) \le y + dy \text{ and the individual is still} \atop \text{alive } X(t) = x\} \tag{7.89}$$

Following the same procedures as in the previous section leads us to

$$f_s = \frac{1}{2}(a(y,s)f)_{yy} - (b(y,s)f)_y - \psi(x)f \tag{7.90}$$

with the appropriate delta function as a condition as s approaches t.

Finally, let us consider one more equation, in this case assuming that $X(t)$ represents the population size of a harvested stock and that at time s when stock size is $X(s)$ the economic return is $r(X(s), s)$. If the discount rate is δ, the long-term discounted rate of return given that $X(t) = x$ is

$$u(x,t) = E\left\{\int_t^\infty r(X(s),s)e^{-\delta s}ds \middle| X(t) = x\right\} \tag{7.91}$$

where, as before, the expectation refers to an average over the sample paths that begin at $X(t) = x$. We now break the integral into two pieces

$$u(x,t) = E\left\{\int_t^{t+dt} r(X(s),s)e^{-\delta s}ds + \int_{t+dt}^\infty r(X(s),s)e^{-\delta s}ds \middle| X(t) = x\right\} \tag{7.92}$$

and recognize that the first integral on the right hand side is

$$r(x,t)e^{-\delta t}dt + o(dt)$$

and that the second integral can be conditioned into an average over dX of the average over the new starting point $X(t + dt) = x + dX$:

$$u(x,t) = r(x,t)e^{-\delta t}dt + o(dt)$$
$$+ E_{dX}\left[E\left\{\int_{t+dt}^\infty r(X(s),s)e^{-\delta s}ds \middle| X(t+dt) = x + dX\right\}\right] \tag{7.93}$$

where the inner expectation on the right hand side of Eq. (7.93) is over the sample paths starting at $X(t+dt)=x+dX$. Of course, the inner expectation is also $u(x+dX, t+dt)$ so that we conclude

$$u(x,t) = r(x,t)e^{-\delta t}dt + o(dt) + E_{dX}\{u(x+dX, t+dt)\} \qquad (7.94)$$

Exercise 7.14 (E/M)

Finish the calculation to show that $u(x, t)$ satisfies the differential equation

$$u_t + b(x,t)u_x + \frac{1}{2}a(x,t)u_{xx} + r(x,t)e^{-\delta t} = 0 \qquad (7.95)$$

Next, assume that a, b and r are functions of x but not functions of time. Set $u(x,t) = v(x)e^{-\delta t}$. What equation does $v(x)$ satisfy?

An alternative to Brownian motion: the Poisson increment

This has been a long chapter, and it is nearly drawing to a close. Before closing, however, I want to introduce an alternative to Brownian motion as a model for stochastic effects. Recall that Brownian motion is a continuous process, but many processes in life are discrete. In gambling, one's holdings usually change by a discrete amount with each hand. As we discussed above, offspring come in discrete units and catastrophes may kill a large number of individuals at one time (Mangel and Tier 1993, 1994). Very often energy reserves or position changes in a more or less discrete manner.

To capture such effects, let us consider the increment of the generalized Poisson process, defined according to

$$d\Pi = \begin{cases} v \text{ with probability } \frac{1}{2}cdt + o(dt) \\ 0 \text{ with probability } 1 - cdt + o(dt) \\ -v \text{ with probability } \frac{1}{2}cdt + o(dt) \end{cases} \qquad (7.96)$$

where v is the intensity of the process, since it describes the size of the jumps, and c is the rate of the process. It would be possible to standardize the intensity to $v=1$, if we wanted to do so, but for this illustration it is better left as it is.

Exercise 7.15 (E)

Compute the mean and variance of $d\Pi$.

We could imagine now a process that satisfies, for example

$$dX = rXdt + d\Pi \qquad (7.97)$$

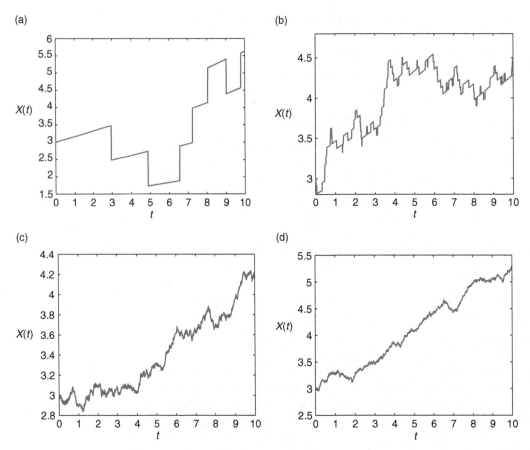

Figure 7.16. Simulated trajectories of Eq. (7.97) for the case in which the product $vc = 1$: (a) $v = 1$, $c = 1$; (b) $v = 0.2$, $c = 5$; (c) $v = 0.1$, $c = 10$; and (d) $v = 0.02$, $c = 50$. Other parameters are $dt = 0.01$, $r = 0.05$.

What interpretation do we give to Eq. (7.97)? If $X(t)$ were the size of a population, then Eq. (7.97) corresponds to deterministic exponential growth with stochastic jumps of $\pm v$ individual, determined by the increment of a Poisson process.

What would we expect trajectories to look like? In Figure 7.16, I show four sample paths, corresponding to different values of v and c, but with their product held constant at 1. When v is large (and thus c is small), transitions occur rarely but when one takes place, it involves a big jump. As c increases (and thus v decreases), the transition is more and more likely, but the size of the transition is smaller and smaller. Indeed, Figure 7.16d almost looks like a deterministic trajectory perturbed by Brownian motion. To make a little bit more sense of this, and to get us ready for the next chapter, let us assume that $X(t)$ satisfies

Eq. (7.97), that there is a population ceiling x_{max} and a critical level x_c corresponding to extinction and define

$$u(x) = \Pr\{X(t) \text{ reaches } x_{max} \text{ before } x_c \,|\, X(0) = x\} \qquad (7.98)$$

We derive the equation that $u(x)$ satisfies in the usual manner

$$u(x) = E_{dX}\{u(x + dX)\}$$

$$= u(x + rxdt - v)\frac{1}{2}cdt + u(x + rxdt)(1 - cdt) \qquad (7.99)$$

$$+ u(x + rxdt + v)\frac{1}{2}cdt + o(dt)$$

Clearly, Eq. (7.99) calls for a Taylor expansion. Note that the first and third terms on the right hand side are already O(dt), so that when we do the Taylor expansion only the first term of the expansion will be appropriate. We thus Taylor expand around $x - v$, x, and $x + v$, respectively, to obtain

$$u(x) = u(x - v)\frac{1}{2}cdt + [u(x) + u_x rxdt](1 - cdt) + u(x + v)\frac{1}{2}cdt + o(dt)$$

$$(7.100)$$

and we now subtract $u(x)$ from both sides, divide by dt and allow dt to approach 0, to be left with a differential-difference equation

$$0 = \frac{cu(x - v)}{2} - cu(x) + u_x rx + \frac{cu(x + v)}{2} \qquad (7.101)$$

which is a formidable equation. But now let us think about the limiting process used in Figure 7.16, in which v becomes smaller and smaller. Then we might consider Taylor expanding $u(x - v)$ and $u(x + v)$ around x according to $u(x - v) = u(x) - u_x v + \frac{1}{2}u_{xx}v^2 + o(v^2)$ and $u(x + v) = u(x) + u_x v + \frac{1}{2}u_{xx}v^2 + o(v^2)$. If we do this, notice that $cu(x)$ will cancel, as will the first derivatives from the Taylor expansion. If we assume that $cv^2 = a$ and that the terms that are $o(v^2)$ approach 0, we are left with

$$0 = \frac{a}{2}u_{xx} + rxu_x \qquad (7.102)$$

which is exactly the equation we would have derived had we started not with Eq. (7.97) but with the stochastic differential equation $dX = rXdt + \sqrt{a}\,dW$. Thus, the limiting result when the rate of transitions increases but their size decreases with cv^2 constant does indeed look very much like a diffusion. Indeed, this is called the diffusion approximation and it too finds common use in population genetics and conservation biology (see Connections).

Connections

Thinking along sample paths and path integrals

Richard Feynman's paper on the formulation of quantum mechanics by thinking along sample paths (Feynman 1948) is still worth reading, even if one does not know a lot about quantum mechanics. From this idea (and Kac's use in the solution of the standard diffusion equation) the notion of path integrals developed, and there are lots of instances of the application of path integrals today (two of my favorites are Schulman (1981) and Friedlin and Wentzell (1984)). For more about Feynman and Kac, see Gleick (1992) and Kac's autobiography (Kac 1985).

Independent increments, Ito and Stratonovich

Our treatment of Brownian motion has been to look from time t to the end of the interval $t + dt$. This is called the Ito calculus, named after the great Japanese mathematician Kiyosi Ito (whose daughter happens to be a faculty member in Department of Linguistics at UCSC). Ito's work has recently been reviewed, in very mathematical form, by Stroock (2003). This is a hard book to read, but the preface is something that everyone can understand. If we think about not the end of the interval, but imagine that the process is truly continuous, then there are changes throughout the interval. This view is called Stratonovich calculus and the effect is to change the form of the diffusion coefficient (see, for example Stratonovich (1963), Wong and Zakai (1965), Wong (1971), and van Kampen (1981a, b)). Engineers sometimes call Stratonovich calculus the Wong–Zakai correction. Karlin and Taylor (1981) give a readable introduction to the different stochastic calculi. Hakoyama et al. (2000) consider a problem in risk of the extinction of populations in which the environmental fluctuations obey Stratonovich calculus and the demographic fluctuations obey Ito calculus; see also Hakoyama and Iwasa (2000).

Fluctuation and dissipation

One of Einstein's great contributions in his 1905 paper was to show how fluctuation could be connected to the Maxwell–Boltzmann distribution; this result has connections to nineteenth-century mechanistic material-ism and the general phenomenon of diffusion (Wheatley and Agutter 1996). The crowning achievement in this area belongs to Uhlenbeck and Ornstein (1930) who showed how to fully connect fluctuation and dissipation (the resistance or friction term in the Ornstein–Uhlenbeck process). Uhlenbeck was a particularly interesting person, who had an enormous effect on twentieth-century physics. When my UC Davis colleague Joel Keizer began his development of non-equilibrium thermodynamics (summarized in Keizer (1987)), Uhlenbeck acted as referee and it took Kac as interpreter of Joel's ideas to convince Uhlenbeck of their validity.

Red, white, and blue noise

There are examples of biological systems with noise that has a spectrum which is far from white (Cohen 1995, White *et al.* 1996, Vasseur and Yodzis 2004). For example, slowly varying environments (as in the North Pacific ocean; see Hare and Francis (1995), Mantua *et al.* (1997)) will have a spectrum that is very red, so that low frequencies are represented more strongly. Some diseases exhibit high frequency fluctuations, so that their spectra are bluer. These can be generated, in discrete time, from a model of the form $Y(t+1) = \alpha Y(t) + \sqrt{1 - \alpha^2} Z(t+1)$, where $Y(t)$ is the environmental noise at time t, α is a parameter with range $-1 \leq \alpha \leq 1$ and $Z(t)$ is a normally distributed random variable. When α is positive, low frequency components dominate (the spectrum is red) and when α is negative the high frequency components dominate. Furthermore, because biological responses are generally nonlinear, they can filter the environmental noise (for examples, see Petchey *et al.* (1997), Petchey (2000) or Laakso *et al.* (2003)).

Stochastic differential equations and stochastic integrals

There is an enormous literature on stochastic differential equation and stochastic integrals. The mathematical levels range from pretty applied, as here, to highly abstract and theoretical. Two older but solid introductions to the material are Arnold (1973) and Gardiner (1983). Another good starting point is Karlin and Taylor (1981), who have a 240 page chapter on diffusion processes. A general discussion of numerical methods for stochastic differential equations is found in Higham (2001). Exact numerical methods for the Ornstein–Uhlenbeck process and its integral are discussed by Gillespie (1996).

Applications in ecology

As we discussed in Chapter 2, diffusion processes arise in a natural way in the study of organismal movement and dispersal (Turchin 1998) and the various connections given there can now take on deeper meaning.

For example, if we were to let (X, Y) denote the position of an animal, we could now write stochastic differential equations to characterize the increments in X and Y. Stochastic differential equations can also be used to describe the dynamics of populations (see, for example, Nisbet and Gurney (1982), Engen *et al.* (2002), Lande *et al.* (2003), and Saether and Engen (2004)). Costantino and Desharnais (1991) use diffusion models to characterize the population dynamics of flour beetles. If $N(t)$ denotes adult numbers at time t, they work with models of the form $dN = N(t)[be^{-cN(t)} - \mu]dt + \sigma N(t)dw$ where the parameters have a natural interpretation. In this case, the stationary distribution of population size is a gamma density (also see Peters *et al.* (1989)), thus connecting us to material in Chapter 3.

Applications in population genetics

Diffusion processes underlie an entire approach to population genetics, with many entrance points to the literature. Some of my favorites are Crow and Kimura (1970), Kimura and Ohta (1971), and Gillespie (1991, 1998). For applications in population genetics, we usually work with the forward equation to specify the evolution of a gene frequency from an initial starting distribution of the frequency, or with the backward equation to describe the time until fixation of an allele.

Chapter 8

Applications of stochastic population dynamics to ecology, evolution, and biodemography

We are now in a position to apply the ideas of stochastic population theory to questions of ecology and conservation (extinction times) and evolutionary theory (transitions from one peak to another on adaptive landscapes), and demography (a theory for the survival curve in the Euler–Lotka equation, which we will derive as review). These are idiosyncratic choices, based on my interests when I was teaching the material and writing the book, but I hope that you will see applications to your own interests. These applications will require the use of many, and sometimes all, of the tools that we have discussed, and will require great skill of craftsmanship. That said, the basic idea for the applications is relatively simple once one gets beyond the jargon, so I will begin with that. We will then slowly work through calculations of more and more complexity.

The basic idea: "escape from a domain of attraction"

Central to the computation of extinction times and extinction probabilities or the movement from one peak in a fitness landscape to another is the notion of "escape from a domain of attraction." This impressive sounding phrase can be understood through a variety of simple metaphors (Figure 8.1). In the most interesting case, the basic idea is that deterministic and stochastic factors are in conflict – with the deterministic ones causing attraction towards steady state (the bottom of the bowl or the stable steady states in Figure 8.1) and the stochastic factors causing perturbations away from this steady state. The cases of the ball

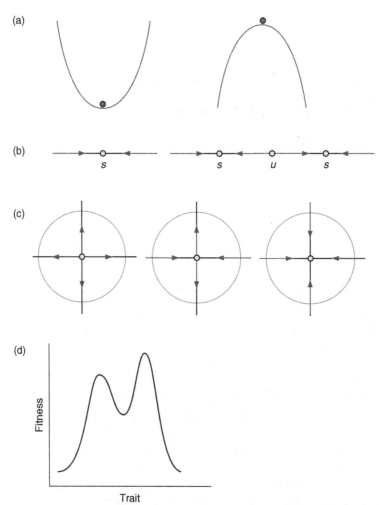

Figure 8.1. Some helpful ways to think about escape from a domain of attraction. (a) The marble in a cup, when slightly perturbed, will return to the bottom of the cup (a domain of attraction). The converse of this is the ball on the hill, in which any small perturbation is going to be magnified and the ball will move either to the right or the left. (b) In one dimension, we could envision a deterministic dynamical system $dX/dt = b(X)$ in which there is a single steady state that is globally stable (as in the Ornstein–Uhlenbeck process), denoted by s. Fluctuations will cause departures from the steady state, but in some sense the stochastic process has nowhere else to go. On the other hand, if the deterministic system has multiple steady states, in which two stable steady states are separated by an unstable one (denoted by u), the situation is much more interesting. Then a starting value near the upper stable steady state might be sufficiently perturbed to cross the unstable steady state and be attracted towards the lower stable steady state. If $X(t)$ were the size of a population, we might think of this as an extinction. (c) For a two dimensional dynamical system of the form $dX/dt = f(X, Y)$, $dY/dt = g(X, Y)$ the situation can be more complicated. If a steady state is an unstable node, for example, then the situation is like the ball at the top of the hill and perturbations from the steady state will be amplified (of course, now there are many directions in which the phase points might move). Here the circle indicates a domain of interest and escape occurs when we move outside of the circle. If the steady state is a saddle point, then the separatrix creates two domains of attraction so that perturbations from the steady state become amplified in one direction but not the other. If the steady state is a stable node, then the deterministic flow is towards the steady state but the fluctuations may force phase points out of the region of interest. (d) If we conceive that natural selection takes place on a fitness surface (Schluter 2000), then we are interested in transitions from one local peak of fitness to a higher one, through a valley of fitness.

on the top of the hill or the steady state being unstable or a saddle point are also of some interest, but I defer them until Connections.

We have actually encountered this situation in our discussion of the Ornstein–Uhlenbeck process, and that discussion is worth repeating, in simplified version here. Suppose that we had the stochastic differential equation $dX = -Xdt + dW$ and defined

$$u(x, t) = \Pr\{X(s) \text{ stays within } [-A, A] \text{ for all } s, 0 \leq s \leq t | X(0) = x\} \quad (8.1)$$

We know that $u(x, t)$ satisfies the differential equation

$$u_t = \frac{1}{2}u_{xx} - x u_x \tag{8.2}$$

so now look at Exercise 8.1.

Exercise 8.1 (M)

Derive Eq. (8.2). What is the subtlety about time in this derivation?

Equation (8.2) requires an initial condition and two boundary conditions. For the initial condition, we set $u(x, 0) = 1$ if $-A < x < A$ and to 0 otherwise. For the boundary conditions, we set $u(-A, t) = u(A, t) = 0$ since whenever the process reaches A it is no longer in the interval of interest. Now suppose we consider the limit of large time, for which $u_t \to 0$. We then have the equation $0 = (1/2)u_{xx} - x u_x$ with the boundary conditions $u(-A) = u(A) = 0$.

Exercise 8.2 (E)

Show that the general solution of the time independent version of Eq. (8.2) is $u(x) = k_1 \int_{-A}^{x} \exp(s^2)ds + k_2$, where k_1 and k_2 are constants. Then apply the boundary conditions to show that these constants must be 0 so that $u(x)$ is identically 0. Conclude from this that with probability equal to 1 the process will escape the interval $[-A, A]$.

We will thus conclude that escape from the domain of attraction is certain, but the question remains: how long does this take. And that is what most of the rest of this chapter is about, in different guises.

The MacArthur–Wilson theory of extinction time

The 1967 book of Robert MacArthur and E. O. Wilson (MacArthur and Wilson 1967) was an absolutely seminal contribution to theoretical ecology and conservation biology. Indeed, in his recent extension of it, Steve Hubbell (2001) describes the work of MacArthur and Wilson as a "radical theory." From our perspective, the theory of MacArthur and Wilson has two major contributions. The first, with which we will not deal, is a qualitative theory for the number of species on an island determined by the balance of colonization and extinction rates and the roles of chance and history in determining the composition of species on an island.

The second contribution concerns the fate of a single species arriving at an island, subject to stochastic processes of birth and death. Three questions interest us: (1) given that a propagule (a certain initial number of individuals) of a certain size arrives on the island, what is the frequency distribution of subsequent population size; (2) what is the chance that descendants of the propagule will successfully colonize the island; and (3) given that it has successfully colonized the island, how long will the species persist, given the stochastic processes of birth and death, possible fluctuations in those birth and death rates, and the potential occurrence of large scale catastrophes? These are heady questions, and building the answers to them requires patience.

The general situation

We begin by assuming that the dynamics of the population are characterized by a birth rate $\lambda(n)$ and a death rate $\mu(n)$ when the population is size n (and for which there are at least some values of n for which $\lambda(n) > \mu(n)$ because otherwise the population always declines on average and that is not interesting) in the sense that the following holds:

Pr{population size changes in the next
$$dt|N(t) = n\} = 1 - \exp(-(\lambda(n) + \mu(n))dt)$$

$$\Pr\{N(t + dt) - N(t) = 1|\text{change occurs}\} = \frac{\lambda(n)}{\lambda(n) + \mu(n)} \quad (8.3)$$

$$\Pr\{N(t + dt) - N(t) = -1|\text{change occurs}\} = \frac{\mu(n)}{\lambda(n) + \mu(n)}$$

Note that Eq. (8.3) allows us to change the population size only by one individual or not at all. Furthermore, since the focus of Eq. (8.3) is an interval of time dt, it behooves us to think about the case in which dt is small. However, also note that there is no term $o(dt)$ in Eq. (8.3) because that equation is precise. For simplicity, we will define $dN = N(t + dt) - N(t)$.

Exercise 8.3 (E)

Show that, when dt is small, Eq. (8.3) is equivalent to

$$\Pr\{dN = 1|N(t) = n\} = \lambda(n)dt + o(dt)$$
$$\Pr\{dN = -1|N(t) = n\} = \mu(n)dt + o(dt) \quad (8.4)$$
$$\Pr\{dN = 0|N(t) = n\} = 1 - (\lambda(n) + \mu(n))dt + o(dt)$$

and note that we implicitly acknowledge in Eq. (8.4) that

$$\Pr\{|dN| > 1 | N(t) = n\} = o(dt)$$

All of this should remind you of the Poisson process. We continue by setting

$$p(n, t) = \Pr\{N(t) = n\} \tag{8.5}$$

and know, from Chapter 3, to derive a differential equation for $p(n, t)$ by considering the changes in a small interval of time:

$$p(n, t + dt) = p(n - 1, t)\lambda(n - 1)dt + p(n, t)(1 - (\lambda(n) + \mu(n))dt)$$
$$+ p(n + 1, t)\mu(n + 1)dt + o(dt) \tag{8.6}$$

which we then convert to a differential-difference equation by the usual procedure

$$\frac{d}{dt}p(n, t) = -(\lambda(n) + \mu(n))p(n, t) + \lambda(n - 1)p(n - 1, t)$$
$$+ \mu(n + 1)p(n + 1, t) \tag{8.7}$$

This equation requires an initial condition (actually, a whole series for $p(n, 0)$) and is generally very difficult to solve (note that, at least thus far, there is no upper limit to the value that n can take, although the lower limit $n = 0$ applies).

One relatively easy thing to do with Eq. (8.7) is to seek the steady state solution by setting the left hand side equal to 0. In that case, the right hand side becomes a balance between probabilities $p(n)$, $p(n - 1)$, and $p(n + 1)$ of population size n, $n - 1$, and $n + 1$. Let us write out the first few cases. When $n = 0$, there are only two terms on the right hand side since $p(n - 1) = 0$, so we have $0 = -\lambda(0)p(0) + \mu(1)p(1)$ where we have made the sensible assumption that $\mu(0) = 0$ and that $\lambda(0) > 0$. How might the latter occur? When we are thinking about colonization from an external source, this condition tells us that even if there are no individuals present now, there can be some later because the population is open to immigration of new individuals. Populations can be open in many ways. For example, if $N(t)$ represents the number of adult flour beetles in a microcosm of flour, then even if $N(t) = 0$ subsequent values can be greater than 0 because adults emerge from pupae, so that the time lag in the full life history makes the adult population "open" to immigration from another life history stage. For example, Peters *et al.* (1989) use the explicit form $\lambda(n) = a(n + \delta)e^{-cn}$ for which $\lambda(0) = a\delta$.

In general, we conclude that $p(1) = [\lambda(0)/\mu(1)]p(0)$. When $n = 1$, the balance becomes $0 = -(\lambda(1) + \mu(1))p(1) + \lambda(0)p(0) + \mu(2)p(2)$, from which we determine, after a small amount of algebra, that $p(2) = [\lambda(1)\lambda(0)/\mu(1)\mu(2)]p(0)$. You can surely see the pattern that will follow from here.

Exercise 8.4 (E)

Show that the general form for $p(n)$ is

$$p(n) = \frac{\lambda(n-1)\lambda(n)\ldots\lambda(0)}{\mu(1)\mu(2)\ldots\mu(n)}p(0)$$

There is one unknown left, $p(0)$. We find it by applying the condition $\sum_n p(n) = 1$, which can be done only after we specify the functional forms for the birth and death rates, and we will do that only after we formulate the general answers to questions (2) and (3).

On to the probability of colonization. Let us assume that there is a population size n_e at which functional extinction occurs; this could be $n_e = 0$ but it could also be larger than 0 if there are Allee effects, since if there are Allee effects, once the population falls below the Allee threshold the mean dynamics are towards extinction (Greene 2000). Let us also assume that there is a population size K at which we consider the population to have successfully colonized the region of interest. We then define

$$u(n) = \Pr\{N(t) \text{ reaches } K \text{ before } n_e | N(0) = n\} \quad (8.8)$$

for which we clearly have the boundary conditions $u(n_e) = 0$ and $u(K) = 1$. We think along the sample paths (Figure 8.2) to conclude that $u(n) = E_{dN}\{u(n + dN)\}$. With dN given by Eq. (8.4), we Taylor expand to obtain

$$u(n) = u(n+1)\lambda(n)dt + u(n-1)\mu(n)dt + u(n)(1 - (\lambda(n) + \mu(n))dt) + o(dt) \quad (8.9)$$

We now subtract $u(n)$ from both sides, divide by dt, and let dt approach 0 to get rid of the pesky $o(dt)$ terms, and we are left with

$$0 = \lambda(n)u(n+1) - (\lambda(n) + \mu(n))u(n) + \mu(n)u(n-1) \quad (8.10)$$

To answer the third question, we define the mean persistence time $T(n)$ by

$$T(n) = E\{\text{time to reach } n_e | N(0) = n\} \quad (8.11)$$

for which we obviously have the condition $T(n_e) = 0$.

Figure 8.2. Thinking along sample paths allows us to derive equations for the colonization probability and the mean persistence time. Starting at population size n, in the next interval of time dt, the population will either remain the same, move to $n+1$, or move to $n-1$. The probability of successful colonization from size n is then the average of the probability of successful colonization from the three new sizes. The persistence time is the same kind of average, with the credit of the population having survived dt time units.

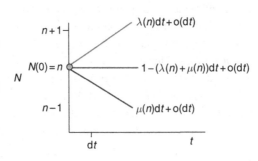

Exercise 8.5 (E)

Use the method of thinking along sample paths, with the hint from Figure 8.2, to show that $T(n)$ satisfies the equation

$$-1 = \lambda(n)T(n+1) - (\lambda(n) + \mu(n))T(n) + \mu(n)T(n-1) \qquad (8.12)$$

which is also Eq. 4-1 in MacArthur and Wilson (1967, p. 70).

We are unable to make any more progress without specifying the birth and death rates, which we now do.

The specific case treated by MacArthur and Wilson

Computationally, 1967 was a very long time ago. The leading technology in manuscript preparation was an electric typewriter with a self-correcting ribbon that allowed one to backspace and correct an error. Computer programs were typed on cards, run in batches, and output was printed to hard copy. Students learned how to use slide rules for computations (or – according to one reader of a draft of this chapter – chose another profession).

In other words, numerical solution of equations such as (8.10) or (8.12) was hard to do. Part of the genius of Robert MacArthur was that he found a specific case of the birth and death rates that he was able to solve (see Connections for more details). MacArthur and Wilson introduce a parameter K, about which they write (on p. 69 of their book): "But since all populations are limited in their maximum size by the carrying capacity of the environment (given as K individuals)" and on p. 70 they describe K as "... a ceiling, K, beyond which the population cannot normally grow." The point of providing these quotations and elaborations is this: in the MacArthur–Wilson model for extinction times (both in their book and in what follows) K is a population ceiling and not a carrying capacity in the sense that we usually understand it in ecology at which birth and death rates balance. In the next section, we will discuss a model in which there is both a carrying capacity in the usual sense and a population ceiling.

For the case of density dependent birth rates, a population ceiling means that

$$\lambda(n) = \begin{cases} \lambda n & \text{if } n \le K \\ 0 & \text{otherwise} \end{cases} \qquad (8.13)$$
$$\mu(n) = \mu n$$

where λ and μ on the right hand sides are now constants. (I know that this is a difficult notation to follow, but it is the one that is used in their book, so I use it in case you choose to read the original, which I strongly

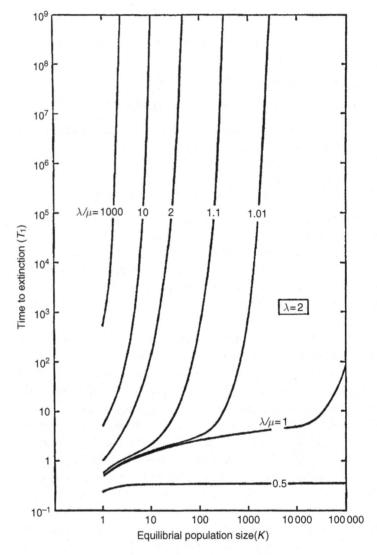

recommend.) For the case of density dependent death rates, MacArthur
and Wilson assume that

$$\lambda(n) = \lambda n$$
$$\mu(n) = \begin{cases} \mu n \text{ for } n \leq K \\ \text{whatever needed to go from } n > K \text{ to } K \text{ otherwise} \end{cases} \qquad (8.14)$$

From these equations it is clear that in neither case is K a carrying
capacity (at which birth rates and death rates are equal); rather it is a
population ceiling in the sense that "the population grows exponentially
to level K, at which point it stops abruptly" (MacArthur and Wilson

1967, p. 70). This point will become important in the next section, when we use modern computational methods to address persistence time. However, the point of Eqs. (8.13) and (8.14) is that they allow one to find the mean time to extinction, which is exactly what MacArthur and Wilson did (see Figure 8.3). The dynamics determined by Eqs. (8.13) or (8.14) will be interesting only if $\lambda \geq \mu$ (preferably strictly greater). Figures such as 8.3 led to the concept of a "minimum viable population" size (Soule 1987), in the sense that once K was sufficiently large (and the number $K = 500$ kind of became the apocryphal value) the persistence time would be very large and the population would be okay.

It is hard to overestimate the contribution that this theory made. In addition to starting an industry concerned with extinction time calculations (see Connections), the method is highly operational. It tells people to measure the density independent birth and death rates and estimate (for example from historical population size) carrying capacity and then provides an estimate of the persistence time. In other words, the developers of the theory also made clear how to operationalize it, and that always makes a theory more popular.

We shall now explore how modern computational methods can be used to extend and improve this theory.

The role of a ceiling on population size

One of the difficulties of the MacArthur–Wilson theory is that the density dependence of demographic interactions and the population ceiling are confounded in the same parameter K. We now separate them. In particular, we will assume that there is a population ceiling N_{max}, in the sense that absolutely no more individuals can be present in the habitat of interest. (My former UC Davis, and now UC Santa Cruz, colleague David Deamer used to make this point when teaching introductory biology by having the students compute how many people could fit into Yolo County, California. You might want to do this for your own county by taking its area and dividing by a nominal value of area per person, perhaps 1 square meter. The number will be enormous; that's closer to the population ceiling, the carrying capacity is much lower.)

We now introduce a steady state population size N_s defined by the condition

$$\lambda(N_s) = \mu(N_s) \tag{8.15}$$

With this condition, N_s does indeed have the interpretation of the deterministic equilibrial population size, or our usual sense of carrying capacity in that birth and death rates balance at N_s. This steady state will be stable if $\lambda(n) > \mu(n)$ if $n < N_s$ and that $\lambda(n) < \mu(n)$ if $n > N_s$. This is the simplest dynamics that we could imagine. There might be many steady states, some stable and some unstable, but all below the population ceiling.

Why bother to contain with a population ceiling? The answer can be seen in Eq. (8.12). In its current form, this is a system of equations that is "open," since each equation involves $T(n-1)$, $T(n)$, and $T(n+1)$. It is closed from the bottom – as we have already discussed – since $\mu(0) = 0$, but introducing the population ceiling is equivalent to $\lambda(N_{max}) = 0$, in which case Eq. (8.12) becomes, for $n = N_{max}$

$$-1 = -(\lambda(N_{max}) + \mu(N_{max}))T(N_{max}) + \mu(N_{max})T(N_{max} - 1) \quad (8.16)$$

and now the system is closed from both the top and the bottom.

Because the system is now closed, and because the population is being measured in number of individuals, the mean extinction time can be viewed as a vector

$$\mathbf{T(n)} = \begin{bmatrix} T(n_e + 1) \\ T(n_e + 2) \\ T(n_e + 3) \\ \cdots \\ T(N_{max} - 1) \\ T(N_{max}) \end{bmatrix} \quad (8.17)$$

and we can write Eq. (8.12) as a product of this vector and a matrix (Mangel and Tier 1993, 1994).

Before doing that, let us expand the framework in Eq. (8.12) to include catastrophic changes in population size. That is, let us suppose that catastrophic changes occur at rate $c(n)$ in the sense that

$\Pr\{\text{population size changes in the next } dt | N(t) = n\} =$
$$1 - \exp(-(\lambda(n) + \mu(n) + c(n))dt)$$

$\Pr\{\text{change is caused by a catastrophe}|\text{change occurs}\} = \dfrac{c(n)}{c(n) + \lambda(n) + \mu(n)}$
$$(8.18)$$

and that, given that a catastrophe occurs, there is a distribution $q(y|n)$ of the number of individuals who die in the catastrophe

$$\Pr\{y \text{ individuals die}|\text{catastrophe occurs}, n \text{ individuals present}\} = q(y|n) \quad (8.19)$$

We now proceed in two steps. First, you will generalize Eq. (8.12); then we will use the population ceiling and matrix formulation to solve the generalization.

Exercise 8.6 (M)

Show that the generalization of Eq. (8.12) is

$$-1 = \lambda(n)T(n+1) - ((\lambda(n) + \mu(n) + c(n))T(n)) + \mu(n)T(n-1)$$
$$+c(n)\sum_{v=0}^{n} q(y|n)T(n-y) \tag{8.20}$$

in which we allow that no individual or all individuals might die in a catastrophe. (This is an unlikely event, chosen mainly for mathematical pleasure of starting the sum from 0, rather than a larger value. In practice, $q(y|n)$ will be zero for small values of y. Although, it is conceivable, I suppose, that a hurricane occurs and there are no deaths caused by it.)

Now we define $s(n)$ by $s(n) = \lambda(n) + \mu(n) + c(n)(1 - q(0|n))$ and a matrix M whose first four rows and five columns are

$$\begin{matrix}
-s(n_e + 1) & \lambda(n_e + 1) & 0 & 0 & 0 \\
\mu(n_e + 2) + c(n_e + 2)q(1|n_e + 2) & -s(n_e + 2) & \lambda(n_e + 2) & 0 & 0 \\
c(n_e + 3)q(2|n_e + 3) & \mu(n_e + 3) + c(n_e + 3)q(1|n_e + 3) & -s(n_e + 3) & \lambda(n_e + 3) & 0 \\
c(n_e + 4)q(3|n_e + 4) & c(n_e + 4)q(2|n_e + 4) & m(n_e + 4) + c(n_e + 4)q(1|n_e + 4) & -s(n_e + 4) & \lambda(n_e + 4)
\end{matrix}$$

$$\tag{8.21}$$

and we define the vector -1 by

$$-1 = \begin{bmatrix} -1 \\ -1 \\ -1 \\ \cdots \\ -1 \\ -1 \end{bmatrix} \tag{8.22}$$

Once we have done this, Eq. (8.20) takes the compact form

$$MT(n) = -1 \tag{8.23}$$

and if we define the inverse matrix M^{-1} then Eq. (8.23) has the formal solution

$$T(n) = -M^{-1}1 \tag{8.24}$$

Now we take advantage of living in the twentyfirst century. Virtually all good software programs have automatic inversion of matrices, so that computation of Eq. (8.24) becomes a matter of filling in the matrix and then letting the computer go at it.

In Figure 8.4, I show the results of this calculation for the flour beetle model (Peters *et al.* 1989) in which $\lambda(n) = b_0(n + \delta)\exp(-b_1 n)$ and $\mu(n) = d_1 n$ for the case in which there are no catastrophes and three different cases of catastrophic declines (Mangel and Tier 1993, 1994). For the parameters $b_0 = 0.13$, $b_1 = 0.0165$, $\delta = 1$, $d_1 = 0.088$ the steady state is at about $n = 26$, so a population ceiling of 50 would be much larger than the steady state. As seen in the figures, whether the

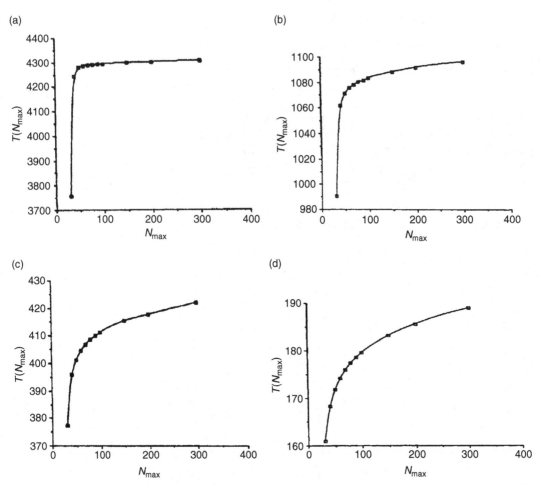

Figure 8.4. Application of Eq. (8.24) to the flour beetle model in which $\lambda(n) = b_0(n+\delta)\exp(-b_1 n)$ and $\mu(n) = d_1 n$ with $b_0 = 0.13$, $b_1 = 0.0165$, $\delta = 1$, $d_1 = 0.088$. (a) No catastrophes. Note the rapid rise in persistence time; (b) rate of catastrophes $c = 0.01$ and $q(y|n)$ following a binomial distribution with probability of death $p = 0.5$; (c) $c = 0.025$, $p = 0.5$; and (d) $c = 0.05$, $p = 0.5$.

population ceiling is 50 or 300 has little effect on the predictions in the absence of catastrophes, but more of an effect in the presence of catastrophes.

This theory is nice, easily extended to other cases (see Connections), reminds us of connections to matrix models, and is easily employed (and easier every day). However, it is also limited because of the assumption about the nature of the stochastic fluctuations that affect population size. In the next two sections, we will turn to a much more general formulation, and investigate both its advantages and its limitations.

A diffusion approximation in the density independent case

We now turn to a formulation in which there is no density dependence and the fluctuations in population size are determined by Brownian motion (Lande 1987, Dennis *et al.* 1991, Foley 1994, Ludwig 1999, Saether *et al.* 2002, Lande *et al.* 2003). As with the method of MacArthur and Wilson, this method is easy to use, but also requires some care in thinking about its application.

When population size is low, density dependent factors are often assumed (rightfully or wrongfully) to be immaterial for the growth of the population. We let $X(t)$ denote the population size at time t and start by assuming discrete dynamics of the form

$$X(t + dt) = \lambda X(t)e^{\zeta(t)} \tag{8.25}$$

where we understand dt to be arbitrary just now (usually people begin with $dt = 1$), λ to be the maximum per capita growth rate, and $\zeta(t)$ to be a Gaussian distributed random variable with mean 0 and variance vdt. If we set $N(t) = \log(X(t))$ then Eq. (8.25) becomes

$$N(t + dt) = N(t) + \log(\lambda) + \zeta(t) \tag{8.26}$$

and if we now define r by $\log(\lambda) = rdt$ and set $\zeta(t) = \sqrt{v}dW$, then Eq. (8.26) becomes a familiar friend

$$dN = rdt + \sqrt{v}dW \tag{8.27}$$

for which we will assume the range of $N(t)$ is 0 (corresponding to 1 individual) to a population ceiling K. (An even simpler case would be to assume that $r = 0$, so that the logarithm of population size simply follows Brownian motion; see Engen and Saether (2000) for an example). The notation is a little bit tricky – in the previous section N represented population size, but here it represents the logarithm of population size; I am confident, however, that you can deal with this switch.

The great advantage of Eq. (8.27) is that the data requirements for its application are minimal: we need to know the mean and variance in the increments in population size. These can often be obtained by surveys, which need not even be regularly spaced in time (although when they are not, one needs to be careful when estimating r and v).

Associated with Eq. (8.27) is a mean persistence time $T(n)$ for a population starting at $N(0) = n$ and defined according to

$$T(n) = E\{\text{time to reach } N = 0 | N(0) = n\} \tag{8.28}$$

with which we associate the boundary condition $T(0) = 0$ (remember that, because we are in log-population space, $n = 0$ corresponds to one individual). We know that a second boundary condition will be needed and we obtain it as follows. If the population ceiling is very large, then following logic we used previously, we expect that $T(K) \approx T(K + \varepsilon)$, where ε is a small number. If we Taylor expand to first order in ε, the condition is the same as the reflecting condition $(dT/dn)|_{n=K} = 0$.

Before discussing the solution of Eq. (8.28), let us reconsider Eq. (8.27) from two perspectives. The first is an alternative derivation. Recall that $X(t)$ is population size, so that if we assumed that there are no density dependent factors, we have in the deterministic case $dX = rXdt$ or $(1/X)(dX/dt) = rdt$, from which Eq. (8.27) follows if we set $N = \log(X)$ and assume that r has a deterministic and a stochastic component.

The second perspective is that we actually know how to solve Eq. (8.27) by inspection, with the initial condition that $N(0) = n$

$$N(t) = n + rt + \sqrt{v}W(t) \tag{8.29}$$

We can read off directly from Eq. (8.29) the mean and confidence intervals for $N(t)$.

In the course of his work on the endangered Alabama Beach Mouse (*Peromyscus polionotus ammobates*), my student Chris Wilcox developed data appropriate for Eqs. (8.27)–(8.29) and kindly allowed me to use them (Figure 8.5). This mouse is found only along the coast of Alabama, USA, in sand dunes and threats to its persistence include development of the coast and periodic catastrophic storms. In Figures 8.5b and c, I show the projections of the mean and 95% confidence intervals for population size at two different sites in the study area. In one case the mean population size shows an increasing trend with time, in the other a decreasing trend (Chris worked at two other sites, which also showed similar properties). Notice, however, that the confidence intervals quickly become very wide – which means that although we have a prediction, it is not very precise. It is data such as these that caused Ludwig (1999) to ask if it is meaningful to estimate probability of extinction (also see Fieberg and Ellner (2000)).

Let us now return to Eqs. (8.27) and (8.28). We know that $T(n)$ will be the solution of the differential equation

$$\frac{v}{2}\frac{d^2T}{dn^2} + r\frac{dT}{dn} = -1 \tag{8.30}$$

with the boundary conditions that we discussed before ($T(0) = 0$ and $(dT/dn)|_{n=K} = 0$).

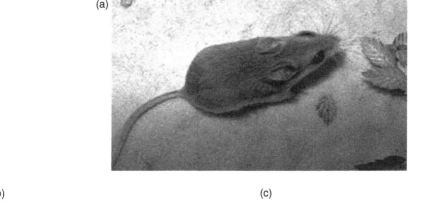

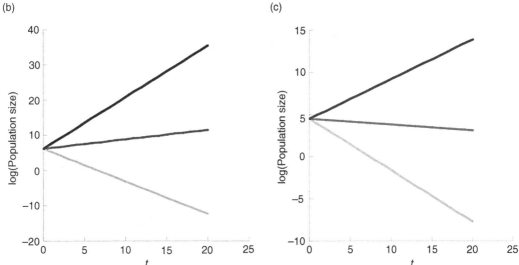

Figure 8.5. (a) The Alabama Beach Mouse, and projections (in 2002) of population size based on Eq. (8.29) at two different sites: (b) the site BPSU, and (c) the site GINS. Photo courtesy of US Fish and Wildlife Service. I show the mean and the upper and lower 95% confidence intervals.

Exercise 8.7 (E/M)

Suppose that $r = 0$. Show that the solution of Eq. (8.30) is

$$T(n) = \frac{2n}{v}\left(K - \frac{n}{2}\right) \tag{8.31}$$

so that if the population starts at the ceiling ($n = K$) the mean persistence time is $T(K) = K^2/v$. Interpret its shape and compare it with the MacArthur–Wilson result (Figure 8.3).

When $r > 0$, we rewrite Eq. (8.30) using subscripts to denote derivatives as

$$T_{nn} + \frac{2r}{v}T_n = -\frac{2}{v} \tag{8.32}$$

and we now recognize that the left hand side is the same as

$$\exp\left(-\frac{2r}{v}n\right)\frac{d}{dn}\left[T_n\exp\left(\frac{2r}{v}n\right)\right]$$

so that Eq. (8.32) can be rewritten as

$$\frac{d}{dn}\left[T_n\exp\left(\frac{2r}{v}n\right)\right] = -\frac{2}{v}\exp\left(\frac{2r}{v}n\right) \tag{8.33}$$

which we integrate once to obtain

$$T_n\exp\left(\frac{2r}{v}n\right) = -\frac{1}{r}\exp\left(\frac{2r}{v}n\right) + c_1 \tag{8.34}$$

where c_1 is a constant of integration. We now apply the boundary condition that the first derivative of $T(n)$ is 0 when $N = K$ to conclude that $c_1 = (1/r)\exp((2r/v)n)$ and we can thus write that

$$T_n = -\frac{1}{r} + \frac{1}{r}\exp\left(\frac{2r}{v}K\right)\exp\left(-\frac{2r}{v}n\right) \tag{8.35}$$

and we now integrate this equation once again to obtain

$$T(n) = -\frac{n}{r} - \frac{v}{2r^2}\exp\left(\frac{2r}{v}K\right)\exp\left(-\frac{2r}{v}n\right) + c_2 \tag{8.36}$$

where c_2 is a second constant of integration and to which we apply the condition $T(0) = 0$ to conclude that $c_2 = (v/2r^2)\exp((2r/v)K)$ so that

$$\begin{aligned} T(n) &= -\frac{n}{r} + \frac{v}{2r^2}\left[\exp\left(\frac{2r}{v}K\right) - \exp\left(\frac{2r}{v}K\right)\exp\left(-\frac{2r}{v}n\right)\right] \\ &= \left(-\frac{n}{r}\right) + \frac{v}{2r^2}\exp\left(\frac{2r}{v}K\right)\left[1 - \exp\left(-\frac{2r}{v}n\right)\right] \end{aligned} \tag{8.37}$$

Note that this solution involves n, r, K, and v in a nonlinear and relatively complicated fashion.

Exercise 8.8 (E/M)

Foley (1994) uses a different method of obtaining the solution (see his Appendix) and also writes it in a different manner by introducing the parameter $s = r/v$:

$$T(n) = \frac{1}{2rs}[\exp(2sK)(1 - \exp(-2sn)) - 2sn] \tag{8.38}$$

Show that Eqs. (8.37) and (8.38) are the same. Now assume that $sK \ll 1$ and show by Taylor expansion of the exponential to third order in K that

$$T(K) \approx \frac{K^2}{v}\left(1 + \frac{2r}{3v}K\right) \tag{8.39}$$

which tells us how the deterministic and stochastic components of the dynamics affect the persistence time. Note, for example, that the mean persistence time now grows as the cube of the population ceiling.

As with the theory of MacArthur and Wilson, this theory is appealing because of its operational simplicity. It tells us to measure the mean and variance of the per capita changes (and, in more advanced form, the autocorrelation of the fluctuations to correct the estimate of variance (Foley 1994, Lande *et al.* 2003) and to estimate the ceiling of the population). From these will come the mean persistence time via Eqs. (8.37) or (8.38). It is reasonable to ask, however, how these predictions might depend upon life history characteristics (see Connections), on more general density dependence, or when we ever might see a population ceiling.

The general density dependent case

We now turn to the general density dependent case, so that, instead of Eq. (8.27), the population satisfies the stochastic differential equation

$$dN = b(N)dt + \sqrt{a(N)}dW \qquad (8.40)$$

where $b(n)$ and $a(n)$ are known functions. We will assume that there is a single stable steady state n_s for which $b(n_s) = 0$, a population size n_e at which we consider the population to be extinct and, although there surely is a true population ceiling, as will be seen we do not need to specify (or use) it.

These ideas are captured schematically in Figure 8.6. We know that $T(n)$ will now satisfy the equation

$$\frac{a(n)}{2}T_{nn} + b(n)T_n = -1 \qquad (8.41)$$

with one boundary condition $T(n_e) = 0$. For the second boundary condition, as before we require that $\lim_{n\to\infty} T_n = 0$, which by analogy with the previous section, indicates that the population ceiling is infinite. Were it not, we would apply the reflecting condition at K.

We solve this equation using the same method as in the previous section, but now in full generality. To begin, we set $W(n) = T_n$, so that Eq. (8.41) can be rewritten as

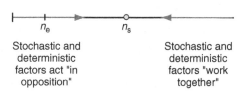

Figure 8.6. A schematic description of the general case for stochastic extinction. The population dynamics are $dN = b(N)dt + \sqrt{a(N)}dW$ with a single deterministic stable steady state n_s and a population size n_e at which we consider the population to be extinct. For starting values of population size smaller than n_s, the factors of stochastic fluctuation toward extinction and deterministic increase towards the steady state are acting in opposition, while for values greater than n_s they are acting in concert in the sense that the deterministic factors reduce population size.

$$W_n + \frac{2b(n)}{a(n)} W = -\frac{2}{a(n)} \tag{8.42}$$

and we now define

$$\Phi(n) = \int_{n_e}^{n} \frac{2b(s)}{a(s)} ds \tag{8.43}$$

which allows us to write Eq. (8.42) as

$$\frac{d}{dn}[We^{\Phi(n)}] = -\frac{2}{a(n)}e^{\Phi(n)} \tag{8.44}$$

and, integrating from n to ∞, we conclude that

$$W(n) = T_n = 2e^{-\Phi(n)} \int_{n}^{\infty} \frac{e^{\Phi(s)}}{a(s)} ds \tag{8.45}$$

and we pause momentarily. Note that Eq. (8.45) automatically satisfies the boundary condition $\lim_{n \to \infty} T_n = 0$. Also note that the function $\Phi(n)$ defined by Eq. (8.43) involves the ratio of the infinitesimal mean and variance. The bigger the variance – thus the stronger the fluctuations – the smaller the ratio (and thus the integral), all else being equal.

We integrate Eq. (8.45) once more, this time from n_e to n (recalling that $T(n_e) = 0$) and end up with the formula for the mean persistence time in the general case

$$T(n) = 2 \int_{n_e}^{n} e^{-\Phi(s)} \int_{s}^{\infty} \frac{e^{\Phi(y)}}{a(y)} dy ds \tag{8.46}$$

Equation (8.46) is our desired result. It gives the mean persistence time for a population starting at size n when the dynamics follow the general stochastic differential equation (8.40). This general formulation tells us actually very little about specific situations, but the literature contains many examples of its application once the functional forms for $b(n)$ and $a(n)$ are chosen according to the biological situation at hand (see Connections for some examples).

Transitions between peaks on the adaptive landscape

Schluter (2000) writes "Natural selection is a surface" (p. 85). When that surface has multiple peaks, we are faced with the problem of understanding how transitions between one adaptive peak to a higher one can occur across a valley of fitness. To my knowledge, there have

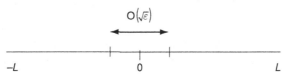

Confidence interval for stationary density

$$O\!\left(\sqrt{\varepsilon}\right)$$

−L   0   L

Figure 8.7. Our understanding of transitions from one fitness peak to another on the adaptive landscape will rely on the metaphor of an Ornstein–Uhlenbeck process $dX = -Xdt + \sqrt{\varepsilon}dW$, for which the stationary density is Gaussian with mean 0 and variance $\varepsilon/2$. We consider an interval $[-L, L]$ that is much larger than the confidence interval for the stationary density, which is $O(\sqrt{\varepsilon})$, as domain of one adaptive peak and values of X outside of this interval another adaptive peak, so that when X escapes from the interval, a transition has occurred. As described in the text, we are interested in three kinds of times: the deterministic time to return from initial value L to $2\sqrt{\varepsilon}$, the mean time to escape from $[-L, L]$, and the mean time to escape from an initial value $X(0) > 0$ without returning to 0.

been just two attempts (Ludwig 1981, Lande 1985) to answer this question. (Gavrilets (2003) has a nice, general review of the topic.) Here, I will walk you through Ludwig's analysis; the problem is highly stylized and the analysis is difficult, but at the end we will have a deepened and sharpened intuition about the general issue. Our starting point is the Ornstein–Uhlenbeck process

$$dX = -Xdt + \sqrt{\varepsilon}dW \tag{8.47}$$

for which we know that the stationary density is Gaussian, with mean 0 and variance $\varepsilon/2$, so that the confidence intervals for the stationary density are $O(\sqrt{\varepsilon})$; for example the 95% confidence interval is approximately $[-\sqrt{\varepsilon}, \sqrt{\varepsilon}]$. Thus the mechanism that we consider consists of deterministic return to the origin with fluctuations superimposed upon that deterministic return.

We shall also consider a larger interval, $[-L, L]$ (Figure 8.7) and metaphorically consider that within this larger interval we have one "fitness peak" and that outside of it we have another "fitness peak," so that escape from the interval $[-L, L]$ corresponds to transition between peaks.

Our first calculation is an easy one. If we replace Eq. (8.47) by the deterministic equation $dx/dt = -x$, we know that the only behavior is attraction towards the origin.

Exercise 8.9 (E)

Show that the deterministic return time $T_d(L)$ to reach $\sqrt{\varepsilon}$, given by the solution of $dx/dt = -x$, with $x(0) = L$, is $T_d(L) = \log(L) - \log(\sqrt{\varepsilon})$. Thus, conclude that the deterministic time to return from initial point L to the vicinity of the origin scales as $\log(L/\sqrt{\varepsilon})$.

Our second calculation is not much more complicated. Suppose that we allow $T(x)$ to denote the mean time to escape from the interval $[-L, L]$, given that $X(0) = x$. We know that $T(x)$ satisfies the equation

$$\frac{\varepsilon}{2}T_{xx} - xT_x = -1 \tag{8.48}$$

with the boundary conditions $T(-L) = T(L) = 0$. The solution of Eq. (8.48) with these boundary conditions is not too difficult, but it is very cumbersome and hard to learn from. Let us think a bit more about the situation. First note that Eq. (8.48) is symmetrical about $x=0$, because if we set $y=-x$, we obtain the same differential equation and same boundary conditions. Second, think about what happens to the stochastic process when it returns, ever so momentarily to $X=0$: at that point there is no deterministic component to the dynamics and the mean of the fluctuations is 0 as well. In other words, we could think of the process at $X=0$ being reflected rather than continuing through to negative values. Thus, we can equivalently consider Eq. (8.48) with the boundary condition $T(L)=0$ and $T_x|_{x=0}=0$, i.e. reflection at the origin.

Exercise 8.10 (M/H)

Show that the solution of Eq. (8.48) satisfying the boundary conditions $T(L)=0$ and $T_x|_{x=0}=0$ is

$$T(x) = \frac{2}{\varepsilon} \int_x^L \exp\left(\frac{y^2}{\varepsilon}\right) \int_0^y \exp\left(-\frac{s^2}{\varepsilon}\right) ds\, dy \qquad (8.49)$$

The solution given by Eq. (8.49) presents some new challenges for analysis, because of the positive exponential in the outer integrals. Let us begin by thinking of the integral over s and making the transformation of variables $u = s/\sqrt{\varepsilon}$ so that the integral over s becomes $\int_0^{y/\sqrt{\varepsilon}} \exp(-u^2)\sqrt{\varepsilon}\,du$. Now if we think that the noise is small ($\varepsilon \ll 1$) then when y gets away from 0 the upper limit is getting large. We recognize then that we are computing the normalization constant for a Gaussian distribution once again.

Exercise 8.11 (E)

Show that $\int_0^\infty \exp(-u^2)du = \sqrt{\pi}$. Here is a hint: remember that $\int_{-\infty}^\infty \exp(-u^2/2\sigma^2)du = \sqrt{2\pi}\sigma$.

If we then approximate the integral over s in Eq. (8.49) by $\sqrt{\pi\varepsilon}$ we can conclude that

$$T(x) \approx 2\sqrt{\frac{\pi}{\varepsilon}} \int_x^L \exp\left(\frac{y^2}{\varepsilon}\right) dy \qquad (8.50)$$

Now the integral in Eq. (8.50) is something new for us, because of the positive exponent. Just looking at this integral suggest that the main contribution to it will come from the vicinity of L, because the integrand is largest there. We can make this more precise. First, let us make the change of variables $v = y/\sqrt{\varepsilon}$ so that the integral we have to consider is

$I = \int_{x/\sqrt{\varepsilon}}^{L/\sqrt{\varepsilon}} \exp(v^2)dv$. We integrate this by parts, much as we did in the expansion of the tail of the Gaussian distribution:

$$\int_a^b \exp(v^2)dv = \int_a^b \frac{d}{dv}(\exp(v^2))\frac{1}{2v}dv = \exp(v^2)\frac{1}{2v}\Big|_a^b + \int_a^b \exp(v^2)\frac{1}{2v^2}dv$$

$$(8.51)$$

Using the right hand side of Eq. (8.51), we conclude that

$$T(x) \approx \sqrt{\pi}\left[\frac{\exp\left(\frac{L^2}{\varepsilon}\right)}{\frac{L}{\sqrt{\varepsilon}}} - \frac{\exp\left(\frac{x^2}{\varepsilon}\right)}{\frac{x}{\sqrt{\varepsilon}}}\right] \qquad (8.52)$$

In other words, the time to escape from $[-L, L]$ when starting at $x > 0$ grows like $T(x) \sim \exp(L^2/\varepsilon)$. Thus, on average it takes a very long time to escape from a domain of attraction.

This conclusion – of a very long average time to escape – accounts for the reality that most trajectories starting at $x > 0$ will be drawn back towards the origin and spend a long time there before ultimately escaping. However, now let us focus on a special subset of trajectories: those which start at $x > 0$ and escape (through L) without ever having returned to the origin. We can thus define

$$u(x,t) = \Pr\{\text{exit } (0,L] \text{ by time } t \text{ without ever having returned to } \qquad (8.53)$$
$$0|X(0) = x > 0\}$$

Now, since $u(x, t)$ is the probability of exiting from $(0, L]$ without having crossed 0 by time t, $u_t(x, t)$ is the probability density for the time of exit. That is

$$u_t(x,t)dt = \Pr\{\text{exit from } (0,L] \text{ in the interval } t, t + dt \qquad (8.54)$$
$$\text{without having crossed } 0|X(0) = x\}$$

and consequently the mean time for trajectories that start at x and exit without having crossed 0 is

$$h(x) = \int_0^\infty t u_t(x,t)dt \qquad (8.55)$$

Now we define

$$w(x) = \lim_{t \to \infty} u(x,t) \qquad (8.56)$$

so that $w(x)$ is the probability of ever escaping from $(0, L]$ without first revisiting L. We recognize that

$$\frac{u(x,t)}{w(x)} = \frac{\Pr\{\text{exiting by time } t \text{ and never hitting } 0|X(0) = x > 0\}}{\Pr\{\text{exiting without hitting } 0|X(0) = x > 0\}} \qquad (8.57)$$

is the conditional probability of exiting by time t without hitting 0. Thus

$$T_c(x) = \frac{h(x)}{w(x)} \tag{8.58}$$

is the mean time to exit $(0, L]$ given that $X(0) = x$ without returning to 0. Our goal is to find this time.

The probability of escape by time t without returning to the origin satisfies the differential equation

$$u_t = \frac{\varepsilon}{2} u_{xx} - x u_x \tag{8.59}$$

with the initial condition $u(x, 0) = 0$ and the boundary conditions $u(0, t) = 0$ and $u(L, t) = 1$. We know that $w(x)$ satisfies the time-independent version of Eq. (8.59) with the same boundary conditions $(w(0) = 0, w(L) = 1)$.

Exercise 8.12 (E)

Show that

$$w(x) = \frac{\displaystyle\int_0^{\frac{x}{\sqrt{\varepsilon}}} \exp(s^2) ds}{\displaystyle\int_0^{\frac{L}{\sqrt{\varepsilon}}} \exp(s^2) ds} \tag{8.60}$$

We now derive an equation for $h(x)$ using what I like to call the Kimura Maneuver, since it was popularized by M. Kimura in his work in population genetics (Kimura and Ohta 1971).

We begin by differentiating Eq. (8.59) with respect to time, multiplying by t and integrating:

$$\int_0^\infty t u_{tt} \, dt = \frac{\varepsilon}{2} \int_0^\infty t u_{txx} \, dt - x \int_0^\infty t u_{tx} \, dt \tag{8.61}$$

We then exchange the order of integration and differentiation on the right hand side of Eq. (8.61), and that, for example $\int_0^\infty t u_{txx} \, dt = (\partial^2/\partial x^2) \int_0^\infty t u_t \, dt = h_{xx}$ and which allows us to rewrite Eq. (8.61) as

$$\frac{\varepsilon}{2} h_{xx} - x h_x = \int_0^\infty t u_{tt} \, dt \tag{8.62}$$

and we now integrate the right hand side of Eq. (8.62) by parts, keeping in mind that both $u(x, 0) = 0$ and that the time derivative of $u(x, t)$ goes to 0 as $t \to \infty$, so that

$$\int_0^\infty t u_{tt} \, dt = t u_t|_0^\infty - \int_0^\infty u_t dt = -[\lim_{t\to\infty} u(x, t) - u(x, 0)] = -w(x) \qquad (8.63)$$

and combining this with Eq. (8.62) we conclude that

$$\frac{\varepsilon}{2} h_{xx} - x h_x = -w(x) \qquad (8.64)$$

We are now going to understand certain properties of $w(x)$, the solution of Eq. (8.60) without actually solving it. To do so, we shall find it handy to employ Dawson's integral (Abramowitz and Stegun (1974); it is also kind of fun to do a web search with key words "Dawson's Integral"):

$$D(y) = \exp(-y^2) \int_0^y \exp(s^2) ds \qquad (8.65)$$

so that we can rewrite $w(x)$ as

$$w(x) = \exp\left(\frac{x^2}{\varepsilon} - \frac{L^2}{\varepsilon}\right) \frac{D\left(\frac{x}{\sqrt{\varepsilon}}\right)}{D\left(\frac{L}{\sqrt{\varepsilon}}\right)} \qquad (8.66)$$

Now recall that the main contribution to the integral component of Dawson's integral will come from the end point (and to leading order is $(1/2y)\exp(y^2)$) so that $D(y) \sim 1/2y$ when y is large. Using this relationship allows us to rewrite Eq. (8.66) as

$$w(x) \sim 2 \frac{L}{\sqrt{\varepsilon}} \exp\left(\frac{x^2}{\varepsilon} - \frac{L^2}{\varepsilon}\right) D\left(\frac{x}{\sqrt{\varepsilon}}\right)$$

and Eq. (8.64) becomes

$$\frac{\varepsilon}{2} h_{xx} - x h_x \sim -2 \frac{L}{\sqrt{\varepsilon}} \exp\left(\frac{x^2}{\varepsilon} - \frac{L^2}{\varepsilon}\right) D\left(\frac{x}{\sqrt{\varepsilon}}\right) \qquad (8.67)$$

Using an integrating factor, we can rewrite Eq. (8.67) as

$$\frac{d}{dx}\left[h_x \exp\left(-\frac{x^2}{\varepsilon}\right)\right] \sim \frac{-4}{\varepsilon} \frac{L}{\sqrt{\varepsilon}} \exp\left(-\frac{L^2}{\varepsilon}\right) D\left(\frac{x}{\sqrt{\varepsilon}}\right) \qquad (8.68)$$

We integrate this equation to obtain

$$h_x \exp\left(-\frac{x^2}{\varepsilon}\right) \sim \frac{-4}{\varepsilon} L \exp\left(-\frac{L^2}{\varepsilon}\right) \left[\int_0^{\frac{x}{\sqrt{\varepsilon}}} D(y) dy - c\right] \qquad (8.69)$$

where c is a constant of integration. Consequently,

$$h_x \sim -\frac{4}{\varepsilon} L \exp\left(\frac{x^2}{\varepsilon} - \frac{L^2}{\varepsilon}\right) \left[\int_0^{\frac{x}{\sqrt{\varepsilon}}} D(y)dy - c\right] \tag{8.70}$$

We are almost there.

To continue the analysis, we set $F(x/\sqrt{\varepsilon}) = \int_0^{x/\sqrt{\varepsilon}} D(y)dy$, so that

$$h_x \sim \frac{4}{\varepsilon} L \exp\left(\frac{x^2}{\varepsilon} - \frac{L^2}{\varepsilon}\right)\left\{c - F\left(\frac{x}{\sqrt{\varepsilon}}\right)\right\} \tag{8.71}$$

which we integrate to obtain

$$h(x) \sim \frac{4}{\varepsilon} L \exp\left(-\frac{L^2}{\varepsilon}\right)\left[\int_0^x \exp\left(\frac{s^2}{\varepsilon}\right)\left\{c - F\left(\frac{s}{\sqrt{\varepsilon}}\right)\right\}ds\right] \tag{8.72}$$

Clearly $h(0) = 0$. To satisfy the other boundary condition, we must have that $\int_0^L \exp(s^2/\varepsilon)\{c - F(s/\sqrt{\varepsilon})\}ds = 0$ from which we conclude that

$$c = \frac{\displaystyle\int_0^L \exp\left(\frac{s^2}{\varepsilon}\right)F\left(\frac{s}{\sqrt{\varepsilon}}\right)ds}{\displaystyle\int_0^L \exp\left(\frac{s^2}{\varepsilon}\right)ds} \tag{8.73}$$

We now recall that $D(y) \sim 1/2y$ for large y, so that $\int_0^{x/\sqrt{\varepsilon}} D(y)dy \sim 1/2\log(x/\sqrt{\varepsilon})$ and consequently, since the main contributions to the integrals in Eq. (8.73) come from the upper limit, we conclude

$$c \sim \frac{1}{2}\log\left(\frac{L}{\sqrt{\varepsilon}}\right) \tag{8.74}$$

We keep this in mind as we proceed to the next, and final, step. Now, since $F(s) > 0$, from Eq. (8.72) we conclude that

$$h(x) < \frac{4}{\varepsilon} L \exp\left(-\frac{L^2}{\varepsilon}\right)c\int_0^x \exp\left(\frac{s^2}{\varepsilon}\right)ds \tag{8.75}$$

so that

$$T_c(x) = \frac{h(x)}{w(x)} < \frac{\frac{4}{\varepsilon} L c \exp\left(-\frac{L^2}{\varepsilon}\right)\displaystyle\int_0^x \exp\left(\frac{s^2}{\varepsilon}\right)ds}{2\frac{L}{\sqrt{\varepsilon}}\exp\left(\frac{x^2}{\varepsilon} - \frac{L^2}{\varepsilon}\right)D\left(\frac{x}{\sqrt{\varepsilon}}\right)} = \frac{\frac{2}{\sqrt{\varepsilon}}c\displaystyle\int_0^x \exp\left(\frac{s^2}{\varepsilon}\right)ds}{\exp\left(\frac{x^2}{\varepsilon}\right)D\left(\frac{x}{\sqrt{\varepsilon}}\right)} \tag{8.76}$$

but, from Eq. (8.65), $\exp(y^2)D(y) = \int_0^y \exp(s^2)ds$. We thus conclude that

$$T_c(x) < \frac{2}{\sqrt{\varepsilon}} c = \frac{1}{\sqrt{\varepsilon}} \log\left(\frac{L}{\sqrt{\varepsilon}}\right) \qquad (8.77)$$

Let us summarize the analysis. The deterministic return time from L to a vicinity of the origin scales as $\log(L/\sqrt{\varepsilon})$, the mean time for all stochastic trajectories to escape from $[-L, L]$ scales as $\exp(L^2/\varepsilon)$ and the mean time to escape without ever returning to 0 scales as $\log(L/\sqrt{\varepsilon})$. These are vastly different times – indeed many orders of magnitude when L is moderate and ε is small. The mean time to escape, conditioned on not returning to the origin, is much, much smaller than the average escape time. Thus, the mean time to escape, conditioned on not returning to the origin appears as a punctuated trajectory. Gavrilets (2003) refers to those trajectories that escape as "lucky" ones and notes that they do it quickly.

That was a lot of hard work. And to some extent, the payoff is in a deeper understanding of the problem, rather than in the details of the mathematical analysis. Indeed, in retrospect, our discussion of the gambler's ruin can shed light on this problem. Recall that, in the gambler's ruin, we decided that in general one is very rarely going to be able to break the bank, but that if it is going to happen, it will happen quickly (with a run of extreme good luck). And the same holds in this case: it is rare for a trajectory starting at $X(0) = x$ to escape without returning to the origin, but when a trajectory does escape, the escape happens quickly.

I feel obligated to end this section with a discussion of punctuated equilibrium. In 1971, Stephen J. Gould and Niles Eldredge (then youngsters aiming to become the Waylon and Willy – the outlaws – of evolutionary biology (see http://en.wikipedia.org/wiki/Outlaw_country if you do not understand the context of this metaphor) coined the phrase "punctuated equilibrium" and offered punctuated equilibria as an alternative to the gradualism of Darwinian theory as it was then understood (Gould and Eldredge 1977; Gould 2002, p. 745 ff.) Writing about it thirty years later, Gould said "First of all, the theory of punctuated equilibrium treats a particular level of structural analysis tied to a particular temporal frame ... Punctuated equilibrium is not a theory about all forms of rapidity, at any scale or level, in biology. Punctuated equilibrium addresses the origin and deployment of species in geological time" (Gould 2002, pp. 765–766). The two key concepts in this theory are stasis and punctuation, which I have illustrated schematically in Figure 8.8; Lande (1985) describes the situation in this manner "species maintain a constant phenotype during most of their existence and that new species originate suddenly in small localized populations" (p. 7641). The question can be put like this: since the geological record

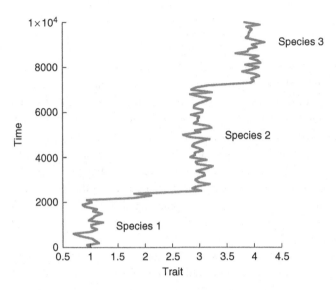

Figure 8.8. The key concepts in Eldredge and Gould's challenge to Darwinian gradualism are stasis and punctuation. These are illustrated here for a hypothetical trajectory of three species characterized by a generic trait. For the first 2000 time units (call one time unit a thousand years if you like) or so, the trait fluctuates around the value 1 (stasis) but then around time 2000 there is a rapid transition (punctuation) to trait value equal to 3, which persists for about another 5000 time units (more stasis) after which another punctuation event occurs. During the periods of stasis, there are fluctuations around trait values. The question, and challenge, is whether this picture is consistent with the notion of gradual modification. Our answer is yes.

does look like the schematic in Figure 8.8, what challenge is posed to the Darwinian notion of gradualism? Indeed, some authors (Margulis and Sagan 2002) have argued that the entire mathematical and technical machinery associated with the gradualist Darwinian paradigm falls apart because of punctuated equilibrium. In her recent and wonderful book, West-Eberhard (2003) emphasizes (pp. 474–475) that "punctuated equilibrium is a hypothesis regarding rates of phenotypic evolution and does not challenge gradualism. The patterns and causes of change in evolutionary rates are at issue, not the relative importance of selection versus development." Gould (2002) makes this clear on p. 756: "Rather, punctuated equilibrium refutes the third and most general meaning of Darwinian gradualism, designated in Chapter 2 (see pp. 152–155) as 'slowness and smoothness (but not constancy) of rate'." Put more simply, we could ask: can a single mechanism account for the pattern shown in Figure 8.8 or does one require multiple mechanisms and processes? There is a flip answer to the question, as there always is a flip answer to any question. In this case, it is that stasis corresponds to a relatively constant environment and microfluctuations around an adaptive peak of fitness and punctuation corresponds to an environmental change in which the current adaptive peak becomes non-adaptive, another peak arises and there is strong selection from the formerly adaptive peak to the new one. But this answer is somewhat dissatisfying since it is flip and it makes key assumptions about the link between the environment and the fitness peaks that are not present in the underlying Darwinian framework. At first it would seem that we have answered the question in this section and to some extent, we have in that we now understand how the pattern of stasis and punctuation might be consistent with gradualism. However, Gavrilets (2003) emphasizes that this kind of analysis is not the full story, which can be found in his paper.

Anderson's theory of vitality and the biodemography of survival

We now turn to the application of diffusion processes to understanding survival. To begin, we will review life tables, the Euler–Lotka equation of population demography and methods for solving it. We will then see how a diffusion model of vitality can be used to characterize survival.

In a life table (Kot 2001, Preston *et al.* 2001), we specify the schedule of survival to age a $\{l(a), a = 0, 1, 2, \ldots, a_{max}\}$ and the expected reproduction at age a $\{m(a), a = 0, 1, 2, \ldots, a_{max}\}$, where a_{max} is the maximum age, for individuals in a population. To characterize the growth of the population, we compute the number of births $B(t)$ at time t. These have two sources: individuals who were born at a time $t - a$ and who are now of age a, and individuals who were present at time 0 and are still contributing to the population. If we denote the births due to the latter individuals by $Q(t)$, we can write that

$$B(t) = \sum_{a=0}^{a_{max}} B(t-a)l(a)m(a) + Q(t) \tag{8.78}$$

Equation (8.78) is called Lotka's renewal equation for population growth. If $t \gg a_{max}$ we assume that $Q(t) = 0$ since none of the individuals present at time 0 will have survived to produce offspring at time t. Let us do that, for convenience drop the upper limit in the summation, and assume that $B(t) = Ce^{rt}$, where the constant C (which actually becomes immaterial) and the population growth rate r are to be determined. Setting $Q(t) = 0$ and substituting into Eq. (8.78) we have

$$Ce^{rt} = \sum_{a=0}^{} Ce^{r(t-a)} l(a)m(a) \tag{8.79}$$

from which we conclude that

$$1 = \sum_{a=0}^{} e^{-ra}l(a)m(a) \tag{8.80}$$

Equation (8.80) is called the Euler–Lotka equation. In the literature, it is usually treated as an equation for r, depending upon $\{l(a), m(a), a = 0, 1, 2, \ldots, a_{max}\}$. Dobzhansky and Fisher recognized that if there are no density-dependent effects, then r is also a measure of fitness for a genotype with schedule of births and survival given by $\{m(a), l(a), a = 0, 1, 2, \ldots, a_{max}\}$. Note also that since $e^{-ra}l(a)m(a)$ sums to 1, we can think of it as the probability density function for the fraction of the population at age a (Demetrius 2001).

According to Eq. (8.80), the solution r is a function $r(l(a), m(a))$. We may ask: how does r change with a change in the schedule of fecundity or survival (Charlesworth 1994)? For example, let us implicitly differentiate Eq. (8.80) with respect to $m(y)$:

$$0 = e^{-ry}l(y) + \sum_{a=0} -e^{-ra}al(a)m(a)\frac{\partial r}{\partial m(y)} \qquad (8.81)$$

from which we conclude that

$$\frac{\partial r}{\partial m(y)} = \frac{e^{-ry}l(y)}{\sum_{a=0} e^{-ra}al(a)m(a)} \qquad (8.82)$$

The denominator of Eq. (8.82) has units of time and in light of our interpretation of $e^{-ra}l(a)m(a)$ as a probability density, we conclude that the denominator is a mean age; indeed it is generally viewed as the mean generation time. Note that the right hand side of Eq. (8.82) declines with age as long as $r > 0$; this observation is one of the foundations of W. D. Hamilton's theory of senescence (Hamilton 1966, 1995).

We can ask the same question about the dependence of r on the schedule of survival. This is slightly more complicated.

Exercise 8.13 (M/H)

Set $l(a) = \prod_{y=0}^{a-1} s(y)$ so that $s(y)$ has the interpretation of the probability of surviving from age y to age $y + 1$. Show that

$$\frac{\partial r}{\partial s(y)} = \frac{\sum_{a=y+1} e^{-ra}l(a)m(a)}{s(y)\sum_{a=0} e^{-ra}al(a)m(a)} \qquad (8.83)$$

Now, to actually employ Eqs. (8.82) or (8.83), we need to know r. In my experience, Newton's method, which I now explain, has always worked to find a solution for the Euler–Lotka equation. Think of Eq. (8.80) as an equation for r, which we write as $H(r) = 0$, where

$$H(r) = \sum_{a=0} e^{-ra}l(a)m(a) - 1 \qquad (8.84)$$

Now suppose that r_T is the solution of this equation (the subscript T standing for True), so that $H(r_T) = 0$. If we Taylor expand $H(r_T)$ around r to first order in $r_T - r$, we have $H(r_T) \sim H(r) + H_r(r_T - r)$, where the derivative is evaluated at r. Now the left hand side of this equation is 0 and if we solve the right hand side for r_T we obtain $r_T \approx r - (H(r)/H_r)$, which suggests an iterative procedure by which we might find the true value of r. Choose an initial value r_0 and then iteratively define r_n by

$$r_n = r_{n-1} - \frac{H(r_{n-1})}{H_r(r_{n-1})} \tag{8.85}$$

Under very general conditions, r_n will converge to the true value. A good starting value is often $r_0 = 0$ or, to be a bit more elaborate, one might write that the expected lifetime reproduction of an individual $R_0 = \sum_{a=0} l(a)m(a)$ as $\exp(rT_g)$, where T_g is the average generation time in a population that is not growing, given by the denominator of Eq. (8.82) when $r = 0$. In that case, a starting value could be $r_0 = \log(R_0)/T_g$.

If the preceding material is new to you, or you feel kind of rusty and would like more familiarity, I suggest that you try the following exercise.

Exercise 8.14 (E)

Waser *et al.* (1995) published the following information on the life history of mongoose in the Serengeti. Some of it is shown in the table below.

Age (a)	$l(a)$	$m(a)$
0	1	0
1	0.41	0
2	0.328	0.21
3	0.252	0.39
4	0.182	0.95
5	0.142	1.32
6	0.085	1.48
7	0.057	2.45
8	0.031	3.78
9	0.021	2.56
10	0.014	4.07
11	0.005	3.76
12	0.005	3
13	0.002	2
14	0.002	0

(a) Compute R_0 and use Newton's method to find r. (b) What do you predict will happen to R_0 and r if the survivorship for age 5 and beyond decreases by just 5%? Now compute the new values. (c) Compute R_0 and r if individuals delay reproduction from year 2 to year 4 because of a food shortage. That is, assume that individuals are now 4 years old when they get the reproduction previously associated with a 2 year old, 5 years old when they get reproduction previously associated with a 3 year old, etc. Interpret your results.

Underlying all of these calculations is the schedule of survival and fecundity and it is the schedule of survival that I now want to investigate, using a theory of organismal vitality determined by Brownian motion due to Jim Anderson at the University of Washington (Anderson 1992, 2000). Survival to any age is the result of internal processes and external processes, so that we write $l(a) = P_e(a)P_v(a)$, where $P_e(a)$ is the probability of survival to age a associated with external causes (random or accidental mortality, we might say), and which we assume to be e^{-ma}, and $P_v(a)$ is the survival to age a associated with internal processes and organismal vitality.

Let us define $V(t)$ to be that vitality, with the notion that $V(t) > 0$ means that the organism is alive and that $V(t) = 0$ corresponds to death. Anderson assumes that $V(t)$ satisfies the following stochastic differential equation

$$dV = -\rho dt + \sigma dW \qquad (8.86)$$

so we see that $V(t)$ declines deterministically at a constant rate and is incremented in a stochastic fashion by Brownian motion. This is clearly the simplest assumption that one can make, but, as the work of Anderson shows, one can go a long way with it.

It may be helpful to think of vitality as the result of a variety of hidden physiological and biochemical processes which, when taken together, determine an overall state of the organism. It may also be that there is no such thing as "external" mortality – that all mortality is vitality driven. For example, the ability to escape a falling tree (a random event in the forest) may depend upon internal state as much as anything else.

The probability density for $V(t)$, defined so that $p(v, t|v_0, 0)dv = \Pr\{v \leq V(t) \leq v + dv|V(0) = v_0\}$, satisfies the forward equation

$$p_t = \rho p_v + \frac{\sigma^2}{2}p_{vv} \qquad (8.87)$$

and from the definition, we know that $p(v, t|v_0, 0) = \delta(v - v_0)$; as before, one boundary condition will be $p(v, t|v_0, 0) \to 0$ as $v \to \infty$. For the second boundary condition, since an organism starting with no vitality is dead $p(v, t|0, 0) = 0$. The solution of Eq. (8.87) satisfying the specified initial and boundary conditions is not exceptionally difficult to find, but this is one of the few cases in this book in which I say "we look it up." Some of the best sources for looking up solutions of the standard diffusion equation are Carslaw and Jaeger (1959), Goel and Richter-Dyn (1974), and Crank (1975) (this particular solution is computed by the "method of images" in which we satisfy the boundary condition at 0 by subtracting an appropriate mirror image quantity). The solution is

$$p(v, t|v_0, 0) = \frac{1}{\sqrt{2\pi\sigma^2 t}} \left[\exp\left(-\frac{(v - v_0 + \rho t)^2}{2\sigma^2 t} \right) \right.$$

$$\left. - \exp\left(-\frac{(v + v_0 + \rho t)^2}{2\sigma^2 t} + \frac{2\rho v_0}{\sigma^2} \right) \right] \qquad (8.88)$$

Exercise 8.15 (E|M)

When one encounters a purported solution in the literature, even if one does not derive the solution, one should check it as much as possible. Do this with Eq. (8.88) by verifying that it satisfies the differential equation, the initial condition and the boundary conditions.

Since the organism survives to age t if it has positive vitality at that age, we conclude that

$$P_v(t) = \int_0^\infty p(v, t|v_0, 0) dv \qquad (8.89)$$

Evidently, $P_v(t)$ will be related to Gaussian cumulative distribution functions, but Anderson chooses to use the error function erf(z) and complementary error function erfc(z) (Abramowitz and Stegun 1974) and since this will give you a new tool, I will do that too (I am also personally very fond of the error function (Mangel and Ludwig 1977)). These functions are defined by

$$\text{erf}(z) = \frac{2}{\sqrt{\pi}} \int_0^z e^{-t^2} dt \qquad \text{erfc}(z) = \frac{2}{\sqrt{\pi}} \int_z^\infty e^{-t^2} dt \qquad (8.90)$$

Exercise 8.16 (E)

Show that $\text{erf}(z) + \text{erfc}(z) = 1$ and that $\text{erf}(z) = 2\Phi(z\sqrt{2}) - 1$, where $\Phi(z)$ is the Gaussian cumulative distribution function, i.e.

$$\Phi(z) = \left(\frac{1}{\sqrt{2\pi}} \right) \int_{-\infty}^z \exp\left(-\left(\frac{u^2}{2} \right) \right) du$$

Anderson next introduces the scaled parameters $r = \rho/v_0$ and $s = \sigma/v_0$ so that both the deterministic loss of vitality and the intensity of the stochastic increments are measured relative to the initial vitality. In terms of these scaled parameters, Eq. (8.90) becomes

$$p(t, v|v_0, 0) = \frac{1}{\sqrt{2\pi s^2 t}} \left[\exp\left(-\frac{(v - 1 + rt)^2}{2s^2 t} \right) - \exp\left(-\frac{(v + 1 + rt)^2}{2s^2 t} + \frac{2r}{s^2} \right) \right]$$

$$(8.91)$$

and the viability related probability of survival is

$$P_v(t) = \frac{1}{2}\left[\operatorname{erfc}\left(\frac{rt-1}{s\sqrt{2t}}\right) - \exp\left(\frac{2r}{s^2}\right)\operatorname{erfc}\left(\frac{rt+1}{s\sqrt{2t}}\right)\right] \qquad (8.92)$$

We are now able to construct the probability of surviving to age t

$$l(t) = e^{-mt}P_v(t) \qquad (8.93)$$

Anderson (2000) explores a number of properties of the model, including the predicted rate of mortality at age, the expected lifespan, maximum likelihood estimates of parameters, and connections between the parameters and physiological variables such as body size or environmental variables such as the dose of a putative toxin (Figure 8.9). I encourage you to read his paper, which is well-written and informative.

Figure 8.9. Comparison of Anderson's theory of vitality (lines) and some experimental results (symbols). (a) Survival of the water flea *Daphnia pulex* at densities of 1, 8, and 32 individuals/ml. (b) Survival of subyearling chinook salmon *Oncorhynchus tshawytscha* at four saturation levels of total dissolved gas. Reprinted with permission.

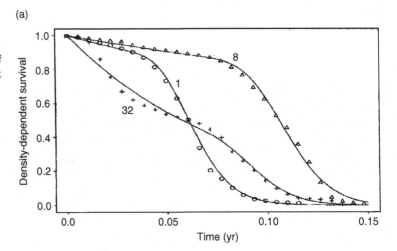

(a)

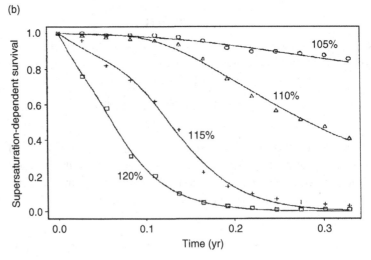

(b)

Connections

Escape from the domain of attraction

Escape from a domain of attraction, or more generally the transition between two deterministic steady states driven by fluctuations, has wide applicability in biology, chemistry, economics, engineering, and physics (Klein 1952, Brinkman 1956, Kubo *et al.* 1973, Arnold 1974, van Kampen 1977, Schuss 1980, Ricciardi and Sato 1990). Many of the ideas go back to Hans Kramers, a master of modern physics (ter Haar 1998) who modeled chemical reactions as Brownian motion in a field of force (Kramers 1940). Other introductions to the problem from the perspective of physics or chemistry can be found in van Kampen (1981b), Gardiner (1983), Gillespie (1992) and Keizer (1987). In biology, the classic paper of Ludwig (1975) brings to bear many of the tools that we have discussed. The more mathematical side of the question is interesting and challenging because the problems involve large deviations (Bucklew 1990). There are ways to use the method of thinking along sample paths to understand the general problem (Freidlin and Wentzell 1984), but the mathematical difficulty rises rapidly.

Extensions of the MacArthur–Wilson theory

The Theory of Island Biogeography spawned an industry (a good place to start is Goel and Richter-Dyn (1974)). Indeed, the late 1960s were heady times for theoretical biology. In the remarkable period of the late 1960s, optimal foraging theory (MacArthur and Pianka 1966, Emlen 1966), island biogeography (MacArthur and Wilson 1967), and metapopulation ecology (Levins 1969) developed. The theories of optimal foraging and island biogeography developed rapidly and led to experiments, and the development of new fields such as behavioral ecology. On the other hand, metapopulation theory languished for quite a while before a phase of development, and the subsequent development in the 1980s was mainly theoretical (see, for example, Hanski (1989)). The rapid success of island biogeography and optimal foraging theory relative to metapopulation ecology teaches us two things. First, the developers of optimal foraging theory and island biogeography provided a prescription: (i) measure a certain set of empirically clear parameters, and (ii) given these parameters, compute a quantity of interest. Levins did not do this as explicitly. For example, in classical rate-maximizing optimal foraging theory as we discussed at the start of the book, one measures handling times of, energy gain from, and encounter rates with, food items, and then is able to predict the diet breadth of a foraging organism. In classical island biogeography, one measures per capita birth and death rates and carrying capacity of an island and is able to predict the mean persistence time. On the other hand, it is not exactly clear what to measure in metapopulation theory or how to apply it. Indeed, authors still revisit the original Levins model trying to operationalize it (Hanski 1999). Second, Levins published his seminal paper in an entomology journal and on biological control. In the heady times of the late 1960s, such "applied" biology was scorned by many colleagues. A very interesting discussion of the role of theory in conservation biology is found in Caughley (1994), which caused an equally interesting rejoinder (Hedrick et al. 1996).

Catastrophes and conservation

Catastrophic changes in population size can occur for many reasons, and in the past decade or so there has been increasing recognition of the role of catastrophes in regulating populations. Connections to the literature can be found in Mangel and Tier (1993, 1994), Young (1994), Root (1998), and Wilcox and Elderd (2003).

Ceilings and the distribution of extinction times

Mangel and Tier (1993) show that the second moment of the persistence time satisfies $S(n) = -2M^{-1}T(n)$ from which the variance and coefficient of variation of the persistence time can be calculated. By using that calculation, they conclude that persistence times are approximately exponentially distributed. Ricciardi and Sato (1990) provide a more general discussion of first-passage times.

The diffusion approximation

In population biology, the most general formulation of the diffusion approximation (Halley and Iwasa 1998, Diserud and Engen 2000, Hakoyama and Iwasa 2000, Lande et al. 2003) takes the form $dX = b(X)dt + \sqrt{a_e}X dW_e + \sqrt{a_d X} dW_d$, where a_e and a_d are the environmental and demographic components of stochasticity and dW_e and dW_d are independent increments in the Brownian motion process, the former interpreted according to Stratonovich calculus (to account for autocorrelation in environmental fluctuation) and the latter according to Ito calculus (to account for demographic stochasticity). Engen et al. (2001) show results similar to those in Figure 8.5, for the decline of the barn swallow, using a model that has both environmental and demographic fluctuations. The main message, however, is the same (see their

Figure 4). It is worthwhile to wonder when the diffusion approximation gives valid conclusions for life histories that do not meet the assumptions of the model (Wilcox and Possingham 2002).

Connecting models and data

In general, we will need to estimate extinction risk and mean time to extinction from time series that may often be short and sparse. This presents new challenges, both conceptually and technically (Ludwig 1999, Hakoyama and Iwasa 2000, Fieberg and Ellner 2000, Iwasa *et al.* 2000).

Punctuated equilibrium

As with some of the other topics in this book, there are probably 1000 papers or more on punctuated equilibrium, what it means, and what it does not mean (Gould and Eldredge 1993). A recent issue of *Genetica* (**112–113** (2001)) was entirely dedicated to the rate, pattern and process of microevolution (see Hendry and Kinnison (2001) for the introduction of the issue). Pigliucci and Murren (2003) have recently wondered if the rate of macroevolution (the escape from a domain of attraction) can be so fast as to pass us by. West-Eberhard (2003) is a grand source of ideas for models (but not of models) in this area. The calculation by Lande (1985) using a very similar approach to the one that we did, with a quantitative genetic framework, warranted a news piece in *Science* (Lewin 1986). Jim Kirchner (Kirchner and Weil 1998, 2000; Kirchner 2001, 2002) has written a series of interesting and excellent papers on the nature of rates in the fossil record.

Biodemography

Demography – generally understood today as a social science – is the statistical study of human populations, especially with respect to size and density, distribution and vital statistics. The goal is to describe patterns, understand pattern and process, and predict the consequences of change on those patterns. The foundations of demography are the life table, the Gompertz mortality model (Gompertz 1825), and the stable age distribution that arises as the solution of the Euler–Lotka renewal equation. Biodemography seeks to merge demography with evolutionary thinking (Gavrilov and Gavrilova 1991, Wachter and Finch 1997, Carey 2001, Carey and Judge 2001, Carey 2003). The result, for example, will be to use the comparative method to explore similarities and differences of patterns across species and to understand the patterns and mechanisms of vital statistics as the result of evolution by natural (and sometimes artificial) selection. The Gavrilovs (Gavrilov and Gavrilova 1991) note that Raymond Pearl actually understood the importance of doing this – and wanted to do it – but lacked the tools. For example, Pearl and Miner (1935) wrote "For it appears clear that there is no one universal 'law' of mortality ... different species may differ in the age distribution of their dying just as characteristically as they differ in their morphology" and that "But what is wanted is a measure of the individual's total activities of all sorts, over its whole life; and also a numerical expression that will serve as a measure of net integrated effectiveness of all the environmental forces that have acted upon the individual throughout its life"; the methods of life history analysis that we have discussed in other chapters allow exactly this kind of calculation. The papers of Pearl are still wonderful reads, and most are easily accessible through *JSTOR*; I encourage you to take a look at them (Pearl 1928, Alpatov and Pearl 1929; Pearl and Parker 1921, 1922a, b, c, d, 1924a, b; Pearl *et al.* 1923, 1927, 1941). As I write the final draft (April 2005) one of the most interesting issues in biodemography, with enormous importance for aging modern societies, is that of mortality plateaus. The Gompertz model can be summarized as

$$\frac{\mathrm{d}N}{\mathrm{d}t} = -m(t)N$$
$$\frac{\mathrm{d}m}{\mathrm{d}t} = km$$

(8.94)

That is, the population declines exponentially and the coefficient of mortality characterizing the decline grows exponentially. However, in the past twenty years many studies of the oldest members of populations (see, for example, Vaupel *et al.* (1998)) have shown that mortality rates may not grow exponentially in the oldest individuals, but may plateau or even decline. Why this is so is not understood and is an area of active and intense research (Mueller and Rose 1996, Pletcher and Curtsinger 1998, Kirkwood 1999, Wachter 1999, Demetrius 2001, Mangel 2001a, Weitz and Fraser 2001, de Grey 2003a, b).

Financial engineering: a different way of thinking

Louis Bachelier developed much of the theory of Brownian motion in the same manner as Einstein did, but five years before Einstein in his (Bachelier's) doctoral thesis "Theory of Speculation". This thesis is translated from the French and published by Cootner (1964). Thus, many of the tools that we have discussed in the previous and this chapter apply to economic problems; this area of research is now called financial engineering (Wilmott 1998) and some readers may decide that this is indeed an attractive career for them. I was very tempted to include a detailed section on these methods, but both one of the referees of the proposal and my wife thought that financial engineering did not belong as an application of this tool kit, so I leave it to Connections. The basic ideas behind the pricing of stock options, due to Merton (1971) and to Black and Scholes (1972), employ stochastic differential equations but in a somewhat different manner than we have used. There are three key components. The first is a stock whose price $S(t)$ follows a log-normal model for dynamics

$$\frac{1}{S}\frac{dS}{dt} = \mu dt + \sigma dW \tag{8.95}$$

where we interpret μ as a measure of the mean rate of return of the stock and σ as a measure of the volatility of the price of the stock. The second component is a riskless investment such as a bank account or bond paying interest rate r, in the sense that if $B(t)$ is the price of the bond then $dB/dt = rB$. Looking backwards from a time T at which we know the value of the bond, $B(T)$, we conclude that the appropriate price at time t is $B(t) = B(T)e^{-r(T-t)}$. The third component is the option, which is a right to buy or sell a stock at a fixed price (called the exercise or strike price k) up to a fixed time T (called the expiration or maturity date). With an American option one can exercise at any time prior to T, with a European option only at time T. A put option is exercised by selling the stock; a call option is exercised by buying the stock. An option does not have to be exercised, but a future has to be exercised (so that a future might better be called a Must). With these definitions, we can compute the values of call and put options. For example, a call option will be exercised only if the price of the stock on day T exceeds the exercise price; for a put option, the reverse is true. Ultimately, the goal is to find the value of the option on days prior to T, so we define

$$W(s, t, T) = E\{\text{European option on day } T | S(t) = s\} \tag{8.96}$$

for which we have the end condition for a call option $W(s, T, T) = \max\{s - k, 0\}$. Unlike evolutionary problems, in which one maximizes a fitness function, option pricing is based on the concepts of hedging and no arbitrage. Hedging consists of buying an amount Δ of actual stock (in addition to the right to buy stock at a later date) in such a manner that whether the price of the stock rises (making the option more valuable) or falls (making it less valuable), the net value of the portfolio, $\Pi(t) = W(s, t, T) - \Delta s$ consisting of the option minus the amount spent on stock, stays the same. The condition of no arbitrage (arbitrage is the general process of profiting from price discrepancies) means that the portfolio grows at the same rate as the riskless investments, so that $d\Pi = r\Pi\,dt$. These conditions are sufficient to allow one to derive the diffusion equation for the price of the option. See Wilmott (1998) for an excellent introduction to such matters. Another terrific book on these topics, and which will seem familiar to you technically if not scientifically, is Dixit and Pindyck (1994).

References

Abrahams, M. V. and Healey, M. C. (1993). Some consequences of variation in vessel density: a manipulative field experiment. *Fisheries Research*, **15**, 315–322.

Abramowitz, M. and Stegun, I. (1974). *Handbook of Mathematical Functions*. New York: Dover Publications.

Acosta, C. A. (2002). Spatially explicit dispersal dynamics and equilibrium population sizes in marine harvest refuges. *ICES Journal of Marine Science*, **59**, 458–468.

Ade, R. (1989). *The Trout and Salmon Handbook*. New York: Facts on File.

Adkison, M. D. and Zhenming, S. (2001). A comparison of salmon escapement estimates using a hierarchical Bayesian approach versus separate maximum likelihood estimation of each year's return. *Canadian Journal of Fisheries and Aquatic Sciences*, **58**, 1663–1671.

Adler, F. R. (1993). Migration alone can produce persistence of host–parasitoid models. *American Naturalist*, **141**, 642–650.

Ahmed, A. M., Baggot, S. L., Maingon, R. and Hurd, H. (2002). The costs of mounting an immune response are reflected in the reproductive fitness of the mosquito *Anopheles gambiae*. *Oikos*, **97**, 371–377.

Allen, J. E. and Maizels, R. M. (1997). Th1-Th2: reliable paradigm or dangerous dogma? *Immunology Today*, **18**, 387–392.

Alpatov, W. W. and Pearl, R. (1929). Experimental studies on the duration of life. XII. Influence of temperature during the larval period and adult life on the duration of the life of the imago of drosophila melanogaster. *American Naturalist*, **63**, 37–67.

Anand, P. (2002). Decision-making when science is ambiguous. *Science*, **295**, 1839.

Anderson, J. J. (1992). A vitality-based stochastic model for organism survival. In D. L. DeAngelis and L. J. Gross (editors) *Individual-Based Models and Approaches in Ecology*. New York: Chapman and Hall, pp. 256–277.

Anderson, J. J. (2000). A vitality-based model relating stressors and environmental properties to organismal survival. *Ecological Monographs*, **70**, 445–470.

Anderson, L. G. (1991a). Efficient policies to maintain total allowable catches in ITQ fisheries with at-sea processing. *Land Economics*, **67**, 141–157.

Anderson, L. G. (1991b). A note on market power in ITQ fisheries. *Journal of Environmental Economics and Management*, **21**, 291–296.

Anderson, R. M. (editor) (1982). *Population Dynamics of Disease: Theory and Applications*. London: Chapman and Hall.

Anderson, R. M. (1991). The Kermack–McKendrick epidemic threshold theorem. *Bulletin of Mathematical Biology*, **53**, 3–32.

Anderson, R. M. and May, R. M. (1978). Regulation and stability of host–parasite population interactions. I Regulatory processes. *Journal of Animal Ecology*, **47**, 219–247.

Anderson, R. M. and May, R. M. (1991). *Infectious Diseases of Humans. Dynamics and Control*. Oxford: Oxford University Press.

Anonymous, (2002). *The State of the World Fisheries and Aquaculture*. Rome: Food and Agriculture Organization of the United Nations.

Anscombe, F. J. (1950). Sampling theory of the negative binomial and logarithmic series distributions. *Biometrika*, **37**, 358–382.

Antia, R. and Lipsitch, M. (1997). Mathematical models of parasite responses to host immune defenses. *Parasitology*, **115**, S115–S167.

Antia, R., Levin, B. R. and May, R. M. (1994). Within-host population dynamics and the evolution and maintenance of microparasite virulence. *American Naturalist*, **144**, 457–472.

Antia, R., Nowak, M. A. and Anderson, R. M. (1996). Antigenic variation and the within-host dynamics of parasites. *Proceedings of the National Academy of Sciences*, **93**, 985–989.

Apollonio, S. (1994). The use of ecosystem characteristics in fisheries management. *Reviews in Fisheries Science*, **2**, 157–180.

Apostolaki, P., Milner-Gulland, E. J., McAllister, M. K. and Kirkwood, G. P. (2002). Modelling the effects of establishing a marine reserve for mobile species. *Canadian Journal of Fisheries and Aquatic Sciences*, **59**, 405–415.

Arnold, L. (1973). *Stochastic Differential Equations*. New York: John Wiley and Sons.

Arnold, L. (1974). *Stochastic Differential Equations: Theory and Applications*. London: Wiley Interscience.

Austin, D. J., White, N. J. and Anderson, R. M. (1998). The dynamics of drug action on the within-host population growth of infectious agents: melding pharmacokinetics with pathogen population dynamics. *Journal of Theoretical Biology*, **194**, 313–339.

Babcock, E. A. and Pikitch, E. K. (2000). A dynamic programming model of fishing strategy choice in a multispecies trawl fishery with trip limits. *Canadian Journal of Fisheries and Aquatic Sciences*, **57**, 357–370.

Bailey, N. T. J. (1953). The total size of a general stochastic epidemic. *Biometrika*, **40**, 177–185.

Bakun, A. (1996). *Patterns in the Ocean. Oceanic Process and Marine Population Dynamics*. La Jolla, CA: California Sea Grant.

Barlow, N. D. (1993). A model for the spread of bovine Tb in New Zealand possum populations. *Journal of Applied Ecology*, **30**, 156–164.

Barlow, N. D. (2000). Non-linear transmission and simple models for bovine tuberculosis. *Journal of Animal Ecology*, **69**, 703–713.

Barnard, G. A. and Bayes, T. (1958). Studies in the history of probability and statistics. IX. Thomas Bayes's essay towards solving a problem in the doctrine of chances. *Biometrika*, **45**, 293–315.

Barrowman, N. J. and Myers, R. A. (2000). Still more spawner-recruitment curves: the hockey stick and its generalizations. *Canadian Journal of Fisheries and Aquatic Sciences*, **57**, 665–676.

Bartlett, B. R. (1964). Patterns in the host-feeding habit of adult parasite hymenoptera. *Annals of the Entomological Society of America*, **57**, 344–350.

Bartlett, M. S. (1957). Measles periodicity and community size. *Journal of the Royal Statistical Society A*, **120**, 48–70.

Basar, T. and Olsder, G. J. (1982). *Dynamic Noncooperative Game Theory*. New York: Academic Press.

Bazykin, A. D. (1998). *Nonlinear Dynamics of Interacting Populations*. Singapore and River Edge, NJ: World Scientific.

Beamish, R. J. (2002). Obituary. William Edwin Ricker OC, FRSC, LLD, DSc. *Journal of Fish Biology*, **60**, 285–287.

Begon, M., Bowers, R. G., Kadianakis, N. and Hodgkinson, D. E. (1992). Disease and community structure: the importance of host self-regulation in a host–host pathogen model. *The American Naturalist*, **139**, 1131–1150.

Begon, M., Feore, S. M., Brown, K., *et al.* (1998). Population and transmission dynamics of cowpox in bank voles: testing fundamental assumptions. *Ecology Letters*, **1**, 82–86.

Begon, M., Hazel, S. M., Baxby, D., *et al.* (1999). Transmission dynamics of a zoonotic pathogen within and between wildlife host species. *Proceedings of the Royal Society of London*, B**266**, 1939–1945.

Behnke, R. J. (1992). *Native Trout of Western North America*. Bethesda, MD: American Fisheries Society.

Behnke, R. J. (2002). *Trout and Salmon of North America*. New York: Free Press.

Belew, R. K. and Mitchell, M. (editors) (1996). *Adaptive Individual in Evolving Populations: Models and Algorithms*. Reading, MA: Addison Wesley.

Bellman, R. E. (1984). *Eye of the Hurricane: An Autobiography*. Singapore: World Scientific.

Bellman, R. and Cooke, K. L. (1963). *Differential-Difference Equations*. New York: Academic Press.

Bellows, T. S. and Hassell, M. P. (1988). The dynamics of age-structured host-parasitoid interactions. *Journal of Animal Ecology*, **57**, 259–268.

Bender, C. M. and Orszag, S. A. (1978). *Advanced Mathematical Methods for Scientists and Engineers*. New York: McGraw Hill.

Berger, R., Quack, M. and Tschumper, G. S. (2000). Electroweak quantum chemistry for possible precursor molecules in the evolution of biomolecular homochirality. *Helvetica Chimica Acta*, **83**, 1919–1950.

Bernoulli, D. (1738/1954). Exposition of a new theory on the measurement of risk. *Economietrica*, **22**, 23–36.

Bertsekas, D. P (1976). *Dynamic Programming and Stochastic Control*. New York: Academic Press.

Bertsekas, D. P. (1995). *Dynamic Programming and Optimal Control. Volumes One and Two*. Belmont, MA: Athena Scientific.

Beverton, R. J. H. (1992). Patterns of reproductive strategy parameters in some marine teleost fishes. *Journal of Fish Biology*, **41** (Supplement B), 137–160.

Beverton, R. J. H. (1994). *Notes on the Use of Theoretical Models in the Study of the Dynamics of Exploited Fish Populations. From the Lectures of R. J. H. Beverton presented at U.S. Fishery Laboratory, Beaufort, North Carolina, Bureau of Commercial Fisheries 1951*. Beaufort, NC: Marine Fisheries Section, American Fisheries Society.

Beverton, R. J. H. and Holt, S. J. (1957). *On the Dynamics of Exploited Fish Populations*. London: HMSO.

Beverton, R. J. H. and Holt, S. J. (1959). A review of the lifespans and mortality rates of fish in nature, and their relation to growth and other physiological characteristics. In G. E. W. Wolstenholme and M. O'Connor (editors) *CIBA Foundation Colloquia on Ageing*. London: CIBA Foundation, pp. 142–174.

Bharucha-Reid, A. T. (1997 (1960)). *Elements of Stochastic Processes*. New York: Dover.

Bjorksted, E. P. (2000). Stock–recruitment relationships for life cycles that exhibit concurrent density dependence. *Canadian Journal of Fisheries and Aquatic Sciences*, **57**, 459–467.

Bjørnstad, O. N. and Grenfell, B. T. (2001). Noisy clockwork: time series analysis of population fluctuations in animals. *Science*, **293**, 638–643.

Black, F. and Scholes, M. (1972). The pricing of options and corporate liabilities. *The Journal of Political Economy*, **81**, 637–654.

Blackmond, D. G., McMillan, C. R., Ramdeechul, S., Schorm, A. and Brown, J. M. (2001). Origins of asymmetric amplifications in autocatalytic alkylzinc additions. *Journal of the American Chemical Society*, **123**, 10 103–10 104.

Bleistein, N. and Handelsman, R. A. (1975). *Asymptotic Expansions of Integrals*. New York: Holt, Rinehart and Winston.

Blythe, S. P., Nisbet, R. M. and Gurney, W. S. C. (1984). The dynamics of population models with distributed maturation periods. *Theoretical Population Biology*, **25**, 289–311.

Bonsall, M. P. and Hastings, A. (2004). Demographic and environmental stochasticity in predator–prey metapopulation dynamics. *Journal of Animal Ecology*, **73**, 1043–1055.

Bonsall, M. B., Hassell, M. P. and Asefa, G. (2002). Ecological trade-offs, resource partitioning, and coexistence in a host–parasitoid assemblage. *Ecology*, **83**, 925–934.

Boreman, J., Nakashima, B. S., Wilson, J. A. and Kendall, R. L. (editors) (1997). *Northwest Atlantic Groundfish: Perspectives on a Fishery Collapse*. Bethesda, MD: American Fisheries Society.

Botsford, L. W., Hastings, A. and Gaines, S. D. (2001). Dependence of sustainability on the configuration of marine reserves and larval dispersal distance. *Ecology Letters*, **4**, 144–150.

Bouloux, C., Langlais, M. and Silan, P. (1998). A marine host-parasite model with direct biological cycle and age structure. *Ecological Modelling*, **107**, 73–86.

Bradford, M., Myers, R. A. and Irvine, J. R. (2000). Reference points for coho salmon (*Oncorhynchus kisutch*) harvest rates and escapement goals based on freshwater production. *Canadian Journal of Fisheries and Aquatic Sciences*, **57**, 677–686.

Brenner, S. (1999). Theoretical biology in the third millennium. *Philosophical Transactions of the Royal Society of London, Series B, Biological Sciences*, **354**, 1963–1965.

Briggs, C. J. (1993). Competition among parasitoid species on a stage-structure host and its effect on host suppression. *American Naturalist*, **141**, 372–397.

Briggs, C. J. and Hoopes, M. F. (2004). Stabilizing effects in spatial parasitoid-host and predator-prey models: a review. *Theoretical Population Biology*, **65**, 299–315.

Brinkman, H. C. (1956). Brownian motion in a field of force and the diffusion theory of chemical reactions, II. *Physica*, **22**, 149–155.

Brodziak, J. K. T., Overholtz, W. J. and Rago, P. J. (2001). Does spawning stock affect recruitment of New England groundfish. *Canadian Journal of Fisheries and Aquatic Sciences*, **58**, 306–318.

Brookhart, M. A., Hubbard, A. E., van der Laan, M. J., Colford Jr., J. M. and Eisenberg, J. N. S. (2002). Statistical estimation of parameters in a disease transmission model: analysis of a *Cryptosporidium* outbreak. *Statistics in Medicine*, **21**, 3627–3638.

Brooks, E. N. (2002). Using reproductive values to define optimal harvesting for multisite density-dependent populations: example with a marine reserve. *Canadian Journal of Fisheries and Aquatic Sciences*, **59**, 875–885.

Brown, R. (1828). A brief account of microscopical observations made in the months of June, July, and August, 1827, on the particles contained in the pollen of plants; and on the general existence of active molecules in organic and inorganic bodies. *The Philosophical Magazine*, **4**, 161–173.

Brunet, L. R., Dunne, D. W. and Pearce, E. J. (1998). Cytokine interaction and immune response during *Schistosoma mansoni* infection. *Parasitology Today*, **14**, 422–427.

Bucklew, J. A. (1990). *Large Deviation Techniques in Decision, Simulation, and Estimation*. New York: Wiley Interscience.

Burnham, K. P. and Anderson, D. R. (1998). *Model Selection and Inference. A Practical Information–Theoretic Approach*. New York: Springer Verlag.

Bynum, W. F. (2002). Mosquitoes bite more than once. *Science*, **295**, 47.

Caddy, J. F. (2002). Limit reference points, traffic lights, and holistic approaches to fisheries management with minimal stock assessment input. *Fisheries Research*, **56**, 133–137.

Calder, W. A. I. (1984). *Size, Function, and Life History*. Cambridge, MA: Harvard University Press.

Callaway, D. S. and Perelson, A. S. (2002). HIV-1 infection and low steady state viral loads. *Bulletin of Mathematical Biology*, **64**, 29–64.

Caraco, T., Duryea, M. C., Glavanakov, S., Maniatty, W. and Szymanski, B. K. (2001). Host spatial heterogeneity and the spread of vector-borne infection. *Theoretical Population Biology*, **59**, 185–206.

Cardinale, M. and Arrhenius, F. (2000). Decreasing weight-at-age of Atlantic herring (*Clupea harengus*) from the Baltic Sea between 1986 and 1996: a statistical analysis. *ICES Journal of Marine Science*, **57**, 882–893.

Cardinale, M. and Modin, J. (1999). Changes in size-at-maturity of Baltic cod (*Gadus morhua*) during a period of large variations in stock size and environmental conditions. *Fisheries Research*, **41**, 285–295.

Carey, J. R. (2001). Insect biodemography. *Annual Review of Entomology*, **46**, 79–110.

Carey, J. R. (2003). *Longevity. The Biology and Demography of Life Span*. Princeton, NJ: Princeton University Press.

Carey, J. R. and Judge, D. S. (2001). Principles of biodemography with special reference to human longevity. *Population: An English Selection*, **13**, 9–40.

Carpenter, S. R., Ludwig, D. and Brock, W. A. (1999). Management of eutrophication for lakes subject to potentially irreversible change. *Ecological Applications*, **9**, 751–771.

Carslaw, H. S. and Jaeger, J. C. (1959). *Conduction of Heat in Solids*. Oxford: Oxford University Press.

Catteruccia, F., Godfray, C. and Crisanti, A. (2003). Impact of genetic manipulation of the fitness of *Anopheles stephensi* mosquitoes. *Science*, **299**, 1225–1227.

Caughley, G. (1994). Directions in conservation biology. *Journal of Animal Ecology*, **63**, 215–244.

Chapman, R. N. (1928). The quantitative analysis of environmental factors. *Ecology*, **9**, 111–122.

Charles, A. T. (1992). Fishery conflicts. A unified framework. *Marine Policy*, **16**, 379–393.

Charles, S., Morand, S., Chasse, J. L. and Auger, P. (2002). Host patch selection induced by parasitism: basic reproduction ratio R_0 and optimal virulence. *Theoretical Population Biology*, **62**, 97–109.

Charlesworth, B. (1990). Optimization models, quantitative genetics, and mutation. *Evolution*, **44**, 520–538.

Charlesworth, B. (1994). *Evolution in Age-Structured Populations*. Cambridge: Cambridge University Press.

Charnov, E. (1993). *Life History Invariants*. New York: Oxford University Press.

Charnov, E. L. (1976). Optimal foraging, the marginal value theorem. *Theoretical Population Biology*, **9**, 129–136.

Charnov, E. L. and Skinner, S. W. (1984). Evolution of host selection and clutch size in parasitoid wasps. *Florida Entomologist*, **67**, 5–21.

Charnov, E. L. and Skinner, S. W. (1985). Complementary approaches to the understanding of parasitoid oviposition decisions. *Environmental Entomology*, **14**, 383–391.

Charnov, E. L. and Skinner, S. W. (1988). Clutch size in parasitoids: the egg production rate as a constraint. *Evolutionary Ecology*, **2**, 167–174.

Chattopadhyay, J. and Sarkar, R. R. (2003). Chaos to order: preliminary experiments with a population dynamics models of three trophic levels. *Ecological Modelling*, **163**, 45–50.

Cheever, J. (1978). *The Stories of John Cheever*. New York: Alfred A. Knopf.

Chen, D. G., Irvine, J. R. and Cass, A. J. (2002). Incorporating Allee effects in fish stock–recruitment models and applications for determining reference points. *Canadian Journal of Fisheries and Aquatic Sciences*, **59**, 242–249.

Chen, Y., Jiao, Y. and Chen, L. (2003). Developing robust frequentist and Bayesian fish stock assessment methods. *Fish and Fisheries*, **4**, 105–120.

Christensen, V., Guénette, S., Heymans, J. J., *et al.* (2003). Hundred-year decline of North Atlantic predatory fishes. *Fish and Fisheries*, **4**, 1–24.

Clark, C. W. (1973). The economics of overexploitation. *Science*, **181**, 630–634.

Clark, C. W. (1985). *Bioeconomic Modelling and Fisheries Management*. New York: Wiley Interscience.

Clark, C. W. (1990). *Mathematical Bioeconomics*, 2nd edn. New York: Wiley Interscience.

Clark, C. W. (2006). *The Worldwide Crisis in Fisheries*. New York: Cambridge University Press.

Clark, C. W. and Mangel, M. (1979). Aggregation and fishery dynamics: a theoretical study of schooling and the purse seine tuna fisheries. *Fishery Bulletin*, **77**, 317–337.

Clark, C. W. and Mangel, M. (2000). *Dynamic State Variable Models in Ecology. Methods and Applications*. New York: Oxford University Press.

Clark, W. G. (1991). Groundfish exploitation rates based on life history parameters. *Canadian Journal of Fisheries and Aquatic Sciences*, **48**, 734–750.

Clark, W. G. (2002). F35% revisited ten years later. *North American Journal of Fisheries Management*, **22**, 251–257.

Cochrane, K. L. (2000). Reconciling sustainability, economic efficiency and equity in fisheries: the one that got away? *Fish and Fisheries*, **1**, 3–21.

Codeço, C. T. (2000). Endemic and epidemic dynamics of cholera: the role of the aquatic reservoir. *BMC Infectious Diseases*, **1**, 1–14.

Cohen, D. (1966). Optimizing reproduction in a randomly varying environment. *Journal of Theoretical Biology*, **12**, 119–129.

Cohen, J. E. (1995). Unexpected dominance of high frequencies in chaotic nonlinear population models. *Nature*, **378**, 610–612.

Cole-King, A. (1993). Marine conservation: a new policy area. *Marine Policy*, May 1993, 171–185.

Comins, H. N. and Hassell, M. P. (1979). The dynamics of optimally foraging predators and parasitoids. *Journal of Animal Ecology*, **48**, 335–351.

Comins, H. N., Hassell, M. P. and May, R. M. (1992). The spatial dynamics of host-parasitoid systems. *Journal of Animal Ecology*, **61**, 735–748.

Congdon, P. (2001). *Bayesian Statistical Modelling*. New York: John Wiley and Sons.

Conover, D. O. and Munch, S. B. (2002). Sustaining fisheries yields over evolutionary time scales. *Science*, **297**, 94–96.

Cook, R. M. and Hubbard, S. F. (1977). Adaptive searching strategies in insect parasites. *Journal of Animal Ecology*, **46**, 115–125.

Cooper, A. B., Hilborn, R. and Unsworth, J. W. (2003). An approach for population assessment in the absence of abundance indices. *Ecological Applications*, **13**, 814–828.

Cootner, P. H. (editor) (1964). *The Random Character of Stock Market Prices*. Cambridge, MA: MIT Press.

Corkett, C. J. (2002). Fish stock assessment as a non-falsifiable science: replacing an inductive and instrumental view with a critical rational one. *Fisheries Research*, **56**, 117–123.

Costantino, R. F. and Desharnais, R. A. (1991). *Population Dynamics and the Tribolium Model: Genetics and Demography*. New York: Springer Verlag.

Cote, I. M., Mosqueira, I. and Reynolds, J. D. (2001). Effects of marine reserve characteristics on the protection of fish populations: a meta-analysis. *Journal of Fish Biology*, **59** (Supplement A), 178–179.

Cottingham, K. L., Chiavelli, D. A. and Taylor, R. K. (2003). Environmental microbe and human pathogen: the ecology and microbiology of *Vibrio cholerae*. *Frontiers in Ecology and the Environment*, **1**, 80–86.

Council, N. R. (editor) (2001). *Marine Protected Areas. Tools for Sustaining Ocean Ecosystems*. Washington, DC: National Academy Press.

Courant, R. and Hilbert, D. (1962). *Methods of Mathematical Physics. Volume II. Partial Differential Equations*. New York: Wiley Interscience.

Crank, J. (1975). *The Mathematics of Diffusion*, 2nd edn. Oxford: Oxford University Press.

Crawley, M. J. (2002). *Statistical Computing. An Introduction to Data Analysis Using S-Plus*. New York: Wiley.

Crow, J. F. and Kimura, M. (1970). *An Introduction to Population Genetics Theory*. Minneapolis, MN: Burgess Publishing Company.

Crowder, L. B. and Murawski, S. A. (1998). Fisheries bycatch: Implications for management. *Fisheries*, **23**, 8–17.

Daskalov, G. M. (2002). Overfishing drives a trophic cascade in the Black Sea. *Marine Ecology – Progress Series*, **225**, 53–63.

Davies, R. (1992). *Reading and Writing*. Salt Lake City: University of Utah Press.

Davis, H. T. (1962). *Introduction to Nonlinear Differential and Integral Equations*. New York: Dover.

Davis, P. J. and Gregerman, R. I. (1995). Occasional Notes: Parse Analysis II: A revised model that accounts for *phi*. *The New England Journal of Medicine*, **332**, 965–966.

Day, T. (2001). Parasite transmission models and the evolution of virulence. *Evolution*, **55**, 2389–2400.

Day, T. (2002a). The evolution of virulence in vector-borne and directly transmitted parasites. *Theoretical Population Biology*, **62**, 199–213.

Day, T. (2002b). On the evolution of virulence and the relationship between various measures of mortality. *Proceedings of the Royal Society of London*, B**269**, 1317–1323.

Day, T. (2002c). Virulence evolution via host exploitation and toxin production in spore-producing pathogens. *Ecology Letters*, **5**, 471–476.

Day, T. (2003). Virulence evolution and the timing of disease life-history events. *Trends in Ecology and Evolution*, **18**, 113–118.

Day, T. and Proulx, S. (2004). A general theory for the evolutionary dynamics of virulence. *American Naturalist*, **164**, E40–E63.

de Bach, P. (1943). The importance of host-feeding by adult parasites in the reduction of host populations. *Journal of Economic Entomology*, **36**, 647–658.

de Bruijn, N. G. (1981). *Asymptotic Methods in Analysis*. New York: Dover Publications.

de Grey, A. D. N. J. (2003a). An engineer's approach to the development of real anti-aging medicine. *Science SAGE KE 2003* (8 January 2003).

de Grey, A. D. N. J. (2003b). Overzealous maximum-likelihood fitting falsely convicts the slope heterogeneity hypothesis. *Experimental Gerontology*, **38**, 921–923.

de Koeijer, A., Diekmann, O. and Reijnders, P. (1998). Modelling the spread of phocine distemper virus among harbour seals. *Bulletin of Mathematical Biology*, **60**, 585–596.

de Koeijer, A., Heesterbeek, H., Schreuder, B., *et al*. (2004). Quantifying BSE control by calculating the basic reproduction ratio R0 for the infection among cattle. *Journal of Mathematical Biology*, **48**, 1–22.

de Roode, J. C. and Read, A. F. (2003). Evolution and ecology, after the malaria genomes. *Trends in Ecology and Evolution*, **18**, 60–61.

de Valpine, P. and Hastings, A. (2002). Fitting population models incorporating process noise and observation error. *Ecological Monographs*, **72**, 57–76.

DeAngelis, D. L. and Gross, L. J. (1992). *Individual-Based Models and Approaches in Ecology*. New York: Chapman and Hall.

DeGroot, M. H. (1970). *Optimal Statistical Decisions*. New York: McGraw-Hill.

Delbrück, M. (1940). Statistical fluctuations in autocatalytic reactions. *Journal of Chemical Physics*, **8**, 120–124.

Demetrius, L. (2001). Mortality plateaus and directionality theory. *Proceedings of the Royal Society of London*, **B268**, 2029–2037.

Denney, N. H., Jennings, S. and Reynolds, J. D. (2002). Life-history correlates of maximum population growth rates in marine fishes. *Proceedings of the Royal Society of London*, **B269**, 2229–2237.

Dennis, B., Munholland, P. L. and Scott, J. M. (1991). Estimation of growth and extinction parameters for endangered species. *Ecological Monographs*, **61**, 115–143.

Dennis, B. and Otten, M. R. M. (2000). The joint effects of density dependence and rainfall on abundance of San Joaquin kit fox. *Journal of Wildlife Management*, **64**, 388–400.

Dick, E. J. (2004). Beyond 'lognormal versus gamma': discrimination among error distributions for generalized linear models. *Fisheries Research*, **70**, 351–366.

Dieckmann, U., Metz, J. A. J., Sabelis, M. W. and Sigmund, K. (2002). Adaptive dynamics of infectious diseases. In *Pursuit of Virulence Management*. Cambridge: Cambridge University Press.

Diserud, O. H. and Engen, S. (2000). A general and dynamic species abundance model, embracing the lognormal and gamma models. *The American Naturalist*, **155**, 497–511.

Dixit, A. K. and Pindyck, R. S. (1994). *Investment Under Uncertainty*. Princeton, NJ: Princeton University Press.

Dobson, A. and Foutopoulos, J. (2001). Emerging infectious pathogens of wildlife. *Philosophical Transactions of the Royal Society London*, **B356**, 1001–1012.

Dobson, A. P. and Hudson, P. J. (1992). Regulation and stability of a free-living host–parasite system: *Trichostrongylus tenius* in red grouse. II. Population models. *Journal of Animal Ecology*, **61**, 487–498.

Dorn, M. W. (2002). Advice on west coast rockfish harvest rates from Bayesian meta-analysis of stock–recruit relationships. *North American Journal of Fisheries Management*, **22**, 280–300.

Dovers, S. R. and Handmer, J. W. (1995). Ignorance, the precautionary principle, and sustainability. *Ambio*, **24**, 92–97.

Dovers, S. R., Norton, T. W. and Handmer, J. W. (1996). Uncertainty, ecology, sustainability and policy. *Biodiversity and Conservation*, **5**, 1143–1167.

Driessen, G. and Hemerik, L. (1992). The time and egg budget of *Leptopilina clavipes*, a parasitoid of *Drosophila*. *Ecological Entomology*, **17**, 17–27.

Dubins, L. E. and Savage, L. J. (1976). *Inequalities for Stochastic Processes: How to Gamble if You Must*. New York: Dover Publications.

Dwyer, G. and Elkinton, J. S. (1993). Using simple models to predict virus epizootics in gypsy moth populations. *Journal of Animal Ecology*, **62**, 1–11.

Dyson, F. (1999). *The Sun, the Genome, and the Internet*. New York: Oxford University Press.

Earn, D. J. D., Dushoff, J. and Levin, S. A. (2002). Ecology and evolution of the flu. *Trends in Ecology and Evolution*, **17**, 334–340.

Easterling, M. R. and Ellner, S. P. (2000). Dormancy strategies in a random environment: comparing structured and unstructured models. *Evolutionary Ecology Research*, **2**, 387–407.

Ebert, D. and Bull, J. J. (2003). Challenging the trade-off model for the evolution of virulence: is virulence management feasible? *Trends in Microbiology*, **11**, 15–20.

Edelstein-Keshet, L. (1988). *Mathematical Models in Biology*. New York: Random House.

Edgar, G. and Barrett, N. S. (1999). Effects of the declaration of marine reserves on Tasmanian reef fishes, invertebrates and plants. *Journal of Experimental Marine Biology and Ecology*, **242**, 107–144.

Edwards, A. W. F. (1992). *Likelihood*, expanded edition. Baltimore: Johns Hopkins University Press.

Edwards, R. L. (1954). The effect of diet on egg maturation and resorption in *Mormoniella vitripennis* (Hymenoptera, Pteromalidae). *Quarterly Journal of Microscopical Science*, **95**, 459–469.

Efron, B. (2005). Bayesians, frequentists, and scientists. *Journal of the American Statistical Association*, **100**, 1–5.

Eigen, M. (1996). Prionics or the kinetic basis of prion diseases. *Biophysical Chemistry*, **63**, A1–A18.

Einstein, A. (1956). *Investigations on the Theory of Brownian Movement*. New York: Dover Publications.

Einum, S. and Fleming, I. A. (2000). Highly fecund mothers sacrifice offspring survival to maximize fitness. *Nature*, **405**, 565–567.

Ellison, A. M. (1996). An introduction to Bayesian inference for ecological research and environmental decision-making. *Ecological Applications*, **6**, 1036–1046.

Ellison, A. M. (2004). Bayesian inference in ecology. *Ecology Letters*, **7**, 509–520.

Emlen, J. M. (1966). The role of time and energy in food preference. *American Naturalist*, **100**, 611–617.

Enderby, J. E. (1998). Sir Frederick Charles Frank (1911–98). *Nature*, **393**, 314.

Engen, S., Lande, R. and Saether, B.-E. (2002). The spatial scale of population fluctuations and quasi-extinction risk. *American Naturalist*, **160**, 439–451.

Engen, S. and Saether, B.-E. (2000). Predicting the time of quasi-extinction for populations far below their carrying capacity. *Journal of Theoretical Biology*, **205**, 649–658.

Engen, S., Saether, B.-E. and Møller, A. P. (2001). Stochastic population dynamics and time to extinction of a declining population of barn swallows. *Journal of Animal Ecology*, **70**, 789–797.

Enquist, B. J., West, G. B., Charnov, E. L. and Brown, J. H. (1999). Allometric scaling of production and life-history variation in vascular plants. *Nature*, **401**, 907–911.

Enserink, M. (2002). How devastating would a smallpox attack really be? *Science*, **296**, 1592–1595.

Enserink, M. (2003). New look at old data irks smallpox-eradication experts. *Science*, **299**, 181.

Esch, G. W., Shostak, A. W., Marcogliese, D. J. and Goater, T. M. (1990). *Parasite Communities: Patterns and Processes*. New York: Chapman and Hall.

Essington, T. E., Kitchell, J. F. and Walters, C. J. (2001). The von Bertalanffy growth function, bioenergetics, and the consumption rates of fish. *Canadian Journal of Fisheries and Aquatic Sciences*, **58**, 2129–2138.

Estes, J. A., Tinker, M. T., Williams, T. M. and Doak, D. F. (1998). Killer whale predation on sea otters linking oceanic and nearshore ecosystems. *Science*, **282**, 473–476.

Esteva, L. and Vargas, C. (2003). Coexistence of different serotypes of dengue virus. *Journal of Mathematical Biology*, **46**, 31–47.

Ewald, P. W. (1994). *Evolution of Infectious Disease*. New York: Oxford University Press.

Fanshawe, S., Vanblaricom, G. R. and Shelly, A. A. (2003). Restored top carnivores as detriments to the performance of marine protected areas intended for fishery sustainability: a case study with red abalones and sea otters. *Conservation Biology*, **17**, 273–283.

Farrow, S. and Sumaila, U. R. (2002). Economics of marine protected areas. *Fish and Fisheries*, **3**, 356–359.

Faruque, S., Chowdhury, N., Kamruzzaman, M. *et al.* (2003). Reemergence of Epidemic *Vibrio cholerae* O139, Bangladesh. *Emerging Infectious Diseases*, **9**, 1116–1122.

Feller, W. (1957). *An Introduction to Probability Theory and Its Applications, Volume I*. New York: John Wiley and Sons.

Feller, W. (1971). *An Introduction to Probability Theory and Its Applications, Volume II*. New York: John Wiley and Sons.

Ferguson, H. M. and Read, A. F. (2002). Genetic and environmental determinants of malaria parasite virulence in mosquitoes. *Proceedings of the Royal Society of London*, B**269**, 1217–1224.

Ferguson, N. M., Ghani, A. C., Donnelly, C. A., Hagenaars, T. J. and Anderson, R. M. (2002). Estimating the human health risk from possible BSE infection of the British sheep flock. *Nature*, **415**, 420–424.

Feynman, R. P. (1948). Space-time approach to non-relativistic quantum mechanics. *Reviews of Modern Physics*, **20**, 367–387.

Feynman, R. P. (1985). *Surely You're Joking Mr. Feynman*. New York: W. W. Norton.

Fieberg, J. and Ellner, S. P. (2000). When is it meaningful to estimate an extinction probability? *Ecology*, **81**, 2040–2047.

Fiksen, Ø. and Slotte, A. (2002). Stock-environment recruitment models for Norwegian spring spawning herring (*Clupea harengus*). *Canadian Journal of Fisheries and Aquatic Sciences*, **59**, 211–217.

Fisher, A. C., Hanemann, W. M. and Keeler, A. G. (1991). Integrating fishery and water resource management: a biological model of a California salmon fishery. *Journal of Environmental Economics and Management*, **20**, 234–261.

Fisher, R. A. (1930) (1958 reprint). *The Genetical Theory of Natural Selection*, 2nd revised edn. New York: Dover.

Fishman, M. A. and Perelson, A. (1999). Th1/Th2 differentiation and cross-regulation. *Bulletin of Mathematical Biology*, **61**, 403–436.

Flanders, S. E. (1950). Regulation of ovulation and egg disposal in the parasitic hymenoptera. *The Canadian Entomologist*, **82**, 134–140.

Flather, C. H. (1996). Fitting species-accumulation functions and assessing regional land use impacts on avian diversity. *Journal of Biogeography*, **23**, 155–168.

Fletcher, J. P., Hughes, J. P. and Harvey, I. F. (1994). Life expectancy and egg load affect oviposition decision of a solitary parasitoid. *Proceedings of the Royal Society of London*, B**258**, 163–167.

Foley, P. (1994). Predicting extinction times from environmental stochasticity and carrying capacity. *Conservation Biology*, **8**, 124–137.

Francis, R. I. C. C. and Shotton, R. (1997). "Risk" in fisheries management: a review. *Canadian Journal of Fisheries and Aquatic Sciences*, **54**, 1699–1715.

Frank, F. C. (1953). On spontaneous asymmetric synthesis. *Biochimica et Biophysica Acta*, **11**, 459–463.

Frank, S. A. (1996). Models of parasite virulence. *The Quarterly Review of Biology*, **71**, 37–78.

Frank, S. A. (1998). *Social Evolution*. Princeton, NJ: Princeton University Press.

Frank, S. A. (2002). *Immunology and Evolution of Infectious Diseases*. Princeton, NJ: Princeton University Press.

Freedman, D., Pisani, R. and Purves, R. (1998). *Statistics*, 3rd edn. New York: W. W. Norton.

Friedlin, M. I. and Wentzell, A. D. (1984). *Random Perturbations of Dynamical Systems*. New York: Springer Verlag.

Frisk, M. G., Miller, T. J. and Fogarty, M. J. (2001). Estimation and analysis of biological parameters in elasmobranch fishes: a comparative life history study. *Canadian Journal of Fisheries and Aquatic Sciences*, **58**, 969–981.

Fuerth, R. (1956). Notes. In *Investigations on the Theory of the Brownian Movement by Albert Einstein, Ph.D.* New York: Dover Publications, pp. 86–119.

Fulford, G. R., Roberts, M. G. and Heesterbeek, J. A. P. (2002). The metapopulation dynamics of an infectious disease: tuberculosis in possums. *Theoretical Population Biology*, **61**, 15–29.

Galvani, A. P. (2003). Epidemiology meets evolutionary ecology. *Trends in Ecology and Evolution*, **18**, 132–139.

Gandon, S., Mackinnon, M. J., Nee, S. and Read, A. F. (2001). Imperfect vaccines and the evolution of pathogen virulence. *Nature*, **414**, 751–755.

Gani, R. and Leach, S. (2001). Transmission potential of smallpox in contemporary populations. *Nature*, **414**, 748–751.

Ganusov, V. V., Bergstrom, C. T. and Antia, R. (2002). Within-host population dynamics and the evolution of microparasites in a heterogeneous host population. *Evolution*, **56**, 213–223.

Gardiner, C. W. (1983). *Handbook of Stochastic Methods for Physics, Chemistry and the Natural Sciences*. Berlin: Springer Verlag.

Gardner, S. N. (2000). Scheduling chemotherapy: Catch 22 between cell kill and resistance evolution. *Journal of Theoretical Medicine*, **2**, 21–232.

Gardner, S. N. (2001). Modeling multi-drug chemotherapy: tailoring treatment to individuals. *Journal of Theoretical Biology*, **214**, 181–207.

Gardner, S. N. and Agrawal, A. A. (2002). Induced plant defense and the evolution of counter-defenses in herbivores. *Evolutionary Ecology Research*, **4**, 1131–1151.

Gardner, S. N. and Thomas, M. B. (2002). Costs and benefits of fighting infection in locusts. *Evolutionary Ecology Research*, **4**, 109–131.

Gavrilets, S. (2003). Models of speciation: what have we learned in 40 years? *Evolution*, **57**, 2197–2215.

Gavrilov, L. A. and Gavrilova, N. S. (1991). *The Biology of Life Span: A Quantitative Approach*. London: Harwood Academic Publishers.

Gelman, A., Carlin, J. B., Stern, H. S. and Rubin, D. B. (1995). *Bayesian Data Analysis*. London: Chapman and Hall.

George, E. (2004). *Write Away*. New York: Harper Collins.

Gillespie, D. T. (1992). *Markov Processes. An Introduction for Physical Scientists*. Boston: Academic Press.

Gillespie, D. T. (1996). Exact numerical simulation of the Ornstein–Uhlenbeck process and its integral. *Physical Review E*, **54**, 2084–2091.

Gillespie, J. H. (1991). *The Causes of Molecular Evolution*. New York: Oxford University Press.

Gillespie, J. H. (1998). *Population Genetics. A Concise Guide*. Baltimore: Johns Hopkins University Press.

Gillig, D., Griffin, W. L. and Ozuna, T. J. (2001). A bioeconomic assessment of Gulf of Mexico red snapper management policies. *Transactions of American Fisheries Society*, **130**, 117–129.

Gillis, D. M. (1999). Behavioral inferences from regulatory observer data: catch rate variation in the Scotian Shelf silver hake (*Merluccius bilinearis*) fishery. *Canadian Journal of Fisheries and Aquatic Sciences*, **56**, 288–296.

Gillis, D. M. (2003). Ideal free distributions in fleet dynamics: a behavioral perspective on vessel movement in fisheries analysis. *Canadian Journal of Zoology*, **81**, 177–187.

Gillis, D. M. and Peterman, R. M. (1998). Implications of interference among fishing vessels and the ideal free distribution to the interpretation of CPUE. *Canadian Journal of Fisheries and Aquatic Sciences*, **55**, 37–46.

Gillis, D. M., Pikitch, E. K. and Peterman, R. M. (1995). Dynamic discarding decisions: foraging theory for high-grading in a trawl fishery. *Behavioral Ecology*, **6**, 146–154.

Giron, D., Rivero, A., Mandon, N., Darrouzet, E. and Casas, J. (2002). The physiology of host feeding in parasitic wasps: implications for survival. *Functional Ecology*, **16**, 750–757.

Giske, J., Mangel, M., Jakobsen, P., *et al.* (2002). Explicit trade-off rules in proximate adaptive agents. *Evolutionary Ecology Reseach*, **5**, 835–865.

Gislason, H. (1994). Ecosystem effects of fishing activities in the North Sea. *Marine Pollution Bulletin*, **29**, 520–527.

Gleick, J. (1988). *Chaos*. London: William Heinneman Ltd.

Gleick, J. (1992). *Genius. The Life and Science of Richard Feynman*. New York: Pantheon.

Godfray, H. C. J. (1994). *Parasitoids. Behavioral and Evolutionary Ecology*. Princeton, NJ: Princeton University Press.

Godfray, H. C. J. and Hassell, M. P. (1989). Discrete and continuous insect populations in tropical environments. *Journal of Animal Ecology*, **58**, 153–174.

Goel, N. S. and Richter-Dyn, N. (1974). *Stochastic Models in Biology*. New York: Academic Press.

Gompertz, B. (1825). On the nature of the function expressive of the law of human mortality, and on a new mode of determining the value of life contingencies. *Philosophical Transactions of the Royal Society*, **115**, 513–583.

Gordon, D. M., Nisbet, R. M., de Roos, A., Gurney, W. S. C. and Stewart, R. K. (1991). Discrete generations in host–parasitoid models with contrasting life cycles. *Journal of Animal Ecology*, **60**, 295–308.

Gordon, H. S. (1954). The economic theory of a common property resource: the fishery. *Journal of Political Economy*, **62**, 124–142.

Gotelli, N. J. (2001). *A Primer of Ecology*, 3rd edn. Sunderland, MA: Sinauer Associates.

Gould, S. J. (2002). *The Structure of Evolutionary Theory*. Cambridge, MA: The Belknap Press of Harvard University Press.

Gould, S. J. and Eldredge, N. (1977). Punctuated equilibria: the tempo and mode of evolution reconsidered. *Paleobiology*, **3**, 115–151.

Gould, S. J. and Eldredge, N. (1993). Punctuated equilibrium comes of age. *Nature*, **366**, 223–227.

Grafen, A. and Hails, R. S. (2002). *Modern Statistics for the Life Sciences. Learn How to Analyse your Experiments*. Oxford: Oxford University Press.

Graham, A. L. (2001). Use of an optimality model to solve the immunological puzzle of concomitant infection. *Parasitology*, **122**, S61–S64.

Graham, A. L. (2002). When T-helper cells don't help: immunopathology during concomitant infection. *Quarterly Review of Biology*, **77**, 409–434.

Gray, R. D. (1987). Faith and foraging: a critique of the "paradigm argument from design". In A. C. Kamil, J. R. Krebs and H. R. Pulliam (editors) *Foraging Behavior*. New York: Plenum, pp. 69–140.

Greenberg, R. (2001). Lecture 38: Nineteenth-Century Italian Opera – Giuseppe Verdi. In *How to Listen to and Understand Great Music*. The Teaching Company.

Greene, C. (2000). Habitat selection reduces extinction of populations subject to Allee effects. *Theoretical Population Biology*, **64**, 1–10.

Greenwood, M. and Yule, G. U. (1920). An inquiry into the nature of frequency distributions representative of multiple happenings with particular references to the occurrence of multiple attacks of disease or of repeated accidents. *Journal of the Royal Statistical Society*, **83**, 255–279.

Greer, R. (1995). *Ferox Trout and Arctic Charr. A Predator, its Pursuit and its Prey*. Shrewsbury, UK: Swan Hill Press.

Grenfell, B. T. (1988). Gastrointestinal nematode parasites and the stability and productivity of intensive ruminant grazing systems. *Philosophical Transactions of the Royal Society of London*, B**321**, 541–563.

Grenfell, B. T. (1992). Parasitism and the dynamics of ungulate grazing systems. *American Naturalist*, **139**, 907–929.

Grindrod, P. (1996). *The Theory and Applications of Reaction–Diffusion Equations. Patterns and Waves*. Oxford: Oxford University Press.

Groot, C. and Margolis, L. (editors) (1991). *Pacific Salmon Life Histories*. Vancouver, BC: University of British Columbia Press.

Groot, C., Margolis, L. and Clarke, W. C. (editors) (1995). *The Physiological Ecology of Pacific Salmon*. Vancouver, BC: University of British Columbia Press.

Guedj, D. (2000). *The Parrot's Theorem*. New York: St. Martin's Griffin.

Guenette, S. and Pitcher, T. J. (1999). An age-structured model showing the benefits of marine reserves in controlling overexploitation. *Fisheries Research*, **39**, 295–303.

Guenette, S., Lauck, T. and Clark, C. (1998). Marine reserves: from Beverton and Holt to the present. *Reviews in Fish Biology and Fisheries*, **8**, 251–272.

Gulland, J. A. (editor) (1977). *Fish Population Dynamics*. Chichester: John Wiley and Sons.

Gulland, J. A. (1988). *Fish Population Dynamics. Implications for Management*, 2nd edn. Chichester: John Wiley & Sons.

Gunderson, D. R. (1997). Trade-off between reproductive effort and adult survival in oviparous and viviparous fishes. *Canadian Journal of Fisheries and Aquatic Sciences*, **54**, 990–998.

Gurney, W. S. C., Nisbet, R. M. and Lawton, J. H. (1983). The systematic formulation of tractable single-species population models incorporating age structure. *Journal of Animal Ecology*, **52**, 479–495.

Haberman, R. (1998). *Elementary Applied Partial Differential Equations with Fourier Series and Boundary Value Problems*. Upper Saddle River, NJ: Prentice Hall.

Haccou, P., de Vlas, S. J., van Alhen, J. J. M. and Visser, M. E. (1991). Information processing by foragers: effects of intra-patch experience on the leaving tendency of *Leptopilina heterotoma*. *Journal of Animal Ecology*, **60**, 93–106.

Hadamard, J. (1954). *The Psychology of Invention in the Mathematical Field.* New York: Dover.

Hakoyama, H. and Iwasa, Y. (2000). Extinction risk of a density-dependent population estimated from a time series of population size. *Journal of Theoretical Biology*, **204**, 337–359.

Hakoyama, H., Iwasa, Y. and Nakanishi, J. (2000). Comparing risk factors for population extinction. *Journal of Theoretical Biology*, **204**, 327–336.

Halley, J. M. and Iwasa, Y. (1998). Extinction rate of a population under both demographic and environmental stochasticity. *Theoretical Population Biology*, **53**, 1–15.

Halpern, B. S. and Warner, R. R. (2002). Marine reserves have rapid and lasting effects. *Ecology Letters*, **5**, 361–366.

Ham, K. D. and Pearsons, T. N. (2000). Can reduced salmonid population abundance be detected in time to limit management impacts? *Canadian Journal of Fisheries and Aquatic Sciences*, **57**, 17–24.

Hamilton, W. D. (1966). The moulding of senescence by natural selection. *Journal of Theoretical Biology*, **12**, 12–45.

Hamilton, W. D. (1967). Extraordinary sex ratios. *Science*, **156**, 477–488.

Hamilton, W. D. (1995). *Narrow Roads of Gene Land. Volume 1. Evolution of Social Behaviour.* New York: W. H. Freeman and Company.

Hammersley, J. M. (1974). Statistical tools. *The Statistician*, **23**, 89–106.

Hammond, T. R. and O'Brien, C. M. (2001). An application of the Bayesian approach to stock assessment model uncertainty. *ICES Journal of Marine Science*, **58**, 648–656.

Hanski, I. (1989). Metapopulation dynamics: does it help to have more of the same? *Trends in Ecology and Evolution*, **4**, 113–114.

Hanski, I. (1999). *Metapopulation Ecology.* Oxford: Oxford University Press.

Hare, S. R. and Francis, R. C. (1995). Climate change and salmon production in the Northeast Pacific Ocean. *Canadian Special Publications of Fisheries and Aquatic Sciences*, **121**, 357–372.

Harley, S. J. and Myers, R. A. (2001). Hierarchial Bayesian models of length-specific catchability of research trawl surveys. *Canadian Journal of Fisheries and Aquatic Sciences*, **58**, 1569–1584.

Harley, S. J., Myers, R. A. and Dunn, A. (2001). Is catch-per-unit effort proportional to abundance? *Canadian Journal of Fisheries and Aquatic Sciences*, **58**, 1760–1772.

Harris, P. J. and Dean, J. M. (1998). Characterization of king mackerel and Spanish mackerel bycatches of South Carolina shrimp trawlers. *North American Journal of Fisheries Management*, **18**, 439–453.

Harrison, L. G. (1993). *Kinetic Theory of Living Pattern.* Cambridge: Cambridge University Press.

Hart, P. J. B. and Reynolds, J. D. (editors) (2002). *Handbook of Fish Biology and Fisheries, Volume 2. Fisheries.* Oxford: Blackwell.

Harte, J. (1988). *Consider a Spherical Cow. A Course in Environmental Problem Solving.* Sausalito, CA: University Science Books.

Harte, J. (2001). *Consider a Cylindrical Cow. More Adventures in Environmental Problem Solving.* Sausalito, CA: University Science Books.

Hassell, M. P. (1978). *The Dynamics of Arthropod Predator Prey Systems*. Princeton, NJ: Princeton University Press.

Hassel, M. P. (2000a). *The Spatial and Temporal Dynamics of Host–Parasitoid Interactions*. Oxford: Oxford University Press.

Hassell, M. P. (2000b). Host–parasitoid population dynamics. *Journal of Animal Ecology*, **69**, 543–566.

Hassell, M. P. and May, R. M. (1988). Spatial heterogeneity and the dynamics of parasitoid–host systems. *Annals Zoologia Fennici*, **25**, 55–61.

Hassell, M. P., Waage, J. K. and May, R. M. (1983). Variable parasitoid sex ratios and their effect on host-parasitoid dynamics. *Journal of Animal Ecology*, **52**, 889–904.

Hastings, A. and Botsford, L. W. (1999). Equivalence in yield from marine reserves and traditional fisheries management. *Science*, **284**, 1537–1538.

Hastings, I. M. (1997). A model for the origins and spread of drug-resistant malaria. *Parasitology*, **115**, 133–141.

Hastings, I. M. (2001). Modelling parasite drug resistance: lessons for management and control strategies. *Tropical Medicine and International Health*, **6**, 883–890.

Hastings, I. M. and D'Allesandro, U. (2000). Modelling a predictable disaster: the rise and spread of drug-resistant malaria. *Parasitology Today*, **16**, 340–347.

Hastings, I. M. and MacKinnon, M. J. (1998). The emergence of drug-resistant malaria. *Parasitology*, **117**, 411–417.

Hastings, I. M., Bray, P. G. and Ward, S. A. (2002). A requiem for chloroquine. *Science*, **297**, 74–75.

Haupt, R. L. and Haupt, S. E. (1998). *Practical Genetic Algorithms*. New York: John Wiley and Sons.

He, J. X. and Stewart, D. J. (2001). Age and size at first reproduction of fishes: predictive models based only on growth trajectories. *Ecology*, **82**, 784–791.

Healey, M. C. (1985). Influence of fishermen's preference on the success of commercial fishery management regimes. *North American Journal of Fisheries Management*, **5**, 173–180.

Healey, M. C. (1990). Implications of climate change for fisheries management policy. *Transactions of American Fisheries Society*, **119**, 366–373.

Healey, M. C. and Morris, J. F. T. (1992). The relationship between the dispersion of salmon fishing vessels and their catch. *Fisheries Research*, **15**, 135–145.

Heath, D. D., Heath, J. W., Bryden, C. A., Johnson, R. M. and Fox, C. W. (2003). Rapid evolution of egg size in captive salmon. *Science*, **299**, 1738–1740.

Hedrick, P. W., Lacy, R. C., Allendorf, F. W. and Soule, M. E. (1996). Directions in conservation biology. *Conservation Biology*, **10**, 1312–1320.

Heimpel, G. E. and Collier, T. R. (1996). The evolution of host-feeding behaviour in parasitoids. *Biological Reviews*, **71**, 373–400.

Heimpel, G. E. and Rosenheim, J. A. (1995). Dynamic host feeding by the parasitoid *Aphytis melinus*: the balance between current and future reproduction. *Journal of Animal Ecology*, **64**, 153–167.

Heimpel, G. E. and Rosenheim, J. A. (1998). Egg limitation in parasitoids: a review of the evidence and a case study. *Biological Control*, **11**, 160–168.

Heimpel, G. E., Mangel, M. and Rosenheim, J. A. (1998). Effects of time limitation and egg limitation on lifetime reproductive success of a parasitoid in the field. *American Naturalist*, **152**, 273–289.

Heimpel, G. E., Rosenheim, J. A. and Adams, J. A. (1994). Behavioral ecology of host feeding in *Aphytis* parasitoids. *Norwegian Journal of Agricultural Sciences*, Supplement, **16**, 101–115.

Heimpel, G. E., Rosenheim, J. A. and Mangel, M. (1996). Egg limitation, host quality, and dynamic behavior by a parasitoid in the field. *Ecology*, **77**, 2410–2420.

Heimpel, G. E., Rosenheim, J. A. and Kattari, D. (1997). Adult feeding and lifetime reproductive success in the parasitoid *Aphytis melinus*. *Entomologia Experimentalis et Applicata*, **83**, 305–315.

Hellriegel, B. (2001). Immunoepidemiology – bridging the gap between immunology and epidemiology. *Trends in Parasitology*, **17**, 102–106.

Helser, T. E., Thunberg, E. M. and Mayo, R. K. (1996). An age-structured bioeconomic simulation of U.S. Silver hake Fisheries. *North American Journal of Fisheries Management*, **16**, 783–794.

Hemerik, L., Driessen, G. and Haccou, P. (1993). Effects of intra-patch experiences on patch time, search time and searching efficiency of the parasitoid *Leptopilina clavipes*. *Journal of Animal Ecology*, **62**, 33–44.

Hemerik, L., van der Hoeven, N. and van Alphen, J. J. M. (2002). Egg distributions and the information a solitary parasitoid has and uses for its oviposition decisions. *Acta Biotheoretica*, **50**, 167–188.

Henderson, D. A. (1999). The looming threat of bioterrorism. *Science*, **283**, 1279–1282.

Hendry, A. P. and Kinnison, M. T. (2001). An introduction to microevolution: rate, pattern, process. *Genetica*, **112–113**, 1–8.

Hennemuth, R. C., Palmer, J. E. and Brown, B. E. (1980). A statistical description of recruitment in eighteen selected fish stocks. *Journal of Northwest Atlantic Fishery Science*, **1**, 101–111.

Hernandez, M.-J. and Barradas, I. (2003). Variation in the outcome of population interactions: bifurcations and catastrophes. *Journal of Mathematical Biology*, **46**, 571–594.

Hershatter, G. (1997). *Dangerous Pleasures*. Berkeley, CA: University of California Press.

Hess, G. (1996). Disease in metapopulation models: implications for conservation. *Ecology*, **77**, 1617–1632.

Hethcote, H. W. (2000). The mathematics of infectious diseases. *SIAM Review*, **42**, 599–653.

Higham, D. J. (2001). An algorithmic introduction to numerical simulation of stochastic differential equations. *SIAM Review*, **43**, 525–546.

Highman, N. J. (1998). *Handbook of Writing for the Mathematical Sciences*, 2nd edn. Philadelphia, PA: SIAM (Society for Industrial and Applied Mathematics).

Hilborn, R. and Mangel, M. (1997). *The Ecological Detective. Confronting Models with Data*. Princeton, NJ: Princeton University Press.

Hilborn, R., Parama, A. and Maunder, M. (2002). Exploitation rate reference points for west coast rockfish: are they robust and are there better alternatives? *North American Journal of Fisheries Management*, **22**, 365–375.

Hines, W. G. S. (1987). Evolutionarily stable strategies: a review of basic theory. *Theoretical Population Biology*, **31**, 195–272.

Hinrichsen, R. A. (2002). The accuracy of alternative stochastic growth rate estimates for salmon populations. *Canadian Journal of Fisheries and Aquatic Sciences*, **59**, 1014–1023.

Hochberg, M. E. and Ives, A. R. (editors) (2000). *Parasitoid Population Biology*. Princeton, NJ: Princeton University Press.

Hofbauer, J. and Sigmund, K. (1998). *Evolutionary Games and Population Dynamics*. Cambridge: Cambridge University Press.

Holden, A. (editor) (2001). *The New Penguin Opera Guide*. London: Penguin Books.

Holland, D. S. and Sutinen, J. G. (1999). An empirical model of fleet dynamics in New England trawl fisheries. *Canadian Journal of Fisheries and Aquatic Sciences*, **56**, 253–264.

Holoman, D. K. (1992). *Evenings with the Orchestra*. New York: Norton.

Horwood, J. W., Nichols, J. H. and Milligan, S. (1998). Evaluation of closed areas for fish stock conservation. *Journal of Applied Ecology*, **35**, 893–903.

Hotelling, H. (1951). The impact of R. A. Fisher on statistics. *Journal of the American Statistical Association*, **46**, 35–46.

Houston, A. I. and McNamara, J. M. (1999). *Models of Adaptive Behaviour*. Cambridge: Cambridge University Press.

Hubbard, S. F. and Cook, R. M. (1978). Optimal foraging by parasitoid wasps. *Journal of Animal Ecology*, **47**, 593–604.

Hubbard, S. F., Marris, G. and Reynolds, D. M. (1987). Adaptive patterns in the avoidance of superparasitism by solitary parasitic wasps. *Journal of Animal Ecology*, **56**, 387–401.

Hubbell, S. (2001). *The Unified Neutral Theory of Biodiversity and Biogeography*. Princeton, NJ: Princeton University Press.

Hudson, D. J. (1971). Interval estimation from the likelihood function. *Journal of the Royal Statistical Society*, Series B, **33**, 256–262.

Hudson, P. J. (1986). The effect of a parasitic nematode on the breeding production of red grouse. *Journal of Animal Ecology*, **55**, 85–92.

Hudson, P. J. and Dobson, A. P. (1997). Transmission dynamics and host–parasite interactions of *Trichostrongylus tenuis* in red grouse (*Lagopus lagopus scoticus*). *Journal of Parasitology*, **83**, 194–202.

Hudson, P. J., Dobson, A. P. and Newborn, D. (1992a). Do parasites make prey vulnerable to predation – red grouse and parasites. *Journal of Animal Ecology*, **61**, 681–692.

Hudson, P. J., Newborn, D. and Dobson, A. P. (1992b). Regulation and stability of a free-living host–parasite system – *Trichostrongylus tenuis* in red grouse. 1. Monitoring and parasite reduction experiments. *Journal of Animal Ecology*, **61**, 477–486.

Hudson, P. J., Dobson, A. P. and Newborn, D. (1998). Prevention of population cycles by parasite removal. *Science*, **282**, 2256–2258.

Hudson, P. J., Dobson, A. P., Cattadori, I. M., *et al.* (2002). Trophic interactions and population growth rates: describing patterns and identifying mechanisms. *Philosophical Transactions of the Royal Society of London*, B357, 1259–1271.

Huelsenbeck, J. P. and Ronquist, F. (2001). MRBAYES: Bayesian inference of phylogenetic trees. *Bioinformatics*, 17, 754–755.

Huelsenbeck, J. P., Ronquist, F., Nielsen, R. and Bollback, J. P. (2001). Bayesian inference of phylogeny and its impact on evolutionary biology. *Science*, 294, 2310–2314.

Hughes, B. D. (1995). *Random Walks and Random Environments. Volume 1: Random Walks*. Oxford: Clarendon Press.

Hur, K., Kim, J.-I., Choi, S.-I., *et al.* (2002). The pathogenic mechanisms of prion diseases. *Mechanisms of Ageing and Development*, 123, 1637–1647.

Huse, G., Strand, E. and Giske, J. (1999). Implementing behaviour in individual-based models using neural networks and genetic algorithms. *Evolutionary Ecology*, 13, 469–483.

Hutchings, J. A. (2000). Collapse and recovery of marine fishes. *Nature*, 406, 882–886.

Hutchings, J. A. (2001). Influence of population decline, fishing, and spawner variability on the recovery of marine fishes. *Journal of Fish Biology*, 59 (Supplement A), 306–322.

Hutchings, J. A. and Myers, R. A. (1994). What can be learned from the collapse of a renewable resource? Atlantic cod, *Gadus morhua*, of Newfoundland and Labrador. *Canadian Journal of Fisheries and Aquatic Sciences*, 51, 2126–2146.

Hutchings, J. A., Walters, C. and Haedrich, R. L. (1997). Is scientific inquiry incompatible with government information control? *Canadian Journal of Fisheries and Aquatic Sciences*, 54, 1198.

Iwasa, Y., Hakoyama, H., Nakamaru, M. and Nakanishi, J. (2000). Estimate of population extinction risk and its application to ecological risk management. *Population Ecology*, 42, 73–80.

Jacobson, L. D. and MacCall, A. D. (1995). Stock–recruitment models for Pacific Sardine (*Sardinops sagax*). *Canadian Journal of Fisheries and Aquatic Sciences*, 52, 566–577.

Jacobson, L. D., De Oliveira, J. A. A., Barange, M., *et al.* (2001). Surplus production, variability, and climate change in the great sardine and anchovy fisheries. *Canadian Journal of Fisheries and Aquatic Sciences*, 58, 1891–1903.

Jacobson, L. D., Cadrin, S. X. and Weinberg, J. R. (2002). Tools for estimating surplus production and FMSY in any stock assessment model. *North American Journal of Fisheries Management*, 22, 326–338.

James, A., Pitchford, J. W. and Brindley, J. (2003). The relationship between plankton blooms, the hatching of fish larvae, and recruitment. *Ecological Modelling*, 160, 77–90.

Janssen, A. (1989). Optimal host selection by *Drosophila* parasitoids in the field. *Functional Ecology*, 3, 469–479.

Janssen, A., Driessen, G., de Haan, M. and Roodbol, N. (1988). The impact of parasitoids on natural populations of temperature woodland *Drosophila*. *Netherlands Journal of Zoology*, 38, 61–73.

Jaynes, E. T. (2003). *Probability Theory. The Logic of Science*. Cambridge: Cambridge University Press.

Jenkins, R. (2001). *Churchill*. New York: Farrar, Straus and Grioux.

Jennings, S. (2000). Patterns and prediction of population recovery in marine reserves. *Reviews in Fish Biology and Fisheries*, **10**, 209–231.

Jennings, S. and Polunin, N. V. C. (1996). Fishing strategies, fishery development and socioeconomics in traditionally managed Fijian fishing grounds. *Fisheries Management and Ecology*, **3**, 335–347.

Jennings, S., Reynolds, J. D. and Mills, S. C. (1998). Life history correlates of responses to fisheries exploitation. *Proceedings of the Royal Society of London*, B**265**, 333–339.

Jennings, S., Kaiser, M. and Reynolds, J. D. (2001). *Marine Fisheries Ecology*. Oxford: Blackwell Science.

Jervis, M. A. and Kidd, N. A. C. (1986). Host-feeding strategies in hymenopteran parasitoids. *Biological Reviews*, **61**, 395–434.

Jervis, M. A., Heimpel, G. E., Ferns, P. N., Harvey, J. A. and Kidd, N. A. C. (2001). Life-history strategies in parasitoid wasps: a comparative analysis of 'ovigeny'. *Journal of Animal Ecology*, **70**, 442–458.

Jonzen, N., Cardinale, M., Gardmark, A., Arrhenius, F. and Lundberg, P. (2002). Risk of collapse in the eastern Baltic cod fishery. *Marine Ecology – Progress Series*, **240**, 225–233.

Kac, M. (1985). *Enigmas of Chance*. New York: Harper and Row.

Kailath, T. (1980). *Linear Systems*. Englewood Cliffs, NJ: Prentice Hall.

Kaplan, E. H. and Wein, L. M. (2003). Smallpox eradication in west and central Africa: surveillance-containment or herd immunity? *Epidemiology*, **14**, 90–92.

Kaplan, E. H., Craft, D. L. and Wein, L. M. (2002). Emergency response to a smallpox attack: the case for mass vaccination. *Proceedings of the National Academy of Sciences*, **99**, 10 935–10 940.

Karlin, S. and Taylor, H. M. (1981). *A Second Course in Stochastic Processes*. New York: Academic Press.

Keeling, M. J. and Grenfell, B. T. (1997). Disease extinction and community size: modeling the persistence of measles. *Science*, **275**, 65–67.

Keeling, M. J. and Grenfell, B. T. (2000). Individual-based perspectives on R_0. *Journal of Theoretical Biology*, **203**, 51–61.

Kehler, D. G., Myers, R. A. and Field, C. A. (2002). Measurement error and bias in the maximum reproductive rate for the Ricker model. *Canadian Journal of Fisheries and Aquatic Sciences*, **59**, 854–864.

Keizer, J. E. (1987). *Statistical Thermodynamics of Nonequilibrium Processes*. New York: Springer Verlag.

Keller, E. F. (2002). *Making Sense of Life*. Cambridge, MA: Harvard University Press.

Keller, J. B. (1974). Optimal velocity in a race. *American Mathematical Monthly*, **81**, 474–480.

Kendall, M. and Stuart, A. (1979). *The Advanced Theory of Statistics. Volume 2. Inference and Relationship,* 4th edn. London: Charles Griffin and Company Limited.

Kermack, W. O. and McKendrick, A. G. (1927). A contribution to the mathematical theory of epidemics. *Proceedings of the Royal Society of London*, A**115**, 700–721.

Kermack, W. O. and McKendrick, A. G. (1932). Contributions to the mathematical theory of epidemics – II. The problem of endemicity. *Proceedings of the Royal Society of London*, A**138**, 55–83.

Kermack, W. O. and McKendrick, A. G. (1933). Contributions to the mathematical study of epidemics. III. Further studies of the problem of endemicity. *Proceedings of the Royal Society of London*, A**141**, 94–122.

Kimura, M. and Ohta, T. (1971). *Theoretical Aspects of Population Genetics*. Princeton, NJ: Princeton University Press.

Kinas, P. G. (1996). Bayesian fishery stock assessment and decision making using adaptive importance sampling. *Canadian Journal of Fisheries and Aquatic Sciences*, **53**, 414–423.

King, J. R. and McFarlane, G. A. (2003). Marine fish life history strategies: applications to fishery management. *Fisheries Management and Ecology*, **10**, 249–264.

King, M. (1995). *Fisheries Biology, Assessment and Management*. Ames, Iowa: Iowa State University Press.

King, S. (2000). *On Writing*. New York: Scribner.

Kingsland, S. E. (1985). *Modeling Nature*. Chicago: University of Chicago Press.

Kirchner, J. W. (2001). Fractal power spectra plotted upside-down. Comment on "Scaling of power spectrum of extinction events in the fossil record" by V. P. Dimri and M. R. Prakash. *Earth and Planetary Science Letters*, **192**, 617–621.

Kirchner, J. W. (2002). Evolutionary speed limits inferred from the fossil record. *Nature*, **415**, 65–68.

Kirchner, J. W. and Roy, B. A. (1999). The evolutionary advantages of dying young: epidemiological implications of longevity in metapopulations. *American Naturalist*, **154**, 140–159.

Kirchner, J. W. and Weil, A. (1998). No fractals in fossil extinction statistics. *Nature*, **395**, 337–338.

Kirchner, J. W. and Weil, A. (2000). Correlations in fossil extinction and origination rates through geological time. *Proceedings of the Royal Society of London*, B**267**, 1301–1309.

Kirkwood, T. B. L. (1999). Evolution, molecular biology and mortality plateaus. In V. A. Bohr, B. F. C. Clark and T. Stevensner (editors) *Molecular Biology of Aging, Alfred Benzon Symposium 44*. Copenhagen: Munksgaard, pp. 383–390.

Klein, E. K., Lavigne, C., Foueillassar, X., Gouyon, P. and Larédo, C. (2003). Corn pollen dispersal: quasi-mechanistic models and field experiments. *Ecological Monographs*, **73**, 131–150.

Klein, G. (1952). Mean first-passage times of Brownian motion and related problems. *Proceedings of the Royal Society of London*, A**211**, 431–443.

Klyashtorin, L. B. (1998). Long-term climate change and main commercial fish production in the Atlantic and Pacific. *Fisheries Research*, **37**, 115–125.

Knell, R. J., Begon, M. and Thompson, D. J. (1996). Transmission dynamics of *Bacillus thuringiensis* infecting *Plodia interpunctella*: a test of the mass action assumption with an insect pathogen. *Proceedings of the Royal Society of London*, B**263**, 75–81.

Koella, J. C. and Boëte, C. (2003). A model for the coevolution of immunity and immune evasion in vector-borne diseases with implications for the epidemiology of malaria. *The American Naturalist*, **161**, 698–707.

Koella, J. C. and Restif, O. (2001). Coevolution of parasite virulence and host life history. *Ecology Letters*, **4**, 207–214.

Koeller, P. (2003). The lighter side of reference points. *Fisheries Research*, **62**, 1–6.

Kolata, G. B. (1977). Catastrophe theory: the emperor has no clothes. *Science*, **196**, 287 + 350–351.

Koopman, B. O. (1980). *Search and Screening. General Principles with Historical Applications*. Elmsford, NY: Pergamon Press.

Korman, J. and Higgins, P. S. (1997). Utility of escapement time series data for monitoring the response of salmon populations to habitat alteration. *Canadian Journal of Fisheries and Aquatic Sciences*, **54**, 2058–2067.

Kot, M. (2001). *Elements of Mathematical Ecology*. Cambridge: Cambridge University Press.

Kot, M., Lewis, M. A. and van den Driessche, P. (1996). Dispersal data and the spread of invading organisms. *Ecology*, **77**, 2027–2042.

Kramers, H. A. (1940). Brownian motion in a field of force and the diffusion model of chemical reactions. *Physica*, **7**, 284–312.

Kubo, R., Matsuo, K. and Kitahara, K. (1973). Fluctuation and relaxation of macrovariables. *Journal of Statistical Physics*, **9**, 51–96.

Kuikka, S., Hilden, M., Gislason, H. F., *et al.* (1999). Modeling environmentally driven uncertainties in Baltic cod (*Gadus morhua*) management by Bayesian inference diagrams. *Canadian Journal of Fisheries and Aquatic Sciences*, **56**, 629–641.

Laakso, J., Kaitala, V. and Ranta, E. (2003). Non-linear biological responses to disturbance: consequences on population dynamics. *Ecological Modelling*, **162**, 247–258.

Lande, R. (1985). Expected time for random genetic drift of a population between stable phenotypic states. *Proceedings of the National Academy of Sciences*, **82**, 7641–7645.

Lande, R. (1987). Extinction thresholds in demographic models of territorial species. *American Naturalist*, **130**, 624–635.

Lande, R., Engen, S. and Saether, B.-E. (2003). *Stochastic Population Dynamics in Ecology and Conservation*. Oxford: Oxford University Press.

Lander, A. D. (2004). A calculus of purpose. *PLoS Biology*, **2**, 0712–0714.

Laurenson, M. K., Norman, R. A., Gilbert, L., Reid, H. W. and Hudson, P. J. (2003). Identifying disease reservoirs in complex systems: mountain hares as reservoirs of ticks and louping-ill virus, pathogens of red grouse. *Journal of Animal Ecology*, **72**, 177–185.

Law, R. (2000). Fishing, selection, and phenotypic evolution. *ICES Journal of Marine Science*, **57**, 659–668.

Law, R. and Dicekmann, U. (1998). Symbiosis through exploitation and the merger of lineages in evolution. *Proceedings of the Royal Society of London*, B**265**, 1245–1253.

Law, R., Murrell, D. J. and Dieckmann, U. (2003). Population growth in space and time: spatial logistic equations. *Ecology*, **84**, 252–262.

Leggett, W. C. and Deblois, E. (1994). Recruitment in marine fishes: is it regulated by starvation and predation in the egg and larval stages? *Netherlands Journal of Sea Research*, **32**, 119–134.

Lenski, R. E. and May, R. M. (1994). The evolution of virulence in parasites and pathogens: reconciliation between two competing hypotheses. *Journal of Theoretical Biology*, **169**, 253–265.

Leonard, T. and Hsu, J. S. J. (1999). *Bayesian Methods*. Cambridge: Cambridge University Press.

Levin, S. A. and Segel, L. A. (1985). Pattern generation in space and aspect. *SIAM Review*, **27**, 45–67.

Levins, R. (1966). The strategy of model building in population biology. *American Scientist*, **54**, 421–431.

Levins, R. (1969). Some demographic and genetic consequences of environmental heterogeneity for biological control. *Bulletin of the Entomological Society of America*, **15**, 237–240.

Levins, R. (1970). Extinction. In M. Gerstenhaber (editor) *Some Mathematical Questions in Biology*. Providence, RI: American Mathematical Society, pp. 75–107.

Lewin, R. (1986). Punctuated equilibrium is now old hat. *Science*, **231**, 672–673.

Lewis, W. J., and Takasu, K. (1990). Use of learned odours by a parasitic wasp in accordance with host and food needs. *Nature*, **348**, 635–636.

Lewontin, R. and Cohen, D. (1969). On population growth in a randomly varying environment. *Proceedings of the National Academy of Sciences*, **62**, 1056–1060.

Liermann, M. and Hilborn, R. (1997). Depensation in fish stocks: a hierarchic Bayesian meta-analysis. *Canadian Journal of Fisheries and Aquatic Sciences*, **54**, 1976–1984.

Lighthill, M. J. (1958). *Introduction to Fourier Analysis and Generalised Functions*. Cambridge: Cambridge University Press.

Lima, S. L. (2002). Putting predators back into behavioral predator-prey interactions. *Trends in Ecology and Evolution*, **17**, 70–75.

Lin, C. C. and Segel, L. A. (1988 (1974)). *Mathematics Applied to Deterministic Problems in the Natural Sciences*. Philadelphia: SIAM (Society for Industrial and Applied Mathematics).

Lin, Z.-S. (2003). Simulating unintended effects restoration. *Ecological Modelling*, **164**, 169–175.

Lindholm, J. B., Auster, P. J., Ruth, M. and Kaufman, L. (2001). Modeling the effects of fishing and implications for the design of marine protected areas: juvenile fish responses to variations in seafloor habitat. *Conservation Biology*, **15**, 424–437.

Lindley, S. T. (2003). Estimation of population growth and extinction parameters from noisy data. *Ecological Applications*, **13**, 806–813.

Link, D. R., Natale, G., Shao, R., *et al.* (1997). Spontaneous formation of macroscopic chiral domains in a fluid smectic phase of achiral molecules. *Science*, **278**, 1924–1927.

Link, J. S., Brodziak, J. K. T., Edwards, S. F., *et al.* (2002). Marine ecosystem assessment in a fisheries management context. *Canadian Journal of Fisheries and Aquatic Sciences*, **59**, 1429–1440.

Lipp, E. K., Huq, A. and Colwell, R. K. (2002). Effects of global climate on infectious disease: the cholera model. *Clinical Microbiology Reviews*, **15**, 757–770.

Lochmiller, R. L. and Deerenberg, C. (2000). Trade-offs in evolutionary immunology: just what is the cost of immunity? *Oikos*, **88**, 87–98.

Lockwood, D. R., Hastings, A. M. and Botsford, L. W. (2002). The effects of dispersal patterns on marine reserves: does the tail wag the dog? *Theoretical Population Biology*, **61**, 297–310.

LoGiudice, K., Ostfeld, R. S., Schmidt, K. A. and Keesing, F. (2003). The ecology of infectious disease: effects of host diversity and community composition on Lyme disease risk. *Proceedings of the National Academy of Science*, **100**, 567–571.

Lorenzen, K. (2000). Population dynamics and management. In M. C. M. Beveridge and B. J. McAndrew (editors) *Tilapias: Biology and Exploitation*. Dordrecht: Kluwer Academic Publishers, pp. 163–226.

Ludwig, D. (1975). Persistence of dynamical systems under random perturbations. *SIAM Review*, **17**, 605–640.

Ludwig, D. (1981). Escape from domains of attraction for systems perturbed by noise. In R. H. Enns, B. L. Jones, R. M. Miura and S. S. Rangnekar (editors) *Nonlinear Phenomena in Physics and Biology*. New York: Plenum Press.

Ludwig, D. (1995). Uncertainty and fisheries management. *Lecture Notes in Biomathematics*, **100**, 516–528.

Ludwig, D. (1999). Is it meaningful to estimate a probability of extinction? *Ecology*, **80**, 298–310.

Ludwig, D., Mangel, M. and Haddad, B. (2001). Ecology, conservation and public policy. *Annual Review of Ecology and Systematics*, **32**, 481–517.

MacArthur, R. H. and Pianka, E. R. (1966). On the optimal use of a patchy environment. *American Naturalist*, **100**, 603–609.

MacArthur, R. H. and Wilson, E. O. (1967). *The Theory of Island Biogeography*. Princeton, NJ: Princeton University Press.

MacCall, A. D. (1990). *Dynamic Geography of Marine Fish Populations*. Seattle, WA: University of Washington Press.

MacCall, A. D. (1998). Use of decision tables to develop a precautionary approach to problems in behavior, life history and recruitment variability. In V. R. Restrepo (editor) *Proceedings of the Fifth NMFS Stock Assessment Workshop: Providing Scientific Advice to Implement the Precautionary Approach under the Magnuson–Stevens Fishery Conservation and Management Act*. Key Largo, FL: US Department of Commerce.

MacCall, A. D. (2002). Use of known-biomass production models to determine productivity of West Coast groundfish stocks. *North American Journal of Fisheries Management*, **22**, 272–279.

MacDonald, N. (1989). *Biological Delay Systems: Linear Stability Theory*. Cambridge: Cambridge University Press.

Mackauer, M. and Voelkl, W. (1993). Regulation of aphid populations by aphidiid wasps: does parasitoid foraging behaviour or hyperparasitism limit impact? *Oecologia*, **94**, 339–350.

Malloch, P. D. (1994). *Life-History and Habits of the Salmon, Sea-Trout, Trout and Other Freshwater Fish*. Derrydale Press.

Mangel, M. (1982). Applied mathematicians and naval operators. *SIAM Review*, **24**, 289–300.

Mangel, M. (1985). *Decision and Control in Uncertain Resource Systems*. New York: Academic Press.

Mangel, M. (1992). Descriptions of superparasitism by optimal foraging theory, evolutionary stable strategies, and quantitative genetics. *Evolutionary Ecology*, **6**, 152–169.

Mangel, M. (1998). No-take areas for sustainability of harvested species and a conservation invariant for marine reserves. *Ecology Letters*, **1**, 87–90.

Mangel, M. (2000a). Irreducible uncertainties, sustainable fisheries and marine reserves. *Evolutionary Ecology Research*, **2**, 547–557.

Mangel, M. (2000b). On the fraction of habitat allocated to marine reserves. *Ecology Letters*, **3**, 15–22.

Mangel, M. (2000c). Trade-offs between fish habitat and fishing mortality and the role of reserves. *Bulletin of Marine Science*, **66**, 663–674.

Mangel, M. (2001a). Complex adaptive systems, aging and longevity. *Journal of Theoretical Biology*, **213**, 559–571.

Mangel, M. (2001b). Required reading for (ecological) battles. (Review of N Eldredge. The Triumph of Evolution . . . and the Failure of Creationism). *Trends in Ecology and Evolution*, **16**, 110.

Mangel, M. and Beder, J. H. (1985). Search and stock depletion: theory and applications. *Canadian Journal of Fisheries and Aquatic Sciences*, **42**, 150–163.

Mangel, M. and Clark, C. W. (1988). *Dynamic Modeling in Behavioral Ecology*. Princeton, NJ: Princeton University Press.

Mangel, M. and Ludwig, D. (1977). Probability of extinction in a stochastic competition. *SIAM Journal on Applied Mathematics*, **33**, 256–266.

Mangel, M. and Roitberg, B. D. (1992). Behavioral stabilization of host–parasitoid population dynamics. *Theoretical Population Biology*, **42**, 308–320.

Mangel, M. and Tier, C. (1993). Dynamics of metapopulations with demographic stochasticity and environmental catastrophes. *Theoretical Population Biology*, **44**, 1–31.

Mangel, M. and Tier, C. (1994). Four facts every conservation biologist should know about persistence. *Ecology*, **75**, 607–614.

Mangel, M., Hofman, R. J., Norse, E. A. and Twiss, J. R. (1993). Sustainability and ecological research. *Ecological Applications*, **3**, 573–575.

Mangel, M., Mullan, A., Mulch, A., Staub, S. and Yasukochi, E. (1998). A generally accessible derivation of the golden rule of bioeconomics. *Bulletin of the Ecological Society of America*, **79**, 145–148.

Mangel, M., Fiksen, O. and Giske, J. (2001). Theoretical and statistical models in natural resource management and research. In T. M. Shenk and

A. B. Franklin (editors) *Modeling in Natural Resource Management. Development, Interpretation, and Application*. Washington, DC: Island Press, pp. 57–72.

Mantua, N. J., Hare, S. R., Zhang, Y., Wallace, J. M. and Francis, R. C. (1997). A Pacific interdecadal climate oscillation with impacts on salmon production. *Bulletin of the American Meteorological Society*, **78**, 1069–1079.

Margulis, L. and Sagan, D. (2002). *Acquiring Genomes. A Theory of the Origins of Species*. New York: Basic Books.

Marris, G., Hubbar, S. and Hughes, J. (1986). Use of patchy resources by *Nemeritis canescens* (Hymenoptera: Ichneumonidae). I. Optimal solutions. *Journal of Animal Ecology*, **55**, 631–640.

Marshall, C. T., Kjesbu, O. S., Yaragina, N. A., Solemdal, P. and Ulltang, O. (1998). Is spawner biomass a sensitive measure of the reproductive and recruitment potential of Northeast Arctic cod? *Canadian Journal of Fisheries and Aquatic Sciences*, **55**, 1766–1783.

Martz, H. F. and Waller, R. A. (1982). *Bayesian Reliability Analysis*. New York: John Wiley and Sons.

Masel, J. and Bergman, A. (2003). The evolution of the evolvability properties of the yeast prion [PSI +]. *Evolution*, **57**, 1498–1512.

Matsuda, H. and Nishimori, K. (2003). A size-structured model for a stock-recovery program for an exploited endemic fisheries resource. *Fisheries Research*, **60**, 223–236.

Maunder, M. (2003). Is it time to discard the Schaefer model from the stock assessment scientist's toolbox? *Fisheries Research*, **61**, 145–149.

Maunder, M. N. (2002). The relationship between fishing methods, fisheries management and the estimation of maximum sustainable yield. *Fish and Fisheries*, **3**, 251–260.

May, R. M. (1974). *Stability and Complexity in Model Ecosystems*, 2nd edn. Princeton, NJ: Princeton University Press.

May, R. M. and Anderson, R. M. (1978). Regulation and stability of host-parasite population interactions. II. Destabilizing processes. *Journal of Animal Ecology*, **47**, 249–267.

May, R. M., Conway, G. R., Hassel, M. P. and Southwood, T. R. E. (1974). Time delays, density dependence and single species oscillations. *Journal of Animal Ecology*, **43**, 747–770.

Maynard Smith, J. (1968). *Some Mathematical Ideas in Biology*. Cambridge: Cambridge University Press.

Maynard Smith, J. (1982). *Evolution and the Theory of Games*. Cambridge: Cambridge University Press.

McAllister, M. K. (1996). Applications of Bayesian decision theory to fisheries policy formation: review. In R. Arnason and T. B. Davidsson (editors) *Essays on Statistical Modelling Methodology for Fisheries Management*. Rekjavek, Iceland: The Fisheries Research Institute University Press.

McAllister, M. K. and Ianelli, J. N. (1997). Bayesian stock assessment using catch-age data and the sampling-importance resampling algorithm. *Canadian Journal of Fisheries and Aquatic Sciences*, **54**, 284–300.

McAllister, M. K. and Kirkwood, G. P. (1998a). Bayesian stock assessment: a review and example application using the logistic model. *ICES Journal of Marine Science*, **55**, 1031–1060.

McAllister, M. K. and Kirkwood, G. P. (1998b). Using Bayesian decision analysis to help achieve a precautionary approach for managing developing fisheries. *Canadian Journal of Fisheries and Aquatic Sciences*, **55**, 2642–2661.

McAllister, M. K. and Kirkwood, G. P. (1999). Applying multivariate conjugate priors in fishery-management system evaluation: how much quicker is it and does it bias the ranking of management options. *ICES Journal of Marine Science*, **56**, 884.

McAllister, M. K. and Peterman, R. M. (1992a). Decision analysis of a large-scale fishing experiment designed to test for a genetic effect of size-selective fishing on British Columbia pink salmon (*Oncorhynchus gorbuscha*). *Canadian Journal of Fisheries and Aquatic Sciences*, **49**, 1305–1314.

McAllister, M. K. and Peterman, R. M. (1992b). Experimental design in the management of fisheries: A review. *North American Journal of Fisheries Management*, **12**, 1–18.

McAllister, M. K., Pikitch, E. K., Punt, A. E. and Hilborn, R. (1994). A Bayesian approach to stock assessment and harvest decision using the sampling/importance resampling algorithm. *Canadian Journal of Fisheries and Aquatic Sciences*, **51**, 2673–2687.

McAllister, M. K., Starr, P. J., Restrepo, V. R. and Kirkwood, G. P. (1999). Formulating quantitative methods to evaluate fishery-management systems: what fishery processes should be modelled and what trade-offs should be made? *ICES Journal of Marine Science*, **56**, 900–916.

McAllister, M. K., Pikitch, E. K. and Babcock, E. A. (2001). Using demographic methods to construct Bayesian priors for the intrinsic rate of increase in the Schaefer model and implications for stock rebuilding. *Canadian Journal of Fisheries and Aquatic Sciences*, **58**, 1871–1890.

McCallum, H., Barlow, N. and Hone, J. (2001). How should pathogen transmission be modelled? *Trends in Ecology and Evolution*, **16**, 295–300.

McClanahan, T. R. and Kaunda-Arara, B. (1996). Fishery recovery in a coral-reef marine park and its effect on the adjacent fishery. *Conservation Biology*, **10**, 1187–1199.

McDonald, A. D., Sandal, L. K. and Steinshamn, S. I. (2002). Implications of a nested stochastic/deterministic bio-economic model for a pelagic fishery. *Ecological Modelling*, **149**, 193–201.

McDowall, R. M. (1988). *Diadromy in Fishes. Migrations Between Freshwater and Marine Environments*. Portland, OR: Timber Press.

McGarvey, R. (2003). Demand-side fishery management: integrating two forms of input control. *Marine Policy*, **27**, 207–218.

McGregor, R. (1997). Host-feeding and oviposition by parasitoids on hosts of different fitness value: influences of egg load and encounter rate. *Journal of Insect Behavior*, **10**, 451–462.

McGregor, R. R. and Roitberg, B. D. (2000). Size-selective oviposition by parasitoids and the evolution of life-history timing in hosts: fixed preference vs frequency-dependent host selection. *Oikos*, **89**, 305–312.

Mchich, R., Auger, P. M., de la Parra, R. B. and Raissi, N. (2002). Dynamics of a fishery on two fishing zones with fish stock dependent migrations: aggregation and control. *Ecological Modelling*, **158**, 51–62.

McLeod, P., Martin, A. P. and Richards, K. J. (2002). Minimum length scale for growth-limited oceanic plankton distributions. *Ecological Modelling*, **158**, 111–120.

McNamara, J. M., Houston, A. I. and Collins, E. J. (2001). Optimality models in behavioral biology. *SIAM Review*, **43**, 413–466.

McNeill, J. R. (2000). *Something New Under the Sun: an Environmental History of the Twentieth-Century World*. New York: Norton.

McPhee, J. (2002). *The Founding Fish*. New York: Farrar, Straus and Giroux.

McPhee, M. V. and Quinn, T. P. (1998). Factors affecting the duration of nest defense and reproductive lifespan of female sockeye salmon, *Oncorhynchus nerka*. *Environmental Biology of Fishes*, **51**, 469–475.

Medvinsky, A. B., Petrovskii, S. V., Tikhonova, I. A., Malchow, H. and Li, B.-L. (2002). Spatiotemporal complexity of plankton and fish dynamics. *SIAM Review*, **44**, 311–370.

Meier, C., Senn, W., Hauser, R. and Zimmerman, M. (1994). Strange limits of stability in host-parasitoid systems. *Journal of Mathematical Biology*, **32**, 563–572.

Meltzer, M. (2003). Risks and benefits of preexposure and postexposure smallpox vaccination. *Emerging Infectious Diseases*, **9**, 1363–1370.

Meltzer, M. I., Damon, I., LeDuc, J. W. and Millar, D. J. (2001). Modeling potential responses to smallpox as a bioterrorist weapon. *Emerging Infectious Diseases*, **7**, 959–968.

Merton, R. C. (1971). Optimum consumption and portfolio rules in a continuous-time model. *Journal of Economic Theory*, **3**, 373–413.

Mesnil, B. (2003). The catch-survey analysis (CSA) method of fish stock assessment: an evaluation using simulated data. *Fisheries Research*, **63**, 193–212.

Meyer, R. and Millar, R. B. (1999a). Bayesian stock assessment using a state-space implementation of the delay difference model. *Canadian Journal of Fisheries and Aquatic Sciences*, **56**, 37–52.

Meyer, R. and Millar, R. B. (1999b). BUGS in Bayesian stock assessments. *Canadian Journal of Fisheries and Aquatic Sciences*, **56**, 1078–1086.

Millar, R. B. (2002). Reference priors for Bayesian fisheries models. *Canadian Journal of Fisheries and Aquatic Sciences*, **59**, 1492–1502.

Millar, R. B. and Meyer, R. (2000). Bayesian state-space modeling of age-structured data: fitting a model is just the beginning. *Canadian Journal of Fisheries and Aquatic Sciences*, **57**, 43–50.

Mills, D. (1989). *Ecology and Management of Atlantic Salmon*. London: Chapman and Hall.

Mitchell, W. A. and Valone, T. J. (1990). The Optimization Research Program: studying adaptations by their function. *The Quarterly Review of Biology*, **65**, 43–52.

Moerland, T. S. (1995). Temperature: enzyme and organelle. In P. Hochachka and T. P. Mommsen (editors) *Biochemistry and Molecular Biology of Fishes*. Amsterdam: Elsevier Science B.V.

Moret, Y. and Schmid-Hempel, P. (2000). Survival for immunity: the price of immune system activation for bumblebee workers. *Science*, **290**, 1166–1167.

Morgan, E. R., Milner-Gulland, E. J., Torgerson, P. R. and Medley, G. F. (2004). Ruminating on complexity: macroparasites of wildlife and livestock. *Trends in Ecology and Evolution*, **19**, 181–188.

Morris, W. F., Mangel, M. and Adler, F. R. (1995). Mechanisms of pollen deposition by insect pollinators. *Evolutionary Ecology*, **9**, 304–317.

Morse, P. M. and Kimball, G. E. (1951). *Methods of Operations Research*. Technology Press of Massachusetts Institute of Technology; and Cambridge and New York: Wiley.

Mosmann, T. R. and Sad, S. (1996). The expanding universe of T-cell subsets: Th1, Th2 and more. *Immunology Today*, **17**, 138–146.

Mosquera, J. and Adler, F. R. (1998). Evolution of virulence: a unified framework for coinfection and superinfection. *Journal of Theoretical Biology*, **195**, 293–313.

Mosquera, J., Côté, M., Jennings, S. and Reynolds, J. D. (2000). Conservation benefits of marine reserves for fish populations. *Animal Conservation*, **4**, 321–332.

Mueller, L. D. and Rose, M. R. (1996). Evolutionary theory predicts late-life mortality plateaus. *Proceedings of the National Academy of Sciences*, **93**, 15 249–15 253.

Mullan, M. (1993). *Webs and Scales. Physical and Ecological Processes in Marine Recruitment*. Seattle, WA: University of Washington Press.

Murawski, S. A. (2000). Definitions of overfishing from an ecosystem perspective. *ICES Journal of Marine Science*, **57**, 649–658.

Murdoch, W. W. (1994). Population regulation in theory and practice. *Ecology*, **75**, 271–287.

Murdoch, W. W., Nisbet, R., Blythe, S. P., Gurney, W. S. C. and Reeve, J. D. (1987). An invulnerable age class and stability in delay-differential parasitoid–host models. *American Naturalist*, **129**, 263–282.

Murdoch, W. W., Briggs, C. J. and Nisbet, R. (2003). *Consumer–Resource Dynamics*. Princeton, NJ: Princeton University Press.

Murray, J. D. (1990). Turing's theory of morphogenesis – its influence on modelling biological pattern and form. *Bulletin of Mathematical Biology*, **52**, 119–153.

Murray, J. D. (2002). *Mathematical Biology I: An Introduction*. New York: Springer Verlag.

Murray, J. D. (2003). *Mathematical Biology II: Spatial Models and Biomedical Applications*. New York: Springer Verlag.

Myers, R. A., Mertz, G. and Fowlow, P. S. (1997a). Maximum population growth rates and recovery times for Atlantic cod, *Gadus morhua. Fishery Bulletin*, **95**, 762–772.

Myers, R. A., Hutchings, J. A. and Barrowman, N. J. (1997b). Why do fish stocks collapse? The example of cod in Atlantic Canada. *Ecological Applications*, **7**, 91–106.

Myers, R. A., MacKenzie, B. R., Bowen, K. G. and Barrowman, N. J. (2001). What is the carrying capacity for fish in the ocean? A meta-analysis of population dynamics of North Atlantic cod. *Canadian Journal of Fisheries and Aquatic Sciences*, **58**, 1464–1476.

Myers, R. A., Barrowman, N. J., Hilborn, R. and Kehler, D. G. (2002). Inferring Bayesian priors with limited direct data: Applications to risk analysis. *North American Journal of Fisheries Management*, **22**, 351–364.

Nakken, O., Sandberg, P. and Steinshamn, S. I. (1996). Reference points for optimal fish stock management. *Marine Policy*, **20**, 447–462.

Narasimhan, T. N. (1999). Fourier's heat conduction equation: history, influence, and connections. *Reviews of Geophysics*, **37**, 151–172.

Nasell, I. (2002). Stochastic models of some endemic infections. *Mathematical Biosciences*, **179**, 1–19.

Needle, C. L. (2002). Recruitment models: diagnosis and prognosis. *Reviews in Fish Biology and Fisheries*, **11**, 95–111.

Nesse, R. M. and Williams, G. C. (1994). *Why We Get Sick*. New York: Vintage Books.

Nicholson, A. J. (1933). The balance of animal populations. *Journal of Animal Ecology*, **2**, 131–178.

Nicholson, A. J. (1954). An outline of the dynamics of animal populations. *Australian Journal of Zoology*, **2**, 9–65.

Nicholson, A. J. and Bailey, V. A. (1935). The balance of animal populations. *Proceedings of the Zoological Society of London*, **3**, 551–598.

Nisbet, R. and Gurney, W. S. C. (1982). *Modelling Fluctuating Populations*. New York: John Wiley and Sons.

Nisbet, R. M. and Gurney, W. S. C. (1983). The systematic formulation of population models for insects with dynamically varying instar duration. *Theoretical Population Biology*, **23**, 114–135.

Nisbet, R. M., Blythe, S. P., Gurney, W. S. C. and Metz, J. A. J. (1985). Stage-structure models of populations with distinct growth and development processes. *IMA Journal of Mathematics Applied in Medicine and Biology*, **2**, 57–68.

Norse, E. A. (editor) (1993). *Global Marine Biological Diversity*. Washington, DC: Island Press.

Nowak, M. A. and May, R. M. (2000). *Virus Dynamics. Mathematical Principles of Immunology and Virology*. Oxford: Oxford University Press.

Nuland, S. B. (1993). *How We Die*. New York: Vintage Books.

O'Neill, P. D. (2002). A tutorial introduction to Bayesian inference for stochastic epidemic models using Markov chain Monte Carlo methods. *Mathematical Biosciences*, **180**, 103–114.

Oaks, Jr., S. C., Mitchell, V. S., Pearson, G. W. and Carpenter, C. C. J. (editors) (1991). *Malaria. Obstacles and Opportunities*. Washington, DC: National Academy Press.

Okey, T. A. (2003). Membership of the eight Regional Fishery Management Councils in the United States: are special interests over-represented? *Marine Policy*, **27**, 193–206.

Olson, D. M. and Andow, D. A. (1997). Primary sex allocation in *Trichogramma* (Hymenoptera: Trichogrammatidae) and the effects of sperm on oviposition behavior. *Annals of the Entomological Society of America*, **90**, 689–692.

Olson, D. M., Fadamiro, H., Lundgren, J. G. and Heimpel, G. E. (2000). Effects of sugar feeding on carbohydrate and lipid metabolism in a parasitoid wasp. *Physiological Entomology*, **25**, 17–26.

Olver, C., Shuter, B. J. and Minns, C. K. (1995). Toward a definition of conservation principles for fisheries management. *Canadian Journal of Fisheries and Aquatic Sciences*, **52**, 1584–1594.

Overholtz, W. J. (1999). Precision and uses of biological reference points calculated from stock recruitment data. *North American Journal of Fisheries Management*, **19**, 643–657.

Overholtz, W. J., Edwards, S. F. and Brodziak, J. K. T. (1995). Effort control in the New England groundfish fishery: a bioeconomic perspective. *Canadian Journal of Fisheries and Aquatic Sciences*, **52**, 1944–1957.

Owen-Smith, N. (2002). *Adaptive Herbivore Ecology*. Cambridge: Cambridge University Press.

Pacala, S. W., Hassell, M. P. and May, R. M. (1990). Host–parasitoid associations in patchy environments. *Nature*, **344**, 150–153.

Packer, C., Holt, R., Hudson, P., Lafferty, K. and Dobson, A. (2003). Keeping the herds healthy and alert: implications of predator control for infectious disease. *Ecology Letters*, **6**, 797–802.

Paddack, M. J. and Estes, J. A. (2000). Kelp forest fish populations in marine reserves and adjacent exploited areas of Central California. *Ecological Applications*, **10**, 855–870.

Pandolfini, B. (1989). *Chess Openings: Traps and Zaps*. New York: Simon and Schuster.

Patterson, K., Cook, R., Darby, C., *et al.* (2001). Estimating uncertainty in fish stock assessment and forecasting. *Fish and Fisheries*, **2**, 125–157.

Patterson, K. R. (1998). Assessing fish stocks when catches are misreported: model, simulation tests, and application to cod, haddock and whiting in the ICES area. *ICES Journal of Marine Science*, **55**, 878–891.

Patterson, K. R. (1999). Evaluating uncertainty in harvest control law catches using Bayesian Markov chain Monte Carlo virtual population analysis with adaptive rejection sampling and including structural uncertainty. *Canadian Journal of Fisheries and Aquatic Sciences*, **56**, 208–221.

Pearcy, W. G. (1992). *Ocean Ecology of Pacific Salmonids*. Seattle, WA: University of Washington Press.

Pearl, R. (1928). *The Rate of Living*. New York: Alfred Knopf.

Pearl, R. and Miner, J. R. (1935). Experimental studies on the duration of life. XIV. The comparative mortality of certain lower organisms. *Quarterly Review of Biology*, **10**, 60–79.

Pearl, R. and Parker, S. A. (1921). Experimental studies on the duration of life. I. Introductory discussion of the duration of life in drosophila. *American Naturalist*, **55**, 481–509.

Pearl, R. and Parker, S. A. (1922a). Experimental studies on the duration of life. II. Hereditary differences in duration of life in the line-bred strains of drosophila. *American Naturalist*, **56**, 174–187.

Pearl, R. and Parker, S. A. (1922b). Experimental studies on the duration of life. III. The effect of successive etherizations on the duration of life in drosophila. *American Naturalist*, **56**, 273–280.

Pearl, R. and Parker, S. A. (1922c). Experimental studies on the duration of life. IV. Data on the influence of density of population on duration of life in drosophila. *American Naturalist*, **56**, 312–321.

Pearl, R. and Parker, S. A. (1922d). Experimental studies on the duration of life. V. On the influence of certain environmental factors on the duration of life in drosophila. *American Naturalist*, **56**, 385–405.

Pearl, R. and Parker, S. A. (1924a). Experimental studies on the duration of life. IX. New life tables for drosophila. *American Naturalist*, **58**, 71–82.

Pearl, R. and Parker, S. A. (1924b). Experimental studies on the duration of life. X. The duration of life of drosophila melanogaster in the complete absence of food. *American Naturalist*, **58**, 193–218.

Pearl, R., Miner, J. R. and Parker, S. A. (1927). Experimental studies on the duration of life. XI. Density of population and life duration in drosophila. *American Naturalist*, **56**, 289–318.

Pearl, R., Parker, S. A. and Gonzalez, B. M. (1923). Experimental studies on the duration of life. VII. The Mendelian inheritance of duration of life in crosses of wild type and quintuple stocks of drosophila melanogaster. *American Naturalist*, **57**, 153–192.

Pearl, R., Parker, T. and Miner, J. R. (1941). Experimental studies on the duration of life. XVI. Life tables for the flour beetle *Tribolium confusum* duval. *American Naturalist*, **75**, 5–19.

Pellmyr, O. (2003). Yuccas, yucca moths, and coevolution: a review. *Annals of the Missouri Botanical Garden*, **90**, 35–55.

Perelson, A. S. and Nelson, P. W. (1999). Mathematical analysis of HIV-1 dynamics in vivo. *SIAM Review*, **41**, 3–44.

Perkins, P. C. and Edwards, E. F. (1995). A mixture model for estimating discarded bycatch from data with many zero observations: tuna discards in the eastern tropical Pacific Ocean. *Fishery Bulletin*, **94**, 330–340.

Petchey, O. L. (2000). Environmental colour affects aspects of single-species population dynamics. *Proceedings of the Royal Society of London*, B**267**, 747–754.

Petchey, O. L., Gonzalez, A. and Wilson, H. B. (1997). Effects on population persistence: the interaction between environmental noise colour, intraspecific competition and space. *Proceedings of the Royal Society of London*, B**264**, 1841–1847.

Peterman, R. M. (1989). Application of statistical power analysis to the Oregon coho salmon (*Oncorhynchus kisutch*) problem. *Canadian Journal of Fisheries and Aquatic Sciences*, **46**, 1183–1187.

Peterman, R. M. (1990a). The importance of reporting statistical power: the forest decline and acid deposition example. *Ecology*, **71**, 2024–2027.

Peterman, R. M. (1990b). Statistical power analysis can improve fisheries research and management. *Canadian Journal of Fisheries and Aquatic Sciences*, **47**, 2–15.

Peters, C. S., Mangel, M. and Costantino, R. F. (1989). Stationary distribution of population size in *Tribolium. Bulletin of Mathematical Biology*, **51**, 625–638.

Peters, R. H. (1983). *The Ecological Implications of Body Size*. Cambridge: Cambridge University Press.

Peters, R. H. (1991). *A Critique for Ecology*. Cambridge: Cambridge University Press.

Pezzey, J. C. V., Roberts, C. M. and Urdal, B. T. (2000). A simple bioeconomic model of a marine reserve. *Ecological Economics*, **33**, 77–91.

Pielke Jr., R. A. and Conant, R. T. (2003). Best practices in prediction for decision-making: lessons from the atmospheric and earth sciences. *Ecology*, **84**, 1351–1358.

Pierce, N. E. and Nash, D. R. (1999). The Imperial Blue: *Jalmenus evagoras* (Lycaenidae). In R. L. Kitching, R. E. Jones, N. E. Pierce, and E. Scheermeyer, eds., *Biology of Australian Butterflies*. Victoria: CSIRO Publishing, pp. 277–313.

Pierce, N. E., Braby, M. F., Heath, A., *et al.* (2002). The ecology and evolution of ant association in the Lycaenidae (Lepidoptera). *Annual Review of Entomology*, **47**, 733–771.

Pigliucci, M. and Murren, C. J. (2003). Genetical assimilation and a possible evolutionary paradox: can macroevolution sometimes be so fast as to pass us by? *Evolution*, **57**, 1455–1464.

Pikitch, E. K., Huppert, D. D. and Sissenwine, M. P. (editors) (1997). *Global Trends: Fishery Management*. Bethesda, MD: American Fisheries Society.

Pikitch, E. K., Santora, C., Babcock, E. A., *et al.* (2004). Ecosystem-based fishery management. *Science*, **305**, 346–347.

Pincock, R. E. and Wilson, K. R. (1973). Spontaneous generation of optical activity. *Journal of Chemical Education*, **50**, 455–457.

Pincock, R. E., Perkins, R. R., Ma, A. S. and Wilson, K. R. (1971). Probability distribution of enantiomorphous forms in spontaneous generation of optically active substances. *Science*, **174**, 1018–1020.

Pirsig, R. M. (1974). *Zen and the Art of Motorcycle Maintenance*. New York: William Morrow.

Pitcher, T. J. (2000). Ecosystem goals can reinvigorate fisheries management, help dispute resolution and encourage public support. *Fish and Fisheries*, **1**, 99–103.

Pitcher, T. J., Watson, R., Forrest, R., Valtysson, H. P. and Guénette, S. (2002). Estimating illegal and unreported catches from marine ecosystems: a basis for change. *Fish and Fisheries*, **3**, 317–339.

Pletcher, S. D. and Curtsinger, J. W. (1998). Mortality plateaus and the evolution of senescence: why are old-age mortality rates so low? *Evolution*, **52**, 454–464.

Prager, M. H. (2002). Comparison of logistic and generalized surplus-production models applied to swordfish, *Xiphias gladius*, in the north Atlantic Ocean. *Fisheries Research*, **58**, 41–57.

Prager, M. H., Porch, C. E., Shertzer, K. W. and Caddy, J. F. (2003). Targets and limits for management of fisheries: a simple probability-based approach. *North American Journal of Fisheries Management*, **23**, 349–361.

Preston, S. H., Heuveline, P. and Guillot, M. (2001). *Demography. Measuring and Modeling Population Processes*. Oxford: Blackwell.

Pritchard, G. (1969). The ecology of a natural population of Queensland fruit fly, *Dacus tryoni*. *Australian Journal of Zoology*, **17**, 293–311.

Pritchett, V. S. (1990a). *The Complete Collected Stories*. New York: Random House.

Pritchett, V. S. (1990b). *The Complete Collected Essays*. New York: Random House.

Provine, W. B. (1986). *Sewall Wright and Evolutionary Biology*. Chicago, IL: University of Chicago Press.

Punt, A. E. and Hilborn, R. (1997). Fisheries stock assessment and decision analysis: the Bayesian approach. *Reviews in Fish Biology and Fisheries*, **7**, 35–63.

Puterman, M. L. (1994). *Markov Decision Processes. Discrete Stochastic Dynamic Programming*. New York: John Wiley and Sons.

Pybus, O., Charleston, M. A., Gupta, S., *et al.* (2001). The epidemic behavior of the hepatitis C virus. *Science*, **292**, 2323–2325.

Quinn, T. J. I. and Deriso, R. B. (1999). *Quantitative Fish Dynamics*. New York: Oxford University Press.

Raftery, A. E. (1988). Inference for the binomial N-parameter: A hierarchical Bayes approach. *Biometrika*, **75**, 223–228.

Railsback, S. F. (2001). Concepts from complex adaptive systems as a framework for individual-based modelling. *Ecological Modelling*, **139**, 47–62.

Read, A. F. (2003). Simplicity and serenity in advanced statistics. *Trends in Ecology and Evolution*, **18**, 11–12.

Read, A. F., Anwar, M., Shutler, D. and Nee, S. (1995). Sex allocation and population structure in malaria and related parasitic protozoa. *Proceedings of the Royal Society of London*, **B260**, 359–363.

Read, A. F., Narara, A., Nee, S., Keymer, A. E. and Day, K. P. (1992). Gametocyte sex ratios as indirect measures of outcrossing rates in malaria. *Parasitology*, **104**, 387–395.

Reed, W. J. and Simons, C. M. (1996). Analyzing catch-effort data by means of the Kalman filter. *Canadian Journal of Fisheries and Aquatic Sciences*, **53**, 2157–2166.

Rees, M., Mangel, M., Turnbull, L., Sheppard, A. and Briese, D. (2000). The effects of heterogeneity on dispersal and colonization in plants. In M. J. Hutchings, E. A. John, and A. J. A. Stewart, eds., *The Ecological Consequences of Environmental Heterogeneity*. Oxford: Blackwell Science, pp. 237–265.

Reid, C. (1976). *Courant in Göttingen and New York. The Story of an Improbable Mathematician*. New York: Springer Verlag.

Reidl, J. and Klose, K. E. (2002). *Vibrio cholerae* and cholera: out of the water and into the host. *FEMS Microbiology Reviews*, **26**, 125–139.

Ricciardi, L. M. and Sato, S. (1990). Diffusion processes and first-passage-time problems. In L. M. Ricciardi, ed., *Lectures in Applied Mathematics and Informatics*. New York: St. Martin's Press, pp. 206–284.

Richards, L. J. and Maguire, J.-J. (1998). Recent international agreements and the precautionary approach: new directions for fisheries management science. *Canadian Journal of Fisheries and Aquatic Sciences*, **55**, 1545–1552.

Rickman, S. J., Dulvy, N. K., Jennings, S. and Reynolds, J. D. (2000). Recruitment variation related to fecundity in marine fishes. *Canadian Journal of Fisheries and Aquatic Sciences*, **57**, 116–124.

Rivot, E., Prevost, E. and Parent, E. (2001). How robust are Bayesian posterior inferences based on a Ricker model with regards to measurement errors and prior assumptions about parameters? *Canadian Journal of Fisheries and Aquatic Sciences*, **58**, 2284–2297.

Robb, C. A. and Peterman, R. M. (1998). Application of Bayesian decision analysis to management of a sockeye salmon (*Oncorhynchus nerka*) fishery. *Canadian Journal of Fisheries and Aquatic Sciences*, **55**, 86–98.

Robertson, I. C., Robertson, W. G. and Roitberg, B. D. (1998). A model of mutual tolerance and the origin of communal associations between unrelated females. *Journal of Insect Behavior*, **11**, 265–286.

Roff, D. A. (1984). The evolution of life history parameters in teleosts. *Canadian Journal of Fisheries and Aquatic Sciences*, **41**, 989–999.

Roff, D. A. (1991). The evolution of life-history variation in fishes, with particular reference to flatfishes. *Netherlands Journal of Sea Research*, **27**, 197–207.

Rohani, P., Keeling, M. J. and Grenfell, B. T. (2002). The interplay between determinism and stochasticity in childhood diseases. *American Naturalist*, **159**, 469–481.

Roitberg, B. D. (1990). Optimistic and pessimistic fruit flies: evaluating fitness consequences of estimation errors. *Behaviour*, **114**, 65–82.

Roitberg, B. D. and Friend, W. G. (1992). A general theory for host seeking decisions in mosquitoes. *Bulletin of Mathematical Biology*, **54**, 401–412.

Roitberg, B. D. and Lalonde, R. G. (1991). Host marking enhance parasitism risk for a fruit-infesting fly *Rhagoletis basiola*. *Oikos*, **61**, 389–393.

Roitberg, B. D. and Prokopy, R. J. (1987). Insects that mark hosts. *BioScience*, **37**, 400–406.

Roitberg, B. D., Cailrl, R. S. and Prokopy, R. J. (1984). Oviposition deterring pheromone influences dispersal distance in tephritid fruit flies. *Entomologia Experimentalis et Applicata*, **35**, 217–220.

Roitberg, B. D., Mangel, M., Lalonde, R., *et al.* (1992). Seasonal dynamic shifts in patch exploitation by parasitic wasps. *Behavioral Ecology*, **3**, 156–165.

Roitberg, B. D., Sircom, J., Roitberg, C. A., van Alphen, J. J. M. and Mangel, M. (1993). Life expectancy and reproduction. *Nature*, **364**, 351.

Rolff, J. and Siva-Jothy, M. T. (2002). Copulation corrupts immunity: a mechanism for a cost of mating in insects. *PNAS*, **99**, 9916–9918.

Rolff, J. and Siva-Jothy, M. T. (2003). Invertebrate ecological immunology. *Science*, **301**, 472–475.

Romagnani, S. (1996). TH1 and TH2 in human diseases. *Clinical Immunology and Immunopathology*, **80**, 225–235.

Root, K. V. (1998). Evaluating the effects of habitat quality, connectivity, and catastrophes on threatened species. *Ecological Applications*, **8**, 854–865.

Rosà, R., Pugliese, A., Villani, A. and Rizzoli, A. (2003). Individual-based vs. deterministic models for macroparasites: host cycles and extinction. *Theoretical Population Biology*, **63**, 295–307.

Rosenberg, A. A. (2003). Managing to the margins: the overexploitation of fisheries. *Frontiers in Ecology and the Environment*, **1**, 102–106.

Rosenberg, A. A. and Restrepo, V. R. (1994). Uncertainty and risk evaluation in stock assessment advice for U.S. marine fisheries. *Canadian Journal of Fisheries and Aquatic Sciences*, **51**, 2715–2720.

Rosenheim, J. A. (1996). An evolutionary argument for egg limitation. *Evolution*, **50**, 2089–2094.

Rosenheim, J. A. (1999). Characterizing the cost of oviposition in insects: a dynamic model. *Evolutionary Ecology*, **13**, 141–165.

Rosenheim, J. A. and Heimpel, G. E. (1998). Egg limitation in parasitoids: A review of the evidence and a case study. *Biological Control*, **11**, 160–168.

Rosenheim, J. A. and Heimpel, G. E. (2000). Egg maturation, egg resorption, and the costliness of transient egg limitation in insects. *Proceedings of the Royal Society of London*, B**267**, 1565–1573.

Rosenheim, J. A. and Mangel, M. (1994). Patch-leaving rules for parasitoids with imperfect host discrimination. *Ecological Entomology*, **19**, 374–380.

Rosenheim, J. A. and Rosen, D. (1992). Influence of egg load and host size on host-feeding behaviour of the parasitoid *Aphytis lingnaensis*. *Ecological Entomology*, **17**, 263–272.

Rosenzweig, M. L. (1995). *Species Diversity in Space and Time*. Cambridge: Cambridge University Press.

Ross, R. (1916). An application of the theory of probabilities to the study of a priori pathometry. Part I. *Proceedings of the Royal Society of London*, A**92**, 204–230.

Ross, R. and Hudson, H. P. (1917a). An application of the theory of probabilities to the study of a priori pathometry. Part II. *Proceedings of the Royal Society of London*, A**93**, 212–225.

Ross, R. and Hudson, H. P. (1917b). An application of the theory of probabilities to the study of a priori pathometry. Part III. *Proceedings of the Royal Society of London*, A**93**, 225–240.

Royall, R. (1997). *Statistical Evidence. A Likelihood Paradigm*. London: Chapman and Hall.

Rudd, M. A., Tupper, M. H., Folmer, H. and van Kooten, G. C. (2003). Policy analysis for tropical marine reserves: challenges and directions. *Fish and Fisheries*, **4**, 65–85.

Ruxton, G. D. and Rohani, P. (1996). The consequences of stochasticity for self-organized spatial dynamics, persistence and coexistence in spatially extended host-parasitoid communities. *Proceedings of the Royal Society of London*, B**263**, 625–631.

Sadovy, Y. (2001). The threat of fishing to highly fecund fishes. *Journal of Fish Biology*, **59** (Supplement A), 90–108.

Saether, B.-E., Engen, S., Lande, R., Both, C. and Visser, M. E. (2002). Density dependence and stochastic variation in a newly established population of a small songbird. *Oikos*, **99**, 331–337.

Saether, B. E. and Engen, S. (2004). Stochastic population theory faces reality in the laboratory. *Trends in Ecology and Evolution*, **19**, 351–353.

Sanchez Lizaso, J. L., Goni, R., Renones, O., *et al.* (2000). Density dependence in marine protected populations: a review. *Environmental Conservation*, **27**, 144–158.

Sanchirico, J. N., Stoffle, R., Broad, K. and Talaue-McManus, L. (2003). Modeling marine protected areas. *Science*, **301**, 47–48.

Sauvage, F., Langlais, M., Yoccoz, N. G. and Pontier, D. (2003). Modelling hantavirus in fluctuating populations of bank voles: the role of indirect transmission on virus persistence. *Journal of Animal Ecology*, **72**, 1–13.

Scheffer, M. and Carpenter, S. (2003). Catastrophic regime shifts in ecosystems: linking theory to observation. *Trends in Ecology and Evolution*, **18**, 648–656.

Scheffer, M., Carpenter, S., Foley, J. A., Folke, C. and Walker, B. (2001). Catastrophic shifts in ecosystems. *Nature*, **413**, 591–596.

Schemske, D. W. and Bierzychudek, P. (2001). Perspective: evolution of flower color in the desert annual *Linanthus parryae*: Wright revisited. *Evolution*, **55**, 1269–1282.

Schluter, D. (2000). *The Ecology of Adaptive Radiation*. New York: Oxford University Press.

Schmid-Hempel, P. and Ebert, D. (2003). On the evolutionary ecology of specific immune defense. *Trends in Ecology and Evolution*, **18**, 27–32.

Schnute, J., Cass, A. and Richards, L. J. (2000). A Bayesian decision analysis to set escapement goals for Fraser River sockeye salmon (*Oncorhynchus nerka*). *Canadian Journal of Fisheries and Aquatic Sciences*, **57**, 962–979.

Schnute, J. T. (1993). Ambiguous inferences from fisheries data. In V. Barnett and K. F. Turkman (editors) *Statistics for the Environment*. London: John Wiley and Sons, pp. 293–309.

Schnute, J. T. and Kronlund, A. R. (1996). A management oriented approach to stock recruitment analysis. *Canadian Journal of Fisheries and Aquatic Sciences*, **53**, 1281–1293.

Schnute, J. T. and Kronlund, A. R. (2002). Estimating salmon stock-recruitment relationships from catch and escapement data. *Canadian Journal of Fisheries and Aquatic Sciences*, **59**, 433–449.

Schnute, J. and Richards, L. J. (1998). Analytical models for fishery reference points. *Canadian Journal of Fisheries and Aquatic Sciences*, **55**, 515–528.

Schnute, J. T. and Richards, L. J. (2001). Use and abuse of fishery models. *Canadian Journal of Fisheries and Aquatic Sciences*, **58**, 10–17.

Schrag, S. J. and Wiener, A. P. (1995). Emerging infectious disease: what are the relative roles of ecology and evolution? *Trends in Ecology and Evolution*, **10**, 319–324.

Schulman, L. S. (1981). *Techniques and Applications of Path Integration*. New York: Wiley Interscience.

Schuss, Z. (1980). *Theory and Applications of Stochastic Differential Equations*. New York: Wiley.

Schwartz, M. (2003). *How the Cows Turned Mad*. Berkeley, CA: University of California Press.

Schwarz, G. (1978). Estimating the dimension of a model. *Annals of Statistics*, **6**, 461–464.

Seber, G. A. F. (1982). *The Estimation of Animal Abundance*, 2nd edn. New York: MacMillan.

Settle, W. H. and Wilson, L. T. (1990). Invasion by the variegated leafhopper and biotic interactions: parasitism, competition, and apparent competition. *Ecology*, **71**, 1461–1470.

Sevenster, J. A., Ellers, J. and Driessen, G. (1998). An evolutionary argument for time limitation. *Evolution*, **52**, 1241–1244.

Shaw, D. J. and Dobson, A. P. (1995). Patterns of macroparasite abundance and aggregation in wildlife populations: a quantitative review. *Parasitology*, **111**, S111–S113.

Shears, N. T. and Babcock, R. C. (2003). Continuing trophic cascade effects after 25 years of no-take marine reserve protection. *Marine Ecology – Progress Series*, **246**, 1–16.

Sheldon, B. C. and Verhulst, S. (1996). Ecological immunology: costly parasite defences and trade-offs in evolutionary ecology. *TREE*, **11**, 317–321.

Shelton, P. A. and Healey, B. P. (1999). Should depensation be dismissed as a possible explanation for the lack of recovery of the northern cod (*Gadus morhua*) stock? *Canadian Journal of Fisheries and Aquatic Sciences*, **56**, 1521–1524.

Shepherd, J. G. (1982). A versatile new stock recruitment relationship for fisheries, and the construction of sustainable yield curves. *J. Conseil International Exploration de la Mer*, **40**, 67–75.

Shepherd, J. G. (1999). Extended survivors analysis: An improved method for the analysis of catch-at-age data and abundance indices. *ICES Journal of Marine Science*, **56**, 584–591.

Sherman, K. and Alexander, L. M. (editors) (1989). *Biomass Yields and Geography of Large Marine Ecosystems*. Boulder, CO: Westview Press.

Sherman, K., Alexander, L. M. and Gold, B. D. (editors) (1990). *Large Marine Ecosystems. Patterns, Process and Yields*. Washington, DC: American Association for the Advancement of Science.

Sherman, K., Alexander, L. M. and Gold, B. D. (editors) (1991). *Food Chains, Yields, Models and Management of Large Marine Ecosystems*. Boulder, CO: Westview Press.

Sherman, K., Alexander, L. M. and Gold, B. D. (editors) (1993). *Large Marine Ecosystems. Stress, Mitigation and Sustainability*. Washington, DC: American Association for the Advancement of Science.

Shudo, E. and Iwasa, Y. (2001). Inducible defense against pathogens and parasites: optimal choice among multiple options. *Journal of Theoretical Biology*, **209**, 233–247.

Siegel, J. S. (2002). Shattered mirrors. *Nature*, **419**, 346–347.

Simberloff, D. (1980). A succession of paradigms in ecology: essentialism to materialism and probabilism. *Synthese*, **43**, 3–39.

Sinclair, A. F. and Swain, D. P. (1996). Comment: spatial implications of a temperature-based growth model for Atlantic Cod (*Gadus morhua*) off the eastern coast of Canada. *Canadian Journal of Fisheries and Aquatic Sciences*, **53**, 2909–2911.

Sinclair, A. F., Swain, D. P. and Hanson, J. M. (2002a). Disentangling the effects of size-selective mortality, density, and temperature on length-at-age. *Canadian Journal of Fisheries and Aquatic Sciences*, **59**, 372–382.

Sinclair, M., Arnason, R., Csirke, J., *et al.* (2002b). Responsible fisheries in the marine ecosystem. *Fisheries Research*, **58**, 255–265.

Singleton, D. A. and Vo, L. K. (2002). Enantioselective synthesis without discrete optically active additives. *Journal of the American Chemical Society*, **124**, 10 010–10 011.

Siwoff, S., Hirdt, S., Hirdt, P. and Hirdt, T. (1990). *The 1990 Elias Baseball Analyst.* New York: Collier Books.

Skellam, J. G. (1951). Random dispersal in theoretical populations. *Biometrika*, **38**, 196–218.

Skorokhod, A. V., Hoppensteadt, F. C. and Salehi, H. (2002). *Random Perturbation Methods*. New York: Springer Verlag.

Slepoy, A., Singh, R. R. P., Pázmándi, F., Kulkarni, R. V. and Cox, D. L. (2001). Statistical mechanics of prion diseases. *Physical Review Letters*, **87**, 058101.

Smith, G. C. and Wilkinson, D. (2002). Modelling disease spread in a novel host: rabies in the European badger *Meles meles*. *Journal of Applied Ecology*, **39**, 865–874.

Smith, M. C., Reeder, R. H. and Thomas, M. B. (1997). A model to determine the potential for biological control of *Rottboellia cochinchinensis* with the head smut *Sporisorium ophiuri*. *Journal of Applied Ecology*, **34**, 388–398.

Smith, R. D. (2000). *Can't You Hear Me Callin'. The Life of Bill Monroe*. Boston: Little, Brown and Company.

Smith, T. D. (1994). *Scaling Fisheries*. Cambridge: Cambridge University Press.

Snow, C. P. (1965). *The Two Cultures*. Cambridge: Cambridge University Press.

Soai, K., Niwa, S. and Hori, H. (1990). Asymmetric self-catalytic reaction. Self-production of chiral 1-(3-pyridyl)alkanols as chiral self-catalysts in the enantioselective addition of dialkylzinc reagents to pyridine-3-caraldehyde. *Journal of the Chemical Society, Chemical Communications*, **1990**, 982–983.

Soh, S., Gunderson, D. R. and Ito, D. H. (2000). The potential role of marine reserves in the management of shortraker rockfish (*Sebastest borealis*) and rougheye rockfish (*S. aleutianus*) in the Gulf of Alaska. *Fishery Bulletin*, **99**, 168–179.

Soule, M. E. (editor) (1987). *Viable Populations for Conservation*. Cambridge: Cambridge University Press.

Sounes, H. (2001). *Down the Highway. The Life of Bob Dylan*. New York: Grove Press.

Southwood, T. R. E. (1978). *Ecological Methods*. London: Chapman and Hall.

Souza, P. D. (1998). On hyperbolic discounting and uncertain hazard rates. *Proceedings of the Royal Society of London*, **B265**, 2015–2020.

Spielman, A. and D'Antonio, M. (2001). *Mosquito*. New York: Hyperion.

Stachel, J. (editor) (1998). *Einstein's Miraculous Year. Five Papers that Changed the Face of Physics*. Princeton, NJ: Princeton University Press.

Starr, P., Annala, J. H. and Hilborn, R. (1998). Contested stock assessment: two case studies. *Canadian Journal of Fisheries and Aquatic Sciences*, **55**, 529–537.

Stearns, S. C. (editor) (1999). *Evolution in Health and Disease*. New York: Oxford University Press.

Stearns, S. C. (2000). Daniel Bernoulli (1738): evolution and economics under risk. *Journal of Biosciences*, **25**, 221–228.

Stenseth, N. C., Bjørnstad, O. N., Falck, W., *et al.* (1999). Dynamics of coastal cod populations: intra- and intercohort density dependence and stochastic processes. *Proceedings of the Royal Society of London*, A**266**, 1645–1654.

Stephens, D. W. and Krebs, J. R. (1986). *Foraging Theory*. Princeton, NJ: Princeton University Press.

Stewart, I. (2000). The Lorenz attractor exists. *Nature*, **406**, 948–949.

Stockwell, C. A., Hendry, A. P., Berg, O. K., Quinn, T. P. and Kinnison, M. T. (2003). Contemporary evolution meets conservation biology. *Trends in Ecology and Evolution*, **18**, 94–101.

Stolz, J. and Schnell, J. (editors) (1991). *Trout*. Harrisburg, PA: Stackpole Books.

Strand, E., Huse, G. and Giske, J. (2002). Artificial evolution of life history and behavior. *American Naturalist*, **159**, 624–644.

Stratonovich, R. L. (1963). *Topics in the Theory of Random Noise*. New York: Gordon and Breach.

Stratoudakis, Y., Fryer, R. J. and Cook, R. M. (1998). Discarding practices for commercial gadoids in the North sea. *Canadian Journal of Fisheries and Aquatic Sciences*, **55**, 1632–1644.

Strogatz, S. H. (1994). *Nonlinear Dynamics and Chaos*. Cambridge, MA: Westview.

Stroock, D. W. (2003). *Markov Processes from K. Ito's Perspective*. Princeton, NJ: Princeton University Press.

Strunk, W. J. and White, E. B. (1979). *The Elements of Style*, 3rd edn. New York: MacMillan.

Sumpter, D. J. T. and Martin, J. (2004). The dynamics of virus epidemics in *Varroa*-infested honey bee colonies. *Journal of Animal Ecology*, **73**, 51–63.

Suter, G. W. (1996). Abuse of hypothesis testing statistics in ecological risk assessment. *Human and Ecological Risk Assessment*, **2**, 331–347.

Swansburg, E., Chaput, G., Moore, D., Caissie, D. and El-Jabi, N. (2002). Size variability of juvenile Atlantic salmon: links to environmental conditions. *Journal of Fish Biology*, **61**, 661–683.

Sylvain, G., van Baalen, M. and Jansen, V. A. A. (2002). The evolution of parasite virulence, superinfection, and host resistance. *American Naturalist*, **159**, 658–669.

Tanaka, M. M., Kumm, J. and Feldman, M. W. (2002). Coevolution of pathogens and cultural practices: a new look at behavioral heterogeneity in epidemics. *Theoretical Population Biology*, **62**, 111–119.

Tang, S. and Wang, Y. (2002). A parameter estimation program for the error-in-variable model. *Ecological Modelling*, **156**, 225–236.

Taper, M. L. and Lele, S. R. (2004). *The Nature of Scientific Evidence. Statistical, Philosophical and Empirical Considerations*. Chicago: University of Chicago Press.

Taylor, A. D. (1988a). Large-scale spatial structure and population dynamics in arthropod predator–prey systems. *Annals Zoologia Fennici*, **25**, 63–74.

Taylor, A. D. (1988b). Parasitoid competition and the dynamics of host–parasitoid models. *American Naturalist*, **132**, 417–436.

Taylor, A. D. (1993). Aggregation, competition, and host–parasitoid dynamics: stability conditions don't tell it all. *American Naturalist*, **141**, 501–506.

Taylor, L. R. (1984). Assessing and interpreting the spatial distributions of insect populations. *Annual Review of Entomology*, **29**, 321–357.

Taylor, L. R., Woiwood, I. P. and Perry, J. N. (1978). The density-dependence of spatial behaviour and the rarity of randomness. *Journal of Animal Ecology*, **47**, 383–406.

Taylor, L. R., Woiwood, I. P. and Perry, J. N. (1979). The negative binomial as a dynamic ecological model for aggregation, and the density dependence of k. *Journal of Animal Ecology*, **48**, 289–304.

Taylor, L. R., Woiwood, I. P. and Perry, J. N. (1980). Variance and the large scale spatial stability of aphids, moths and birds. *Journal of Animal Ecology*, **49**, 831–854.

ter Haar, D. (1998). *Master of Modern Physics. The Scientific Contributions of H. A. Kramers.* Princeton, NJ: Princeton University Press.

Thom, R. (1972/1975). *Stabilité structurelle et morphogenese (Structural Stability and Morphogenesis)*. Reading, MA: Benjamin.

Thompson, G. G. (1992). A Bayesian approach to management advice when stock-recruitment parameters are uncertain. *Fishery Bulletin*, **90**, 561–573.

Thompson, G. G. (1993). A proposal for a threshold stock size and maximum fishing mortality rate. In S. J. Smith, J. J. Hunt and D. Rivard (editors) *Risk Evaluation and Biological Reference Points for Fisheries Management* (Canadian Special Publication of Fisheries and Aquatic Sciences 120). Ottawa: National Research Council of Canada.

Toft, C. A. and Mangel, M. (1991). From individuals to ecosystems: the papers of Skellam, Lindeman and Hutchinson. *Bulletin of Mathematical Biology*, **53**, 121–134.

Trumble, R. J. (1998). Northeast Pacific flatfish management. *Journal of Sea Research*, **39**, 167–181.

Turchin, P. (1998). *Quantitative Analysis of Movement. Measuring and Modeling Population Redistribution in Animals and Plans.* Sunderland, MA: Sinauer Associates.

Turelli, M. A., Schemske, D. W. and Bierzychudek, P. (2001). Stable two-allele polymorphisms maintained by fluctuating fitnesses and seed banks: protecting the blues in *Linanthus parryae*. *Evolution*, **55**, 1283–1298.

Turing, A. M. (1952). The chemical basis of morphogenesis. *Proceedings of the Royal Society of London*, **B237**, 37–72.

Tyutyunov, Y., Senina, I., Jost, C. and Arditi, R. (2002). Risk assessment of the harvested pike-perch population of the Azov Sea. *Ecological Modelling*, **149**, 297–311.

Uhlenbeck, G. E. and Ornstein, L. S. (1930). On the theory of Brownian motion. *Physical Review*, **36**, 823–841.

Ulrich, C., Le Gallic, B., Dunn, M. R. and Gascuel, D. (2002a). A multi-species multi-fleet bioeconomic simulation model for the English Channel artisinal fisheries. *Fisheries Research*, **58**, 379–401.

Ulrich, C. and Marchal, P. (2002). Sensitivity of some biological reference points to shifts in exploitation patterns and inputs uncertainty for three North Sea demersal stocks. *Fisheries Research*, **58**, 153–169.

Ulrich, C., Pascoe, D., Sparre, P. J., de Wilde, J.-W. and Marchal, P. (2002b). Influence of trends in fishing power on bioeconomics in the North Sea flatfish fishery regulated by catches or by effort quotas. *Canadian Journal of Fisheries and Aquatic Sciences*, **59**, 829–843.

Valleron, A.-J., Boelle, P.-Y., Will, R. and Cesbron, J.-Y. (2001). Estimation of epidemic size and incubation time based on age characteristics of vCJD in the United Kingdom. *Science*, **294**, 1726–1728.

van Alphen, J. J. M. and Visser, M. E. (1990). Superparasitism as an adaptive strategy. *Annual Review of Entomology*, **35**, 59–79.

van Alphen, J. J. M., Van Dijken, M. J. and Waage, J. K. (1987). A functional approach to superparasitism: host discrimination needs not to be learnt. *Netherlands Journal of Zoology*, **37**, 167–179.

van Alphen, J. J. M., Bernstein, C. and Driessen, G. (2003). Information acquisition and time allocation in insect parasitoids. *Trends in Ecology and Evolution*, **18**, 81–87.

van Baalen, M. (2000). The evolution of parasitoid egg load. In M. E. Hochberg and A. R. Ives (editors). *Parasitoid Population Biology*. Princeton, NJ: Princeton University Press.

van Kampen, N. G. (1977). A soluble model for diffusion in a bistable potential. *Journal of Statistical Physics*, **17**, 71–88.

van Kampen, N. G. (1981a). Ito versus Stratonovich. *Journal of Statistical Physics*, **24**, 175–187.

van Kampen, N. G. (1981b). *Stochastic Processes in Physics and Chemistry*. Amsterdam: North-Holland Publishing Company.

van Oostenbrugge, J. A. E., van Densen, W. L. T. and Machiels, M. A. M. (2001). Risk aversion in allocating fishing effort in a highly uncertain coastal fishery for pelagic fish, Moluccas, Indonesia. *Canadian Journal of Fisheries and Aquatic Sciences*, **58**, 1683–1691.

van Randen, E. J. and Roitberg, B. D. (1996). The effect of egg load on superparasitism by the snowberry fly. *Entomologia Experimentalis et Applicata*, **79**, 241–245.

van Winkle W., Jager, H. I., Railsback, S. F., *et al.* (1998). Individual-based model of sympatric populations of brown and rainbow trout for instream flow assessment: model description and calibration. *Ecological Modelling*, **110**, 175–207.

Vasseur, D. A. and Yodzis, P. (2004). The color of environmental noise. *Ecology*, **85**, 1146–1152.

Vaupel, J. W., Carey, J. R., Christensen, K., *et al.* (1998). Biodemographic trajectories of longevity. *Science*, **280**, 855–859.

Vestergaard, N. (1996). Discard behavior, highgrading and regulation: the case of the Greenland Shrimp Fishery. *Marine Resource Economics*, **11**, 247–266.

Vestergaard, N., Squires, D. and Kirkley, J. (2003). Measuring capacity and capacity utilization in fisheries: the case of the Danish Gill-net fleet. *Fisheries Research*, **60**, 357–368.

Visser, M. E. and Rosenheim, J. A. (1998). The influence of competition between foragers on clutch size decisions in insect parasitoids. *Biological Control*, **11**, 169–174.

von Bertalanffy, L. (1957). Quantitative laws in metabolism and growth. *Quarterly Review of Biology*, **32**, 217–231.

Voronka, R. and Keller, J. (1975). Asymptotic analysis of stochastic models in population genetics. *Mathematical Biosciences*, **25**, 331–362.

Vos, M. and Hemerik, L. (2003). Linking foraging behavior to lifetime reproductive success for an insect parasitoid: adaptation to host distributions. *Behavioral Ecology*, **14**, 236–245.

Vos, M., Hemerik, L. and Vet, L. E. M. (1998). Patch exploitation by the parasitoids *Cotesia rubecula* and *Cotesia glomerata* in multi-patch environments with different host distributions. *Journal of Animal Ecology*, **67**, 774–783.

Wachter, K. W. (1999). Evolutionary demographic models for mortality plateaus. *Proceedings of the National Academy of Sciences*, **96**, 10 544–10 547.

Wachter, K. W. and Finch, C. E. (editors) (1997). *Between Zeus and the Salmon. The Biodemography of Longevity*. Washington, DC: National Academy Press.

Wade, P. R. (2000). Bayesian methods in conservation biology. *Conservation Biology*, **14**, 1308–1316.

Waeckers, F. L. (1994). The effect of food deprivation on the innate visual and olfactory preferences in the parasitoid *Cotesia rubecula*. *Journal of Insect Physiology*, **40**, 641–649.

Wajnberg, E., Fauveergue, X. and Pons, O. (2000). Patch leaving decision rules and the marginal value theorem: an experimental analysis and a simulation model. *Behavioral Ecology*, **11**, 577–586.

Walde, S. J. and Murdoch, W. W. (1988). Spatial density dependence in parasitoids. *Annual Review of Entomology*, **33**, 441–466.

Walter, G. H. (1988). Activity patterns and egg production in *Coccophagus bartletti*, an aphelinid parasitoid of scale insects. *Ecological Entomology*, **13**, 95–105.

Walters, C. J. and Martell, S. J. D. (2004). *Fisheries Ecology and Management*. Princeton, NJ: Princeton University Press.

Waser, P. M., Elliott, L. F., Creel, N. M. and Creel, S. R. (1995). Habitat variation and viverrid demography. In A. R. E. Sinclair and P. Arcese (editors) *Serengeti II: Dynamics, Management, and Conservation of an Ecosystem*. Chicago, IL: University of Chicago Press, pp. 421–447.

Washburn, A. R. (1981). *Search and Detection*. Arlington, VA: Military Applications Section, Operations Research Society of America.

Watson, R. (1993). *The Trout. A Fisherman's Natural History*. Shrewsbury, UK: Swan Hill Press.

Watson, R. (1999). *Salmon, Trout and Charr of the World. A Fisherman's Natural History*. Shrewsbury, UK: Swan Hill Press.

Wegner, K., Kalbe, M., Kurtz, J., Reusch, T. and Milinski, M. (2003). Parasite selection for immunogenetic optimality. *Science*, **301**, 1343.

Weitz, J. S. and Fraser, H. B. (2001). Explaining mortality plateaus. *Proceedings of the National Academy of Sciences*, **98**, 15 383–15 386.

West, G. B., Brown, J. H. and Enquist, B. J. (1997). A general model for the origin of allometric scaling laws in biology. *Science*, **276**, 122–126.

West, G. B., Brown, J. H. and Enquist, B. J. (2001). A general model for ontogenetic growth. *Nature*, **413**, 628–631.

West, M. and Harrison, J. (1997). *Bayesian Forecasting and Dynamic Models*. New York: Springer Verlag.

West-Eberhard, M. J. (2003). *Developmental Plasticity and Evolution*. New York: Oxford University Press.

Wheatley, D. N. and Agutter, P. S. (1996). Historical aspects of the origin of diffusion theory in 19th-century mechanistic materialism. *Perspectives in Biology and Medicine*, **40**, 139–155.

White, A., Bowers, R. G. and Begon, M. (1996). Red/blue chaotic power spectra. *Nature*, **381**, 198.

White, P. C. L. and Harris, S. (1995). Bovine tuberculosis in badger (*Meles meles*) populations in southwest England: the use of a spatial stochastic simulation model to understand the dynamics of the disease. *Philosophical Transactions of the Royal Society of London*, B**349**, 391–413.

Whittle, P. (1955). The outcome of a stochastic epidemic. A note on Bailey's paper. *Biometrika*, **42**, 116–122.

Whittle, P. (1983). *Optimization over Time. Dynamic Programming and Stochastic Control. Volume II*. New York: John Wiley and Sons.

Wilcox, C. and Elderd, B. (2003). The effect of density-dependent catastrophes on population persistence time. *Journal of Applied Ecology*, **40**, 859–871.

Wilcox, C. and Possingham, H. (2002). Do life history traits affect the accuracy of diffusion approximations for mean time to extinction? *Ecological Applications*, **12**, 1163–1179.

Wilkinson, D. and Sherratt, T. N. (2001). Horizontally acquired mutualisms, an unsolved problem in ecology? *Oikos*, **92**, 377–384.

Williams, E. H. (2002). The effects of unaccounted discards and misspecified natural mortality on harvest policies based on estimates of spawners per recruit. *North American Journal of Fisheries Management*, **22**, 311–325.

Williams, G. C. and Nesse, R. M. (1991). The dawn of Darwinian medicine. *Quarterly Review of Biology*, **66**, 1–22.

Williams, J. G. (2003). Sardine fishing in the early 20th century. *Science*, **300**, 2032–2033.

Wilmott, P. (1998). *Derivatives. The Theory and Practice of Financial Engineering*. Chichester: John Wiley and Sons.

Wilson, W. G., Morris, W. F. and Bronstein, J. L. (2003). Coexistence of mutualists and exploiters on spatial landscapes. *Ecological Monographs*, **73**, 397–413.

Wong, E. (1971). *Stochastic Processes in Information and Dynamical Systems*. New York: McGraw-Hill.

Wong, E. and Zakai, M. (1965). On the relation between ordinary and stochastic differential equations. *International Journal of Engineering Science*, **3**, 213–229.

Wooster, W. S. (editor) (1988). *Fishery Science and Management. Objectives and Limitations*. New York: Springer Verlag.

Wouk, H. (1962). *Youngblood Hawke*. New York: Doubleday.

Wouk, H. (1994). *The Glory*. New York: Little, Brown.

Wynia, M. K. and Gostin, L. (2002). The bioterrorist threat and access to health care. *Science*, **296**, 1613.

Yates, F. (1951). The influence of *Statistical Methods for Research Workers* on the development of the science of statistics. *Journal of the American Statistical Association*, **46**, 19–34.

Yildiz, F. H. and Schoolnik, G. K. (1999). *Vibrio cholerae* O1 El Tor: Identification of a gene cluster required for the rugose colony type, exopolysaccharide production, chlorine resistance, and biofilm formation. *Proceedings of the National Academy of Sciences*, **96**, 4028–4033.

Young, T. P. (1994). Natural die-offs of large mammals: implications for conservation. *Conservation Biology*, **8**, 410–418.

Zabel, R. W., Harvey, C. J., Katz, S. L., Good, T. P. and Levin, P. S. (2003). Ecologically sustainable yield. *American Scientist*, **91**, 150–157.

Zinnser, W. (1989). *Writing to Learn*. New York: Harper and Row.

Zinsser, W. (2001). *On Writing Well*. New York: Harper Collins.

Zuk, M. and Stoehr, A. M. (2002). Immune defense and host life history. *American Naturalist*, **160**, S9–S22.

Index

Printed in the United States
by Baker & Taylor Publisher Services